GLOBAL DIMENSIONS OF
INTELLECTUAL PROPERTY RIGHTS
IN SCIENCE AND TECHNOLOGY

Office of International Affairs
National Research Council

Edited by

Mitchel B. Wallerstein
Mary Ellen Mogee
Roberta A. Schoen

NATIONAL ACADEMY PRESS
Washington, D.C. 1993

National Academy Press • 2101 Constitution Avenue, NW • Washington, D.C. 20418

NOTICE: The project that is the subject of this report was approved by the Governing Board of the National Research Council, whose members are drawn from the councils of the National Academy of Sciences, the National Academy of Engineering, and the Institute of Medicine. The members of the advisory committee responsible for the project were chosen for their special competencies and with regard for appropriate balance.

Publication of signed work signifies that it is judged a competent and useful contribution worthy of public consideration, but it does not imply endorsement of conclusions or recommendations by the governing bodies, officers, or staffs of the National Research Council, or the NAS, NAE, or IOM.

The program described in this report was supported by the Alfred P. Sloan Foundation, the U.S. Agency for International Development, the U.S. Department of Commerce, the Ford Aerospace Corporation, the Industrial Biotechnology Association, the National Academy of Engineering, the National Science Foundation, and the Pharmaceutical Manufacturers Association. The views expressed in this volume are those of the authors and are not presented as the views of the conference sponsors or organizers.

Library of Congress Cataloging-in-Publication Data

The Global dimensions of intellectual property rights in science and technology : a conference, January 7-8, 1992, National Academy of Sciences, Washington, D.C. / Office of International Affairs, National Research Council ; edited by Mitchel B. Wallerstein, Mary Ellen Mogee, Roberta A. Schoen.

p. cm.

Includes bibliographical references and index.

ISBN 0-309-04833-8

1. Intellectual property—Congresses. 2. Technology and law—Congresses. I. Wallerstein, Mitchel B. II. Mogee, Mary Ellen. III. Schoen, Roberta A. IV. National Academy of Sciences (U.S.) V. National Research Council (U.S.). Office of International Affairs.

K1401.A55 1992 — 93-720

341.7'58—dc20 — CIP

B-119

Printed in the United States of America.

ADVISORY COMMITTEE FOR CONFERENCE ON THE GLOBAL DIMENSIONS OF INTELLECTUAL PROPERTY RIGHTS IN SCIENCE AND TECHNOLOGY

ALBERT R.C. WESTWOOD, *Chair*, Vice President, Research and Technology, Martin Marietta Corporation, Bethesda, Maryland
ARDEN L. BEMENT, JR., Vice President, Technical Resources, TRW, Inc., Lyndhurst, Ohio
HARVEY J. BERGER, Chairman and Chief Executive Officer, ARIAD Pharmaceuticals, Inc., Devon, Pennsylvania
ANNE W. BRANSCOMB, Research Affiliate, Program on Information Resource Policy, Harvard University, Cambridge, Massachusetts
JACQUES J. GORLIN, President, The Gorlin Group, Washington, D.C.
ZVI GRILICHES, Professor of Economics, Harvard University, Cambridge, Massachusetts
KARL F. JORDA, David Rines Professor of Intellectual Property Law and Industrial Innovation, Franklin Pierce Law Center, Concord, New Hampshire
JAMES L. MERZ, Professor, Department of Electrical Engineering, University of California, Santa Barbara
JOHN T. PRESTON, Director, Technology Licensing, Massachusetts Institute of Technology, Cambridge
GUSTAV RANIS, Frank Altschul Professor of International Economics, Yale University, New Haven, Connecticut
HERBERT C. WAMSLEY, Executive Director, Intellectual Property Owners, Inc., Washington, D.C.

Staff

MITCHEL B. WALLERSTEIN, Deputy Executive Officer
ROBERTA A. SCHOEN, Staff Associate
MARY ELLEN MOGEE, Consultant
MONIQUE GREGORY, Administrative Secretary

The National Academy of Sciences (NAS) is a private, nonprofit self-perpetuating society of distinguished scholars engaged in scientific and engineering research, dedicated to the furtherance of science and technology and to their use for the general welfare. Upon the authority of the charter granted to it by the Congress in 1863, the Academy has a mandate that requires it to advise the federal government on scientific and technical matters. Dr. Frank Press is president of the National Academy of Sciences.

The National Academy of Engineering (NAE) was established in 1964, under the charter of the National Academy of Sciences, as a parallel organization of outstanding engineers. It is autonomous in its administration and in the selection of its members, sharing with the National Academy of Sciences the responsibility for advising the federal government. The National Academy of Engineering also sponsors engineering programs aimed at meeting national needs, encourages education and research, and recognizes the superior achievement of engineers. Dr. Robert M. White is president of the National Academy of Engineering.

The Institute of Medicine (IOM) was established in 1970 by the National Academy of Sciences to secure the services of eminent members of appropriate professions in the examination of policy matters pertaining to the health of the public. The Institute acts under the responsibility given to the National Academy of Sciences by its congressional charter to be an adviser to the federal government and, upon its own initiative, to identify issues of medical care, research, and education. Dr. Kenneth Shine is the president of the Institute of Medicine.

The National Research Council was organized by the National Academy of Sciences in 1916 to associate the broad community of science and technology with the Academy's purposes of furthering knowledge and advising the federal government. Functioning in accordance with general policies determined by the Academy, the Council has become the principal operating agency of both the National Academy of Sciences and the National Academy of Engineering in providing services to the government, the public, and the scientific and engineering communities. The Council is administered jointly by both Academies and the Institute of Medicine. Dr. Frank Press and Dr. Robert M. White are chairman and vice-chairman, respectively, of the National Research Council.

Preface

Intellectual property rights (IPRs) have become an area of international interest and controversy as the rate and cost of technological progress have increased, and as national borders have become ever more transparent. Disagreements have arisen not only over the mechanics of granting such rights, but even over the validity and merits of certain fundamental concepts concerning IPRs. For example, there are those who argue that the existence of robust IPR laws catalyzes innovation, and beneficially influences the economic future of companies and nations. Others argue that such laws are economically inefficient and exploitative, and that they are detrimental to the development of emerging nations.

IPRs are not a recent invention, and the word "right" may not be particularly well chosen. As Paul David notes in Chapter 2 of this volume, patents were used as early as the 14th century by English monarchs to protect the knowledge base of foreign craftsmen imported to enhance the state of the domestic technology. In those days, patents were granted initially for 14 years, which was the time necessary to graduate two generations of apprentices. The fact that rights to exploit advances in technical capability are granted by some governing authority, and are not considered inherent to the creator, is not generally appreciated. In the United States, for example, the government grants rights primarily to promote the public interest, and such rights are formulated so as to balance in some manner the economic benefits to the inventor and to society at large. Thus, it will be appreciated that the center of gravity of this balance may shift with changes in the state of the technology, the market, or social values (U.S. Congress, Office of Technology Assessment, 1992).

Nowadays, the varying laws governing IPRs in different nations play a major role in the strategic thinking of corporations as they attempt to ensure that they receive a sufficient return on their often large and certainly risky investments in research and development. Clearly, a company will not be enthusiastic about doing business in a country unwilling to provide protection for the intellectual content of its products—a concern now facing U.S. businesses as they evaluate opportunities in the former Soviet Union. Moreover, in these times of fiscal constraint, U.S. research universities also are increasingly concerned with exploiting the fruits of their intellectual labors and are encountering problems related to differences in national laws.

Part of the problem is that the United States follows the "first-to-invent" rule and permits an inventor a grace period of one year between the announcement of a discovery in a scientific paper or at a meeting, and filing for a patent. Other nations follow a "first-to-file" rule and do not permit disclosure before filing a patent application. This difference has had unfortunate consequences, for example, in the case of Boyer and Cohen's exploitation of their discoveries associated with rDNA. In this particular case, the recombinant DNA technique was granted patent protection in the United States but not in Europe, thereby causing a considerable loss of royalties to the inventors.

Also of increasing concern is the unauthorized use of intellectual property, which is sometimes referred to as piracy. A recent study by the U.S. International Trade Commission indicated that losses to U.S. companies from unpaid royalties on drugs, software, and electronic technologies, for example, may amount to as much as 2-3 percent of sales (i.e., many billions of dollars per year). Multinational companies thus have had to develop multinational IPR strategies, and these may include the aggressive pursuit of patent royalty income as a means of ensuring profitability.

Governments of developing countries, on the other hand, sometimes condone, either explicitly or implicitly, unauthorized use of IPRs, arguing that all knowledge should be in the public domain, or that some degree of protection from the need to pay IPR royalties is required if industry in an emerging nation is to survive the competition from more advanced and fiscally strong industries in industrialized countries. Indeed, IPR issues have now become sufficiently important that they have appeared on the agenda of recent G-7 Economic Summit meetings and are a principal subject of debate in the current Uruguay Round of the General Agreement on Tariffs and Trade (GATT) negotiations.

The issue of IPR infringement was first addressed by the Academy complex at the annual meeting of the National Academy of Engineering in 1986. Subsequently, in February 1988, a group of experts was convened to identify areas for further study. Mary Ellen Mogee, a consultant with expertise on the IPR issue, was then commissioned to develop a comprehensive background paper. This work was discussed by a larger group of experts and practitioners in June 1988, and led to a recommendation that the National Research Council (NRC) should organize a conference focused on the long-term impact on IPR issues resulting from the accelerating global diffusion of technology and from changes in the nature of technology itself.

Public and private sector sponsors for such a meeting subsequently were sought by Mitchel Wallerstein, who was then the associate executive director of the NRC Office of International Affairs. In April 1991, after funding had been obtained, an oversight committee was appointed to plan and organize a conference on the "Global Dimensions of Intellectual Property Rights in Science and Technology." The principal objectives of the meeting were (1) to examine the mutual impacts of trends in science and technology and in the philosophy and practice of IPRs, and (2) to discuss and define new approaches for resolving emerging conflicts in international IPR policies. The conference was held at the National Academy of Sciences on January 8-9, 1992, and was attended by more than 400 participants. This volume is based, in part, on the proceedings of the meeting. It should provide a valuable compendium of historical facts, current opinions, and options for action for both scholars and practitioners in the field of intellectual property rights.

It is a pleasure to acknowledge the invaluable contributions of the Conference Oversight Committee (Arden L. Bement, Harvey J. Berger, Anne W. Branscomb, Jacques J. Gorlin, Zvi Griliches, Karl F. Jorda, James L. Merz, John T. Preston, Gustav Ranis, and Herbert C. Wamsley); the visionary enthusiasm and energetic persistence of Mitchel B. Wallerstein and his colleagues, Roberta A. Schoen and Mary Ellen Mogee, who served both as the primary organizers of the meeting and as editors of this volume; and the financial support of the National Academy of Engineering, the U.S. Agency for International Development, the National Science Foundation, the U.S. Commerce Department, the Ford Aerospace Corporation, the Industrial Biotechnology Association, the Pharmaceutical Manufacturers Association, and the Alfred P. Sloan Foundation, in making this meeting as timely and valuable as it turned out to be.

A.R.C. Westwood
Chairman, Conference Oversight Committee

REFERENCE

U.S. Congress, Office of Technology Assessment. 1992. Finding a Balance: Computer Software, Intellectual Property, and the Challenge of Technological Change. OTA Report No. 052-003-01278-2 (April). Washington, D.C.: U.S. Government Printing Office.

Contents

I
INTRODUCTION

II
THE CASE FOR AND AGAINST A UNIFORM WORLDWIDE INTELLECTUAL PROPERTY RIGHTS SYSTEM

III
NATIONAL AND INTERNATIONAL APPROACHES TO INTELLECTUAL PROPERTY RIGHTS

IV
SCIENTIFIC AND TECHNOLOGICAL ADVANCE AND ITS IMPACT ON THE ROLE OF INTELLECTUAL PROPERTY RIGHTS

V
ADAPTING INTELLECTUAL PROPERTY RIGHTS TO NEW TECHNOLOGIES

VI
GLOBAL INTELLECTUAL PROPERTY RIGHTS ISSUES IN PERSPECTIVE

VII
APPENDIXES

GLOBAL DIMENSIONS OF
INTELLECTUAL PROPERTY RIGHTS
IN SCIENCE AND TECHNOLOGY

I

Introduction

1

The Global Dimensions of Intellectual Property Rights in Science and Technology

We live today in a world in which the economic health of nations and the competitiveness of firms is determined largely by the ability to develop, commercialize, and most importantly, to *appropriate* (or capture) the economic benefits from scientific and technological (S&T) innovations. Intellectual property rights (IPRs), such as patents and copyrights, are an important means used by firms to help protect their investments in innovation. They are legal instruments that have been used by governments for centuries to encourage industrial development and economic growth.

IPRs protect investments in innovation by granting the innovator a temporary monopoly on the use of the innovation. This prevents rapid imitation that could cut into the innovator's returns and decrease the incentive to innovate. By restricting imitation, however, IPRs arguably raise the cost of the new technology and restrict its availability. This may, in turn, retard further progress in the technology by preventing other firms from developing new innovations or improvements that build on the original innovation in a cumulative way. If the new technology has productivity-enhancing effects when used in economic activity, these too may be retarded by the protection of the original innovation.

Thus, IPRs inherently embody a policy conflict between the objective of providing an incentive to technological innovation and the objective of encouraging the rapid diffusion of new technology and the accumulation of technological knowledge. These competing objectives also represent powerful, competing economic interests—from R&D-intensive and non-R&D-in-

tensive firms at one level, to the industrialized, newly industrialized, and developing countries at another.

Governments have generally recognized, at least implicitly, the trade-offs that are involved in IPR laws, and each nation has established national IPR systems that attempt to strike a balance between competing objectives that is deemed appropriate for its national economic, political, and social context. It is important to note in this regard that IPRs are primarily a matter of *national* jurisdiction (i.e., the protection offered to an innovation is governed by the laws of the nation in which the innovation is made, used, or sold). Thus, for example, a patent obtained from the U.S. Patent and Trademark Office provides protection only within the territory of the United States. If a company is doing business in another country, it must file for and obtain IPR protection in that country. Moreover, the protection offered by that country's laws in many cases is not as strong as U.S. IPR protection. Although international IPR conventions exist, they do not establish specific rights. Instead, the extant international agreements attempt merely to ensure that, in any given country, foreign inventors receive the same rights as those granted to local inventors.

The protection offered by IPRs has never been complete, and for that reason many observers have criticized the idea that they grant even a temporary monopoly. Moreover, there has always been a tendency for some countries to seek to use IPR laws to favor domestic firms over foreign ones. (The major international IPR conventions are aimed at controlling this behavior in the interest of encouraging international trade.) Recent changes in global science, technology, trade, and economic development have, however, strained even further the effectiveness of IPRs in protecting S&T innovations.

This volume focuses on the nature of these changes, the challenges they present for national and international IPR systems, and their implications for science and technology. The Office of International Affairs of the National Research Council undertook an examination of the global dimensions of intellectual property rights in science and technology in response to increasing concern expressed by important segments of U.S. industry—and, to a lesser extent, the U.S. university research community—about the lack of uniform international treatment of IPRs and the difficulty of protecting their innovations from imitation. This examination took the form of a major conference, the proceedings of which are published in this volume.

A report of the U.S. International Trade Commission (1988:viii) estimated that the aggregate losses to U.S. industry from inadequate intellectual property protection in other countries in 1986 were $23.8 billion, or 2.7 percent of total sales. Much of the recent concern has focused on the developing world, particularly on the so-called newly industrializing countries (NICs), where patent and copyright laws have been weak or, in some cases, nonexistent. Unauthorized expropriation of intellectual property in the developing world has resulted from a combination of individual actions (i.e., piracy) and national economic policies (i.e., protectionism).

In an effort to accelerate their rate of economic development and increase their level of wealth in the short term, some governments have, for example, conveniently looked the other way when products or technologies are copied or used without permission. Many of these governments argue that some degree of protection from the need to pay for the use of ideas or technologies developed elsewhere is required if they are successfully to promote the maturation of so-called infant industries, whereas others contend that their countries cannot afford to pay the monopoly prices charged for technology protected by IPRs. There is also concern that some of the United States' major trading partners, notably Japan, may be using their IPR systems to deprive U.S. companies of their intellectual property.

As a result of these concerns, IPR issues have been elevated to high political levels within the Group of Seven (G-7) advanced industrialized countries, as demonstrated by the fact that they were an explicit agenda item at a recent G-7 Economic Summit meeting. IPR issues have also been a major point of discussion (and disagreement) in the so-called Uruguay Round of trade negotiations within the General Agreement on Tariffs and Trade (GATT).

The danger posed by inadequate IPR protection is that economic losses suffered by innovating firms could lead to a reduction in the rate of industrial innovation in the United States and other technologically advanced countries capable of generating innovations. A lower rate of innovation could, in turn, result in slower world economic growth, which would hurt all countries. Despite this argument, however, NICs and less developed countries (LDCs) that might benefit from more robust world economic growth have been reluctant to accept the premise that stronger IPR protection is in their long-term national interest.

This apparent paradox reflects the complexities, conflicts, and uncertainties surrounding IPR issues as they pertain to science and technology. The central purpose of the conference reported here was to identify and illuminate the international IPR issues of concern to the U.S. and international S&T communities and, in so doing, to contribute to the process of public education and debate that must guide policymaking in this area. The primary perspective offered on these issues is that of the United States, and the volume may be judged a success if it contributes to the crafting of a U.S. approach to international IPR issues that will serve the national interest. It is obvious, however, that in today's world of scientific, technological, and economic interdependence, the U.S. national interest in the global IPR system cannot be considered in isolation from the interests of other nations. Therefore, the volume also includes viewpoints of other nations as well.

The rest of this chapter introduces some of the major issues as background for the in-depth discussions that follow. It also suggests a set of themes and questions that may be useful in considering the policy implications of the following chapters.

MAJOR SCIENTIFIC AND TECHNOLOGICAL TRENDS

Today, scientific and technological changes are occurring so rapidly and across such a broad spectrum that they are creating unprecedented pressures for change in intellectual property protection in the United States and abroad. One of the major S&T trends that causes increased demand for effective intellectual property protection is the rising cost of R&D and other innovation-related activities. In the pharmaceutical industry, for example, an average expenditure of more than $231 million is required to discover, test, and secure marketing approval for a new drug in the United States (Pharmaceutical Manufacturers Association, 1992). To recoup such substantial investments, a company must be able to market the resulting products globally, which makes the worldwide intellectual property protection critical.

At the same time, product life cycles in some IPR-sensitive industries are shrinking dramatically, in some cases to 18 months or less. This time compression, in turn, reduces the period in which R&D and innovation costs can be recouped and puts a premium on strong and rapid protection of the innovation. Changes in technology, however, occasionally result in inventions that do not fit the old categories of patentable subject matter or cannot meet other requirements for patentability in certain countries. Biotechnology inventions (particularly microorganisms), for example, do not lend themselves to a written disclosure that enables their reproduction. Computer programs are perceived to have the characteristics of mathematical formulas, which are not patentable in some countries. Semiconductor chip designs are perceived as not meeting the U.S. criteria of novelty and nonobviousness.

Scientific discoveries, which are of increasing economic importance, also face difficulties when it comes to obtaining protection. Patent systems traditionally have denied protection to such discoveries. Moreover, the universities and research institutions in which basic scientific research is performed have traditionally put a premium on early dissemination of results, which is also at odds with the requirements for obtaining patents. For example, in many countries, any disclosure of an invention before a patent application has been filed precludes patentability. In one important case, the Cohen-Boyer patent on the basic recombinant DNA technique was granted in the United States, but the discoveries were denied protection in Europe and thereby suffered a considerable loss of royalties, because of their earlier publication (Benko, 1987:29-30).

Changes in technology also have blurred the distinction between *inventions*, which have been traditionally protected by patents, and *literary works*, which have been traditionally protected by copyrights. A computer program, for example, may be regarded as a literary work and a functional work, because it gives instructions to a computer to make it perform a function. In the United States, computer programs are protected primarily by copyrights, although in recent years the U.S. Patent and Trademark Office has granted an increasing number of patents for computer programs.

Technological changes also can make copying and production of R&D-intensive products cheaper, quicker, and in many cases, harder to detect. Digital audio recording, for example, makes it possible to reproduce thousands of perfect copies of the original. This, in turn, makes it more difficult for owners to assert their right to control their inventions through the traditional self-enforcement mechanisms. In some cases, technological "fixes" may be possible to combat copying (e.g., a piece of software that must be replicated in order to "unlock" a protected use), but these are likely to have only limited effectiveness as a general solution.

These and other challenges to the existing IPR regime that are created by the rapid advance of science and technology are discussed in depth in this volume.

EXAMPLES OF EMERGING TECHNOLOGIES

Some of the most significant emerging technologies—including those in the areas of information, electronics, communications, and the new biotechnology—do not fit neatly within existing categories of intellectual property rights.[1] They may force a reevaluation of current approaches to protection at national and international levels. Although detailed case studies of these technologies—and the adaptation of IPRs in response to them—are presented later in this volume, a brief background summary is provided here.

Computer Software

Computer software is expensive to develop but relatively easy to copy, conditions that make it highly vulnerable to infringement of intellectual property rights. The issue of how to protect computer software was first

[1]The Office of Technology Assessment (OTA, 1986) has done a major study of the implications of information-related technologies for intellectual property rights. Intellectual property issues surrounding the new biotechnology have been the subject of a study done for the Organization for Economic Cooperation and Development (Beier et al., 1985) and were covered in an OTA (1984) assessment and a subsequent OTA (1989) report.

acted on in the United States, where protracted debate and many studies considered such alternatives as patents, copyrights, and sui generis intellectual property rights.

Proponents of copyright protection for computer software argue that it is simply another form of writing brought about by technical change, as were sound recordings and motion pictures. Proponents of patent protection argue that the mental and financial effort required to produce software and the functional uses of software more closely resemble inventive activity rather than artistic creation, and thus patents are the more appropriate analogy. Still others argue for a third approach, some form of sui generis protection (i.e., a unique form of protection for IPRs in computer programs), which might possibly have characteristics of both copyright and patent law.

In the United States, the debate resulted in the decision to protect computer software primarily under the copyright laws. In 1980, the Copyright Act of 1976 was amended explicitly to grant copyright protection for software. The United States also has been encouraging other nations to protect computer software under copyright laws. Important questions remain, however, about the adequacy of copyright protection because of the fundamental limitations of copyright, which protects the form of expression of an idea but *not* the idea itself. Perhaps for this reason, as pointed out in Chapter 12 by Pamela Samuelson, the precise nature of protection for computer programs is still not certain in some countries, even though it may be covered under their copyright laws. Moreover, to complicate matters even further, the U.S. Patent and Trademark Office has been granting a growing number of patents for computer programs.

Semiconductor Chips

Semiconductor chips pose somewhat similar problems with respect to existing forms of intellectual property protection. Like computer software development, designing and preparing masks for chip manufacture are expensive, but reproducing masks is relatively simple and inexpensive. The basic technology for manufacturing chips is well established, so it is difficult to establish novelty or nonobviousness as is generally necessary for patent protection. Yet, a chip design is usually too functional to meet the requirements for copyright.

In the case of semiconductor chips, the U.S. Congress approved a sui generis form of protection in the Semiconductor Chip Protection Act of 1984. The legislation borrows from copyright law in its protection of reproduction, importation, and distribution rights, and from patent law in granting the right to exclude others from manufacturing and selling. The term of protection is only 10 years, which is shorter than for either patents or copyrights.

The New Biotechnology

The main forms of intellectual property that are relevant to inventors and companies working in the new biotechnology—for example, rDNA, cell fusion, and novel bioprocessing techniques—are patents, plant breeders' rights, and trade secrets. Biological and medical inventions are excluded from patent protection in many countries. Whereas U.S. patent law, as a result of *Diamond* v. *Chakrabarty*, allows for the patenting of a broad range of subject matter, including plants and animals, the laws in Europe and Japan generally do not. The European Patent Convention, for example, prohibits patent protection for "plant or animal varieties or essentially biological processes for the production of plants and animals" (Benko, 1987:44).

The new biotechnology is affected particularly by the patent doctrine that excludes scientific discoveries, because of the increasing frequency with which such discoveries can be turned quickly into commercial products. Moreover, because academic scientists are playing such an important role in its development, biotechnology is affected particularly by differing criteria of novelty as reflected in the variable length of grace periods. Some universities, however, require research results to be published within 6 months of completion of the research, which may not be long enough to allow a patent search and application process to be completed. On the other hand, some of the universities that have developed major research support relationships with private industry now permit longer delays between discovery and publication. Previous publication is a bar to patentability in many countries, as in the case of the Cohen-Boyer patent mentioned earlier.

In many countries, a deposit of the microorganism is required to obtain protection.[2] In countries that publish unexamined patent applications (the majority of developed countries), deposited cultures are usually released to the public at the same time—generally 18 months after the application is filed. The culture thus can become publicly available before any patent rights have been granted.[3] This effectively negates the potential for protection under trade secret law (which requires the inventor to take measures to keep the invention secret) if a patent is not issued. As a recent Office of Technology Assessment (OTA, 1984:389) report noted: "In essence . . . the

[2]The United States is party to the Budapest Treaty on the International Recognition of the Deposit of Micro-organisms for the Purpose of Patent Procedure, under which member states recognize in their own patent procedures a microorganism deposit made in another country if the deposit is made in a depository that meets the requirements of the treaty (Office of Technology Assessment, 1984:389).

[3]In the approach taken by the United States and Japan, release is deferred until the date of grant of the patent, because it is considered inequitable for the depositor to lose control of the microorganism before receiving an enforceable, exclusive right (Beier et al., 1985:91).

holder of a patent on a microorganism that produces a commercially useful polypeptide such as insulin must turn his or her 'factory' (i.e. the microorganism) over to competitors."

In most countries, new plant varieties fall within the domain of plant variety rights, or plant breeders' rights, a right established to provide protection for plant products.[4] The scope of protection of breeders' rights falls short of patent protection, however, because the breeder's prior authorization is not required for some important uses of the new plant variety. New plant varieties arising from genetic engineering methods would appear to have some characteristics of patentability—for example, they may be the subject of written description and repeatable, as required by patent law. However, double protection (e.g., plant variety protection and patent protection) is prohibited under the International Union for the Protection of New Varieties of Plants (Benko, 1987:44-45).

INTERNATIONAL INFRINGEMENT OF INTELLECTUAL PROPERTY RIGHTS

Increasing levels of infringement have made IPR issues highly visible in recent years; not only does the level of infringement appear to be rising, but there has also been a change in the industries affected. Whereas, in the past, manufacturing of fashion types of consumer goods was primarily affected, today producers of a broad range of industrial goods, including products and processes in such high-technology areas as computer hardware and software, biotechnology, and pharmaceuticals, are suffering significant economic losses due to infringement (U.S. International Trade Commission, 1984:ix). The losses threaten the incentive of firms in the involved industries to develop and introduce technological innovations.

The rise and spread of infringement of intellectual property rights have a number of causes. First, there is significant profit to be obtained from counterfeiting. Second, in many cases there are only limited risks because of weak intellectual property laws, weak enforcement, or both, and it is difficult to detect infringement. Third, infringement is also becoming significantly easier and cheaper in many instances, often because of technological changes that place the means for copying and producing in the hands of many. This has happened in the case of audio- and videotapes, for example, and in software for personal computers. Finally, the governments of some developing countries apparently allow infringement to flourish within

---

[4]In the United States, ownership rights in new varieties of plants are specifically granted by two federal statutes: the Plant Patent Act of 1930 (35 U.S.C. 161-164) and the Plant Variety Protection Act of 1970 (7 U.S.C. 2321 et seq.) (Office of Technology Assessment, 1984:392).

their borders because such activities contribute to their national economic development.

The definition of infringement varies from nation to nation, depending on a nation's intellectual property laws. Thus, what constitutes infringement in the United States may not be infringement under another country's laws. Often, what U.S. companies find objectionable is not really infringement, but rather the low level of protection afforded by some countries' intellectual property laws and enforcement, compared with those of the United States and other industrialized countries.

The terms *infringement, piracy*, and *counterfeiting* are often used interchangeably, but the policy debate would be better served by observing the distinctions among them. *Infringement*, which refers to the transgression of a legally recognized right that is usually litigable in the courts, is the term generally used in relation to the violation of most forms of intellectual property rights, except for trade secrets, in which case violation is termed *misappropriation.* In this country, patent infringement is defined as the unauthorized making, using, or selling of any patented invention within the United States (OTA, 1984:390). *Piracy*, although not a legal term of art, refers primarily to unauthorized reproduction for commercial gain of literary, musical, artistic, and other copyright works,[5] but may also be used in some cases in the context of trademarked or patented works. *Counterfeiting* is a term used most often to refer to unauthorized duplication of a product's trademark to give a similar appearance to a specific product, but it may also be said to exist when a clone[6] of a legitimate product is produced even without the use of the trademark.

Reliable estimates of the losses due to infringement are necessary to evaluate the severity of the problem and to determine what policy actions are warranted. It is quite difficult to develop such estimates, however, because definitions of infringement vary among nations, and it is difficult to detect infringing activities or products. Infringement problems are specific to certain industries or products, countries, and particular forms of intellectual property rights. To study the problem rigorously, it is necessary to focus on selected products in selected countries. Such analysis does not, of course, provide a basis for extrapolating to worldwide infringement losses.

Even if the dollar volume of infringing sales is known, which it often is not, that is not necessarily the same as the dollar amount of sales lost to the

[5]There remains some question about the legal status of *private copying*, which generally refers to home copying of intellectual property purely for individual consumption.

[6]The term *clone* is generally used to describe the replication of a protected product (e.g., certain pieces of computer hardware)—with or without the use of a trademark—which may or may not be litigable in the courts.

legitimate producer. Moreover, what really counts to companies is not the total amount of sales lost, but the net amount of *profits* lost. Although estimates of lost sales and lost profits can be made, they are subject to assumptions that are of uncertain validity (U.S. International Trade Commission, 1988:4-11). Further, not all economic harm can be measured directly, as when lack of confidence in intellectual property protection causes a firm to avoid a market altogether. Indirect effects, such as reduced research, development, and related innovation activities and reduced U.S. employment, are even more difficult to estimate.

Some of the best estimates of the economic impact of infringement are probably those in two reports by the U.S. International Trade Commission (ITC, 1984, 1988). Even these reports are based on questionnaire responses from nonrandom samples of U.S. companies and thus cannot be extrapolated to all firms. The 1988 ITC report estimated aggregate worldwide losses of $23.8 billion in 1986 for key U.S. industrial sectors due to inadequate intellectual property protection.

The 1988 ITC study also provides some of the best information on the quantitative effects of inadequate intellectual property protection on specific American industries. The scientific and photographic goods industry reported the greatest aggregate worldwide loss as a result of inadequate intellectual property protection in 1986—$5.1 billion, or 21 percent of the total for all industries. This was followed by the computer hardware and software industries ($4.1 billion, or 17 percent), the electronics industry ($2.3 billion, or 10 percent), the motor vehicle and motor vehicle parts industries ($2.2 billion, or 9 percent), the entertainment industry ($2.1 billion, or 9 percent), and the pharmaceutical industry ($1.9 billion, or 8 percent). The largest losses due to infringing imports (not including gray market goods) were reported in the industrial and farm equipment industry, electronics, and textiles and apparel. The largest export losses to infringing products were reported in motor vehicles and parts and in electronics (U.S. International Trade Commission, 1988:viii, 4-3).

POLICY THEMES AND QUESTIONS

During the conference and the preparation of this volume, a discernible pattern of policy themes and questions emerged that should be kept in mind while reviewing the material that follows. The most fundamental of these questions is, How can economic incentives for technological innovation be provided, while ensuring rapid and widespread diffusion of new technology? This is a dilemma for all firms and nations, not just the United States. At a higher level of analysis the question becomes, What kind of international IPR regime would be in the long-term national interest of the United States?

Fundamental Trade-offs

The primary rationale for intellectual property rights has been that such rights are necessary to provide incentives for inventors to invent and disclose their ideas to society. Once an invention has been introduced to the market, implemented in practice, or otherwise disclosed, it can often be imitated easily. In the absence of intellectual property protection, an inventor or company may invest in developing and introducing an invention only to find that imitators rapidly enter the market, thereby reducing the economic return to the originator of the idea. Moreover, because they have not incurred the development costs, imitators may be in a superior economic position relative to the original inventor. In this situation, there may not be much incentive—indeed, there may be a *disincentive*—for inventors to invest resources, even if the return to society as a whole is potentially large. On the other hand, as pointed out by Paul David in Chapter 2, overprotection of an invention can discourage subsequent innovation and diffusion, because in some cases the economic rents demanded by the inventor are too high.

Thus, a tension exists between the financial return necessary to provide an incentive to invest in invention and the rapid, widespread diffusion of new technology. Both are necessary for technology to contribute to economic growth and social welfare. The policy question facing nations has been how to achieve the appropriate balance between incentives to innovate and the diffusion of new technologies, such that the economic costs of granting the rights do not outweigh the benefits of increased innovation.

The appropriate trade-off between incentives to innovate and the diffusion of new technologies is also the fundamental policy question at the international level. Here, a way must be found to balance the diverse interests of nations. Proponents of stronger worldwide intellectual property protection argue that it is necessary to ensure adequate economic returns in order to sustain continued investments in innovation. Opponents argue, on the other hand, that the negative effects of monopolistic restrictions on trade and economic development inherent in IPRs outweigh any benefits from increased innovation and, further, that the worldwide extension of such systems is disadvantageous to the developing countries.

These fundamental policy questions can never be answered completely. At the national level, ongoing political and policy processes ensure that the relevant dimensions of the problems are considered and that some degree of balance is achieved. At the international level, however, there does not appear to be sufficient commitment, or even a consensus on the need to balance these competing objectives. This situation may continue to make it difficult to achieve the kind of international agreement necessary to extend a strong intellectual property rights system worldwide and thereby promote equitable, global economic growth.

Policy Vision

In seeking changes to the current system, it is important to consider the kind of international IPR regime that would be in the long-term U.S. national interest. Two characteristics of such a regime might be that it would (1) foster continued global economic development and (2) accommodate the imperatives of newly emerging technologies. Many of the complaints about inadequate intellectual property protection focus on the NICs and the developing countries. The economic development policies of many of the developing countries have been based on imitation and expropriation of the intellectual property of firms from industrialized nations. To date, such policies apparently have been quite successful for many of the developing countries that have now reached the status of NICs.

For this reason, the NICs have been much less supportive of strong intellectual property protection than industrialized countries. Strong intellectual property rights have been regarded as an obstacle to acquiring and diffusing the advanced technology necessary to fuel economic growth in developing countries. The NICs and LDCs often argue that intellectual property rights do little to stimulate indigenous innovation in their countries because they lack the necessary scientific and technological infrastructure. Rather, they view such rights as beneficial primarily to foreign companies that enter their markets. Moreover, until recently, few nationals of NICs and LDCs wished to obtain intellectual property protection in foreign countries. However, as more and more of the NICs begin to have indigenous technology to protect, their view of the IPR "problem" begins to change.

Two important policy questions emerge in this regard: Can economic growth and industrialization in the LDCs and the NICs be sustained on the basis of existing weak protection for intellectual property? When firms from industrialized countries suffer economic losses due to weak intellectual property protection in developing countries, are those losses large enough to endanger their investment in innovation or their ability to survive? Expropriation of intellectual property reduces the economic return to innovation obtained by firms in the industrialized nations and, if of a large enough magnitude, could lead to a lower rate of innovation. In the past, these firms have been the source of the new technologies that fuel world economic growth. Thus, continued reliance on such policies could lead to slower economic growth for all countries.

Clearly, as noted above, indigenous innovations developed in the NICs will require protection, not only in domestic markets but in foreign markets as well. As intellectual property protection becomes more important to indigenous innovators, the governments of these countries will face more pressures to enact and enforce strong protection. Again, the overarching issue is one of balance.

Advances in computer software, semiconductor chips, and biotechnology have set off major debates over how to protect the innovator's rights in those new technologies. It is not yet possible to determine the adequacy of the solutions reached. Emerging technologies, such as artificial intelligence and biochips, raise even more difficult intellectual property issues for the future. The rapidity of technological change in these fields means that actions taken now to deal with IPR problems, whether on the national or the international level, will have to be reevaluated continually and in all likelihood revised in the years to come.

One of the major questions posed as new technologies emerge is whether existing rights can provide adequate protection or whether a new form of rights is needed. Countries typically have dealt with this issue at the national level, and the resulting divergent national approaches to protection have made international agreement more difficult to achieve. This raises, in turn, a related question about the kinds of institutional structures and processes that can facilitate the development of international norms for protecting new technologies and the continuing review that will be necessary.

Intellectual Property Rights as a Trade Issue

The U.S. government has taken a multifaceted, trade-oriented approach to the international IPR issue, an approach that consists of multilateral and bilateral negotiations, as well as unilateral trade measures. General policy questions concern the effectiveness and long-term implications of this overall approach and its various components. For example, what kinds of trade-offs between IPRs and other trade policy objectives will result? How can the United States develop a consistent policy for worldwide protection of intellectual property when actions are being taken in many different forums?

An international code on patents, trademarks, and copyrights currently is being negotiated as part of the Uruguay Round of GATT. The GATT is viewed by some as having several advantages for achieving worldwide IPR protection. It represents a significant shift in approach, away from the World Intellectual Property Organization (WIPO), the U.N. agency that administers most major international IPR conventions, where strong intellectual property protection has been effectively opposed by the developing countries.

As the premier world trade forum, GATT places intellectual property issues in a trade context and links them to other trade and investment issues, thereby potentially bringing enormous bargaining power to bear. On the other hand, linkages with other trade issues may work to the detriment of those concerned with IPRs. Either way, it is likely that a GATT agreement on IPRs, if one is achieved, would include strengthened minimum standards of protection and procedures for settling disputes.[7]

Policy questions with respect to GATT negotiations include whether a satisfactory IPR agreement can be achieved under the Uruguay Round, whether a significant number of countries will sign such an agreement, and whether effective enforcement procedures can be agreed upon and implemented. If international IPR issues become subject to the GATT, what then will be the role of WIPO? Can WIPO be used to deal with *new* technology issues? Can either GATT or WIPO assume the important balancing role needed to resolve international IPR disputes?

Interactions with Other Policies

Intellectual property rights issues also interact importantly with other economic and health policies. The recent initiative of the National Institutes of Health concerning patentability of genetic sequences has, for example, created the possibility of setting off a frantic race among private companies to "stake out" rights to certain gene sequences before it is even clear how they are commercially useful. At the other extreme, laws that bar patentability for inventions that have been previously disclosed are often at odds with university research policies that stress early and free dissemination of research results. Antitrust policies in the past have sought to place narrow limits on the legally permissible exploitation of intellectual property rights. In the area of health policy, delays in new drug approval can reduce the effective period of patent protection, and price controls on medicines in many countries reduce the economic return on those products, which are highly R&D intensive. Each of these interactions, many of which are addressed in this volume, raises important issues at both the national and the international levels.

ORGANIZATION OF THE VOLUME

The challenge of an edited volume based on the proceedings of a conference is to present the material in an interesting and coherent fashion, while avoiding the tendency to try to recreate the agenda of the meeting itself. We have endeavored in the following pages, therefore, to provide an

---

[7]As of this writing, the uncertainties of the potential trade-offs between various issue areas encompassed within the negotiating framework of the Uruguay Round are still being explored and no final agreement has been reached.

intellectual framework for the salient issues raised during the conference and to include only the most cogent points from various discussion sessions. After an interesting and insightful analysis in Chapter 2 of the economic theory and historical development of intellectual property rights by economic historian Paul David, the remainder of the volume is divided into five major sections.

Section II presents the basic cases *for* and *against* a uniform, worldwide system of intellectual property rights. As success in the global economy turns increasingly on access to information and technical know-how, there are few areas in which the differences in approach between rich and poor countries are more clearly exemplified. Section III supports the analysis in the previous chapters with a comparative examination of national approaches, using the experience and practice of representative countries. This section also provides an up-to-date examination of ongoing efforts to negotiate new international IPR agreements.

Section IV takes an entirely different cut at global IPR issues by considering the impact of scientific and technological advance on the modern-day application of IPRs. After Chapter 8 considers how industries use—and seek to protect—their advanced technology to achieve and sustain global competitiveness, a series of different sectoral views of the problem is presented. Chapter 9 provides views of the problem from the standpoint of government, the university research community, and entrepreneurial business. Chapter 10 presents the perspective of multinational firms. Section V then takes up and examines, in turn, the adaptation of existing IPRs to particular, often problematic, new technologies, such as computer software or biotechnology, some of which do not conform well to existing IPR safeguards.

Section VI provides a distillation of some of the most salient unresolved IPR issues, as addressed in the closing panel discussion of the conference. A menu of interesting research questions is presented in a final coda. Biographical sketches of the conference speakers and contributors are included in the appendixes in Section VII.

REFERENCES

Beier, F.K. et al. 1985. Biotechnology and Patent Protection. Paris: Organization for Economic Cooperation and Development.

Benko, Robert P. 1987. Protecting Intellectual Property Rights. Washington, D.C.: American Enterprise Institute.

Pharmaceutical Manufacturers Association. 1992. The cost of innovation in the pharmaceutical industry: New drug R&D cost estimates. Industry Issue Brief (July).

U.S. Congress, Office of Technology Assessment. 1984. Commercial Biotechnology: An International Analysis. Washington, D.C.: U.S. Government Printing Office.

U.S. Congress, Office of Technology Assessment. 1986. Intellectual Property Rights in an Age of Electronics and Information. Washington, D.C.: U.S. Government Printing Office.

U.S. Congress, Office of Technology Assessment. 1989. Patenting Life. Washington, D.C.: U.S. Government Printing Office.

U.S. International Trade Commission. 1984. The Effects of Foreign Product Counterfeiting on U.S. Industry. Final Report on Investigation No. 332-158 under Section 332(b) of the Tariff Act of 1930, Publication 1479. Washington, D.C.

U.S. International Trade Commission. 1988. Foreign Protection of Intellectual Property Rights and Its Effect on U.S. Industry and Trade. Report to the U.S. Trade Representative, Investigation No. 332-245, Publication No. 2065. Washington, D.C.

2

Intellectual Property Institutions and the Panda's Thumb: Patents, Copyrights, and Trade Secrets in Economic Theory and History

PAUL A. DAVID

INTELLECTUAL PROPERTY ISSUES, ECONOMICS, AND HISTORY

The laws and administrative procedures concerned with intellectual property have once again emerged as a topic of widespread and intense discussion in this country and abroad. Many forces have converged to thrust the subject into the spotlight (see, e.g., Office of Technology Assessment, 1986; Benko, 1987; World Intellectual Property Organization, 1988; Rushing and Brown, 1990). Investment in R&D, for example, has become a central aspect of corporate and national strategies of global competition. The shortening of product life cycles, and the advance of techniques that make "reverse engineering" and outright copying of novel products easier, have made it more difficult for firms to reap the benefits of innovation simply by guarding new technologies as trade secrets while quickly moving along their production learning curves to seize a cost advantage over potential imitators. Also, many awkward ambiguities and widening areas of legal dispute have been created by the application to new technological developments of laws pertaining to patents, copyrights, and trade secrets, particularly in regard to biotechnology and to computer and information technologies.

In addition to the forces being generated within the sphere of scientific and engineering research itself, national economic policy has contributed to the renewed interest in intellectual property. During the 1960s and 1970s, developing countries successfully resisted conforming to a regime of strong international protection of intellectual property (see, e.g., Mody, 1990; Siebeck, 1990). However, during the 1980s, the U.S. government responded to the concerns of American producers—especially chemical, pharmaceutical, electronic, and information technology industries—by working vigorously to reverse the trend of the preceding two decades. Acting with some encouragement from other industrially advanced countries, the United States pursued a direct, unilateral course of action. It did not make any major effort to renegotiate agreements within the framework of the Paris Convention for the Protection of Industrial Property (patents and trademarks), the Berne Convention for the Protection of Literary and Artistic Works (copyrights), or other international conventions, nor did it offer some quid pro quo to developing nations that would agree to sign such conventions. Instead, by threatening within the context of bilateral trade negotiations to impose sanctions on developing and newly industrialized nations whose retaliatory leverage was quite limited, the United States achieved considerable success in convincing foreign governments to acquiesce to its position on the treatment of various forms of intellectual property. The pressures generated by the U.S. campaign, however, and the widening international markets for R&D-intensive goods and services have stirred a profound reconsideration of the merits and drawbacks of global "harmonization" of protections for intellectual property and of the desirability of achieving such uniformity at a strong, rather than weak, standard of enforcement.

Unlike the debates over intellectual property institutions in earlier eras, which had captured the attention of such great political and social philosophers as Thomas Jefferson, the current discussions reflect relatively slight interest in philosophical questions. Little attention is being paid to such issues as the "natural rights" of inventors and authors to the fruits of their creative efforts or the justice of claims advanced on behalf of all humanity to benefit from the collective, social processes through which new scientific and technological ideas arise (for exceptions see, e.g., Dworkin, 1981; Davis, 1989; Berg, 1991). Rather, in keeping with the more pervasively utilitarian spirit of the times, the statutes, legal rulings, administrative regulations, and other institutional arrangements affecting patents, copyrights, and trade secrets are widely regarded as public policy instruments that should be designed to enhance economic welfare by stimulating technological progress.

Even if the rhetoric of argument occasionally appeals to notions of justice and equity, modern economic analysis, and its characteristic preoccupation with questions of efficiency, now set the terms for policy discussions about the protection of intellectual property. On the one hand, economic analysis provides the most widely accepted, overarching interpretation and supporting rationale for public interventions aimed at channeling economic resources into invention and innovation. On the other hand, in continuation of a long tradition, economic analysis yields fundamental criticisms of the systems that have been established to achieve that purpose by securing rights in intellectual property. Thus, it is instructive to begin by taking the economist's approach in discussing U.S. intellectual property legislation and national policies to enforce rights in such property internationally. At the very least, this approach provides a framework for identifying the major problems of allocative efficiency and the distributional issues that are at stake—*from the viewpoint of society as a whole* rather than from the perspective of the various private (and national) interests involved.

Economists as a body, however, have been unable to formulate much in the way of straightforward, practical advice to guide lawyers, jurists, and policymakers in these matters (see Priest, 1986). The fundamental cause of their inconclusiveness is not so much the tendency of economists to engage in theoretical speculations as it is their inability to achieve consensus on the answer to two difficult empirical questions. First, will faster growth in the stock of scientific and technological knowledge always be an unambiguously "good thing" for a particular industrial sector or national economy and, therefore, warrant the sacrifice of other, lesser societal goals? Second, how responsive is the supply of socially useful discoveries and inventions to the creation of greater private economic incentives? For policy analysts not to know the policy goal with any precision is a considerable handicap, just as it is for them to remain unsure about the incentives and constraints that would be required to achieve any particular goal, were one to be agreed on.

Unfortunately, however, the two questions cannot be answered any better by lawyers on the basis of their having delved more deeply into the details of existing or proposed intellectual property regimes. Nonetheless, those who from practice are most at ease applying the logic of microeconomic reasoning to intellectual property rights issues, must pay heed to the skepticism voiced by legal experts. They should take more pains not to allow familiar, simplifying abstractions to obscure a central fact about the nature of the world for which they would prescribe institutional reforms, namely, that the complex body of law, judicial interpretation, and administrative practice that one has to grapple with in this field was not created by some rational, consistent, social welfare-maximizing public agency. What one is faced with, instead, is a mixture of the intended and unintended consequences of an undirected historical process on which the varied interests of many parties, acting at different points (some widely separated in time and space), have left an enduring mark. So, it would be really quite remarkable if the evolution of legal institutions concerning patents, copyrights, and trade secrets had somehow resulted in a set of instruments optimally designed to serve either public policy purposes or the private economic interests of individuals and firms seeking such protections.

Agreement with the above does not deny the general notion of an evolutionary drift toward social optimality in the effects of the law on resource allocation. Clever, modern Panglossians have come up with the proposition that the increasing likelihood that laws resulting in inefficient resource use will be exposed to economically motivated litigation, thereby creating "selective pressure" to remold property law in ways that tend to render it more efficient; that this pressure can work even if the outcome of the litigation is random; and that some beneficent "invisible hand" thus guides the evolution of legal institutions affecting economic performance. These ingenious but nonetheless dubious arguments are confined, even by their most ardent proponents, to the supposed workings of the common law system of judge-made law (see, e.g., Priest, 1977; Rubin, 1977; Goodman, 1978; Cooter and Kornhauser, 1980; and Cooter and Ulen, 1988, for discussions of deficiencies in the selective litigation thesis). The modern "law" of intellectual property, however, consists of statutory and administrative laws pertaining to patents and copyrights, even though the common law roots of the law of trade secrets create a complicating exception.

Thus, it is difficult to find even a speculative, theoretical justification for conceptualizing intellectual property statutes, and the administrative procedures they authorize, as institutional tools that were forged perfectly to "Promote the Progress of Science and Useful Arts." The latter is the specific purpose identified by the framers of the Constitution (Article 1, Section 8, Clause 8) when they granted to Congress the power of "securing for limited Times to Authors and Inventors the exclusive Right to their respective Writings and Discoveries." Nor should the prevailing statutes enacted under that authority be regarded as policy instruments designed to attain a social optimum defined more broadly in terms of economic welfare. Indeed, the first step toward understanding many of the policy dilemmas that arise today in regard to intellectual property would seem to be to acknowledge just this elementary point.

If intellectual property arrangements are to be viewed as utilitarian appendages of the body politic, it would be far more illuminating to recognize their essential nature as most closely akin to the "thumb" of the giant panda. The panda's thumb has been justly celebrated by Stephen Jay Gould (1980:Ch.1) as a striking example of evolutionary improvisation yielding an appendage that is inelegant yet serviceable. Although the panda can grasp and strip the leaves from the stalks of the bamboo plant, its thumb is not anatomically a finger at all, much less an opposable, manipulating digit. In actuality, it is a complex structure formed by the marked enlargement of a bone that otherwise would be a component of the animal's wrist—but for the effect of some genetic mutation—and the related extensive rearrangement of supporting musculature. It is, as Gould says, "a contraption, not a lovely contrivance," and one whose obvious mechanical limitations stem from its remote accidental origins.

Evolutionary processes in biology work largely with the materials that are readily available. So does institutional evolution, especially the processes of incremental change and adaptation in legal and other rule systems that give great weight to precedent (see, e.g., North, 1991). Accordingly, even though the legal provisions and administrative rules that make up the "patent system" and "copyright system" have changed considerably in form and function over their long history, they appear remarkably resistant to rapid and radical reform.

As the nature of technologies changes, however, it is increasingly evident that the familiar legal "contraptions" of patents and copyrights are rather ill-suited to some of the situations to which they are being applied (see, e.g., Office of Technology Assessment, 1986; World International Property Organization, 1989). They continue to be looked to as stimuli for the generation of useful innovations, but while enabling the private appropriation of economic benefits from new scientific and engineering knowledge, they have a variety of untoward side effects that may be distorting and even impeding the progress of technology. Moreover, the problems are not confined to those that might be solved by readjusting old and still serviceable legal tools or forging novel statutes to fit special technical circumstances. The process of more finely articulating and more vigorously enforcing private rights in intellectual property is certainly worth pursuing in some situations, but it cannot be looked to for optimal solutions to all of society's problems in designing institutional mechanisms affecting the production, distribution, and utilization of knowledge.

Identifying the limitations as well as the strengths of the private property approach is a central part of my task in introducing the subject of the global dimensions of intellectual property in science and technology. Setting out the basics of modern economic theory of intellectual property and reviewing the historical development of specific legal institutions that define and protect private rights in such property are also major aspects of my task. This assignment is a daunting one, for any of several reasons.

First, as noted, there is no settled body of economic theory on the subject that can be stated briefly without doing serious injustice to the sophisticated insights that have emerged over many decades of debate. Instead, the relevant economic literature is extensive, convoluted, and characterized by subtle points of inconclusive controversy concerning the appropriate course for public policy. Second, intellectual property law is an intricate, highly specialized area of legal scholarship and one to which I make no pretensions of expertise. Third, the historical development in Western societies of the patent system, the statutory protection of copyright, and the body of law governing trade secrets is a subject area that, unfortunately, has remained all too separated from economic and legal analyses of contemporary intellectual property issues. To link them satisfactorily would be no small undertaking.

Recognition of these difficulties should have been sufficient to dissuade me from accepting the assignment. In the event, as one can see, they were unavailing. Some considerable indulgence and forbearance on the part of the reader will therefore be required if my discussion oversimplifies complex matters of economic reasoning concerning intellectual property and the production and distribution of knowledge, points out only the most salient and early developments in the long history of these western European institutional arrangements, glosses over crucial distinctions and subtle points of modern law, and indulges in some provocative concluding comments on the current U.S.-led campaign for an international regime of uniformly strong intellectual property protection, as that appears from this economic historian's perspective.

KNOWLEDGE, PUBLIC POLICY ECONOMICS, AND INTELLECTUAL PROPERTY

The economist approaches the subject of protection of intellectual property rights, like many other issues, by trying to fit it into the generic formula for public policy decisions (see, e.g., Besen and Raskind, 1991:5). Somewhat loosely stated, the policy objective is to maximize the surplus of social benefits of the new information assets over the social costs of their production, that is, to maximize the "net social benefits." A further objective is to push the allocation of public and private resources in the direction of equalizing the social net rate of return on investments in knowledge and in other kinds of productive assets. This formulation gives rise to the following three classes of questions.

First, will the right amount of new information be created, and at the right times? The concern here is whether, in the absence of public intervention, private incentives would be sufficient to generate the optimal flow of additions to the stock of scientific and technological knowledge. Modern economic analysis recognizes that the peculiar, "public good" nature of information as a commodity creates serious resource allocation problems for competitive market systems. Further, it identifies the institution of private rights in intellectual property as one among a number of countervailing measures that the state may take to rectify the deficiencies of market competition.

Second, will the new information that is created be used productively, that is, in a way that yields the maximum flow of social benefits for the producers and consumers of goods and services? Unless intellectual creations are disseminated for others to enjoy as items of consumption or are used directly and indirectly in producing other goods and services, they cannot be expected to yield improvements in productivity and economic welfare. Consequently, a central set of issues for discussion among economists has been the likely effects that public policy measures meant to stimulate additions to knowledge would have on the diffusion of knowledge into commercial uses.

Third, will the conditions under which new knowledge is created be such that the social costs entailed in its production are minimized? There are opportunity costs to devoting resources to the advancement of knowledge through scientific discovery, inventive activity, and the novel expression of ideas. The goal of public policy cannot be simply that of causing private agents or governmental agencies to conduct these pursuits on an ever-grander scale, without regard to whether they are conducted efficiently. Intellectual property institutions must be evaluated in terms of their implications for the social costs of producing new knowledge, as well as for the utilization of the existing stock of knowledge.

Information, Public Goods, and Competitive Market Failures

The argument most generally offered in support of public policy interventions to enforce patents, copyrights, and trade secrecy is that there is a "market failure." In the absence of governmental protection of private property rights, the argument goes, competitive markets would not give individuals and organizations sufficient incentives to induce the socially optimal amount of investment in public goods in the form of new scientific and technological knowledge. It does not necessarily follow, however, that the best remedy for market failure is to create valuable private rights in intellectual property. In fact, the problem is more complicated than even that of arranging for the right amount of some classic public good, such as national defense or lighthouses.

Knowledge may be viewed as a commodity, but it is not a commonplace commodity. It is highly differentiated and has no obvious natural units of measurement. It can have utility as a pure consumption good or as a capital good, and often as both. Knowledge is unusual in that as a pure capital good yielding a stream of material benefits when combined with other kinds of assets, it possesses an intrinsic value. Such is the case, for example, with information about the operation of a cost-saving manufacturing process or the design of a product with better quality attributes. Still more remarkable is information's extreme indivisibility and durability. Once a bit of knowledge has been obtained, there is no value to acquiring it a second time, or a third. There is no societal need to repeat the same discovery or invention because a piece of information can be used again and again without exhausting it. Karl Marx (1867-1894; 1970: Vol. 1, Ch. XXV:386), among others, was struck by the fact that scientific knowledge could be freely appropriated to productive processes, as are the physical forces found in nature:

> Once discovered, the law of the deviation of the magnetic needle in the field of an electric current, or the law of the magnetization of iron, around which an electric current circulates, costs never a penny.

Related to this, and of even greater importance, knowledge differs from ordinary "private" commodities in being what economists refer to as a *nonrival* good; that is, it can be possessed and enjoyed jointly by as many as care to make use of it. This observation forms the point of departure for the classic analysis of the economics of R&D by Arrow (1962), but it is not a modern insight. Consider the following passage in a letter written in 1813 to Isaac McPherson, a Baltimore inventor, by Thomas Jefferson (reprinted in Koch and Peden, 1972:629-630):

> If nature has made any one thing less susceptible than all others of exclusive property, it is the action of the thinking power called an idea, which an individual may exclusively possess as long as he keeps it to himself; but the moment it is divulged, it forces itself into the possession of every one, and the receiver cannot dispossess himself of it. Its peculiar character, too, is that no one possesses the less, because every other possesses the whole of it. . . . That ideas should freely spread from one to another over the globe, for the moral and mutual instruction of man, and improvement of his condition, seems to have been peculiarly and benevolently designed by nature, when she made them, like fire, expansible over all space, without lessening their density in any point, and like the air in which we breathe, move, and have our physical being, incapable of confinement or exclusive appropriation.

Jefferson grasped the essential point that the cost of transmitting useful knowledge in codified form is negligible compared with the cost of creating it, and that, but for society's need to encourage the pursuit of ideas, such information should be distributed freely. Indeed, on these grounds Jefferson proceeded immediately to reject the argument of the French philosophers that inventors and authors had a natural rights claim to property in their creations (reprinted in Koch and Peden, 1972:630):

> Inventions then cannot, in nature, be a subject of property. Society may give an exclusive right to the profits arising from them, as an encouragement to men to pursue ideas which may produce utility, but this may or may not be done, according to the will and convenience of the society, without claim or complaint from anybody....

This does not mean that all types of knowledge can be transmitted at negligibly low marginal costs or that the private and social costs of filtering, interpreting, and utilizing information are insignificant. Recent discussions of the economics of R&D and technology transfers (see, e.g., Pavitt, 1987; Rosenberg, 1990; Arora, 1991) have recognized the importance of *tacit* components of technological knowledge and emphasized that the information contained in scientific papers, patents, blueprints, and other forms of codified knowledge often will not be sufficient to implement the technology in question; complementary know-how is required, and its acquisition is often costly.

Nelson (1990) makes the same point and goes on to associate codified knowledge with the "generic" parts of technological information—those that can be transferred readily and tend, thus, to move quickly into the public domain—and tacit knowledge with the "specific" bits of information that lend themselves better to being held privately. These particular identifications, however, do not seem either necessary or especially helpful. What is held secret and what becomes publicly disclosed are determined not so much by the inherent nature of the information as by the expected costs and rewards associated with each course of action for the agents involved (see Dasgupta and David, 1990). This much is obvious from considering the factors that enter into a firm's decision whether to file for a patent on a new process of manufacture or to protect it as a trade secret.

Nonrival possession, low marginal cost of reproduction and distribution (which makes it difficult to exclude others from access), and substantial fixed costs of original production—these are the three properties familiarly associated with the definition of a *public good*. When these characteristics are present, competitive markets—in which price tends to be driven down to the cost of supplying the marginal unit of the commodity—generally perform quite badly; competitive producers' revenues will not even cover their full costs of production, much less anything approaching the use-value of the goods to the public. Indeed, the attempt to make the beneficiaries pay for value received would so reduce demand as to result in an inefficiently low level of its consumption. In the literature of public finance economics, therefore, alternative allocative mechanisms are proposed as solutions to "the public goods problem." There are three principal alternatives. One is that society should give independent producers publicly financed subsidies and require that the goods be made available to the public freely or at a nominal charge. A second mechanism would have the state levy general taxes to finance its direct participation in production and distribution of the good, furnish and manage the requisite facilities, and contract when necessary with private agents to carry out the work. Here, again, the objective is to supply the good without having to charge prices for it. The third solution is to create a publicly regulated private monopoly authorized to charge consumers prices that will secure a "normal" rate of profit. This does not guarantee, however, that consumers will line up to purchase the goods and services in question. The legal right to exclude other producers from the market for a product does not, of itself, create a profitable monopoly of that line of business.

Although the nonexcludable and nonrivalrous nature of information qualifies it as a public good, information differs in two respects from the mass of conventional public goods, such as traffic lights, flood control systems, and airport beacons or radar landing beams. The first difference is that the attributes of the commodity—typically, the complete contents of the information itself—will not be known beforehand. Indeed, they are not automatically known to all the interested parties even when the new knowledge becomes available. This asymmetry in the distribution of information greatly complicates the process of arranging contracts for the production and use of new knowledge.

The second differentiating feature of knowledge is its cumulative and interactive nature. The stock of scientific and technological knowledge grows by increments, with each advance building on and sometimes altering the significance of previous findings in complicated and often unpredictable ways. As Thomas Jefferson remarked (reprinted in Koch and Peden, 1972:686),

> The fact is, that one new idea leads to another, that to a third, and so on through a course of time until someone, with whom no one of these ideas was original, combines all together, and produces what is justly called an new invention.

On these same grounds, Michael Polanyi (1944:70-71), a British sociologist of science, maintained that patent law was essentially deficient because it sought to "parcel up a stream of creative thought" into a series of distinct claims each of which could constitute the basis of a separately owned monopoly, whereas "incremental progress interacts at every stage with the whole network of human knowledge and draws at every moment on the most varied and dispersed stimuli."

The same kind of creative recombinant process does not operate when one stockpiles weapons for defense or erects another set of airport landing lights. The light signal from an airport or lighthouse is a form of information, but it is the emission of the signal—rather than the bricks or metal or glass—that imparts the public goods character to those structures. This form of information, however, has no capacity for internal growth and elaboration. Unlike scientific and technological knowledge, light signals just do not evolve and acquire new utility through cumulation and interaction. As discussed below, legal and other institutional arrangements may be imposing high costs on research-intensive firms, and society more generally, by restricting access to some elements in those streams of creative thought and thereby making it less likely that the elements will be rapidly rearranged and recombined in new and fruitful ways.

Imperfect Institutional Solutions and Trade-offs

The importance of the foregoing differentiating features of knowledge notwithstanding, there is a striking correspondence between the three solutions for the standard public goods problem—subsidies, direct governmental production, and regulated monopoly—and the three main institutional arrangements that have been devised to deal with allocational problems in the production of knowledge and pure information goods (see Dasgupta and David, 1988). I refer to the latter arrangements as "the three P's," because they can be described in highly idealized forms as patronage, procurement, and property, respectively.

Patronage stands for the system of awarding publicly financed prizes, research grants based on the submission of competitive proposals, and other subsidies to private individuals and organizations engaged in intellectual discovery and invention, in exchange for full public disclosure of their creative achievements. In Western democratic societies, patronage characterizes the pursuit of "open" scientific inquiry and the dominant institutional and social mode of organization associated with the conduct of academic science (see David, 1991).

Procurement is associated with government's contracting for intellectual work, the products of which it will control and devote to public purposes. Whether the information produced will be made available for public use is a secondary issue, although an important matter for public policy. "Sensitive" defense-related research is usually conducted under governmental auspices in secure, closed laboratories, but much publicly contracted R&D and the scientific work of governmentally managed laboratories and agricultural experiment stations are undertaken with the intention of disseminating the findings widely.

Property refers to society's granting private producers of new knowledge exclusive rights to the use of their creations, thereby forming conditions for the existence of markets in intellectual property and enabling the originators to collect fees for the use of their work by others. The specific legal contrivances of the patent, copyright, and somewhat more problematically, the trade secret fall within the property rubric.

The Intellectual Property System

Patents convey the most potent rights in the intellectual property system, for the patentee may exclude everyone else from making, selling, or using the subject matter of a valid patent throughout its term. Under the current U.S. Patent Act (35 U.S.C., Sec. 1-376), the usual term is 17 years, extendable by 5 years for pharmaceutical and medical device patents and by 14 years for design patents. The conditions that must be satisfied to secure the award are also the most stringent (see, e.g., Chisum, 1989). In addition to being potentially useful to society, a patentable invention must pass three tests: originality (originating with the inventor in question), novelty (not having been invented independently by another), and nonobviousness (not already obvious to a person having ordinary skill in the pertinent "art").

Copyright, as defined under the terms of the U.S. Copyright Act (17 U.S.C., Sec. 101-801, 1982), subsists in an "original" work of authorship fixed in a "tangible medium of expression" from which it can be perceived. The boundary separating copyrights from patents is usually seen by intellectual property lawyers in the restriction of copyright to protecting its holder against the copying of the specific expression, but not against others' use of "the idea, procedure, process, system, method of operation, concept, principle, or discovery, regardless of the form in which it is described, explained, illustrated, or embodied in such work" (17 U.S.C., Sec. 102(b), 1982; quoted in Bender, 1986:920). Novelty, however, is not a requirement, and although statutory protection is provided against copying and other enumerated acts, independent origination is not precluded—more than one author can copyright identical works. Nor is unauthorized reproduction restrained, so long as it is not deemed to have significant adverse effects on the copyright holder's current or future economic interests—as when a work is copied for the purpose of scholarly study or quoted in part in other copyrighted material under the doctrine of "fair use."

Some further points of contrast between patents and copyrights derive from the absence of the novelty requirement. Works can be registered and deposited at the Copyright Office, but the owner's rights under the U.S. Copyright Act—except for the right to injunctions against infringers—exist independently of any formal registration, prior examination, or determination of the validity of the claim to originality. Thus, the scope of copyright protection ultimately must be defined through litigation. Counterbalancing the more restricted nature of the rights conveyed by copyright law, the term of protection provided is much longer than that for patents. In most circumstances, a copyright expires 50 years after the death of the creator, a convention that is now quite standard internationally.

Trade secrets are included here under the property rubric. Technological developments and the recent history of litigation have brought the trade secret closer to the patent and the copyright in its form (as law for the protection of valuable rights in information) and its ostensible social function (strengthening private incentives for R&D expenditures and abetting contractual arrangements for the limited sharing of technological information). However, regarding trade secrets as another type of intellectual property is somewhat problematic. Information that is kept secret can be a source of income, and as a valuable asset, it shares a quality that economists would automatically associate with other forms of tangible and intangible property. Nevertheless, all economically valuable secrets are not afforded protection as trade secrets. Under the American Law Institute's widely cited (1939) definition (see Cheung, 1982:42; Bender, 1986:915),

> A trade secret may consist of any formula, pattern device or compilation of information which is used in one's business, and which gives him an opportunity to obtain an advantage over competitors who do not know or use it. . . . It differs from other secret information in a business . . . in that it is not simply information as to single or ephemeral events. . . . A trade secret is a process or device for continuous use in the operation of the business.

How is it possible for society to recognize and enforce a person's claim to hold property in something that is quite specific but must remain less than fully described and actually be hidden from public view? A further problem is that in the law of real and personal property, as well as in the areas of intellectual property law dealing with patents and copyrights, designating something as "property" has a particular meaning. It usually means that the possessor has the exclusive right to use or enjoy the thing, or to assign it to others for their exclusive use or enjoyment (see, e.g., Friedman et al., 1991:61-62). This special sense of the term is not satisfied in the case of a trade secret, because even when the possessor has taken measures to preserve its secrecy, the law provides no remedy if the information is disclosed by accident or uncovered through deliberate, socially conscionable ("fair") actions of others.

Unlike patent and copyright law, trade secret law (even when given statutory structure) is rooted in principles of common law, including theories of contract and tort, as well as property concepts (see Jager, 1991:49). Indeed, the general tendency is to deemphasize rights to property in the information held secret and to protect its originators indirectly by enforcing relationships of confidentiality that have been established implicitly or through explicit contracts. Trade secrecy, thus, can be viewed as a means of increasing the security of the "default option." It offers a recourse that may be socially as well as privately valuable in circumstances in which patents and copyrights are unavailable, ineffectual, or unattractive means of appropriating the economic benefits deriving from the generation of new knowledge and its reduction to concrete practices. The relevant legacy from the common law of master-servant relations is the recognition of society's interest in the formation of relationships of trust between employers and their employees (and between principals and their agents). Because in many instances it is much more efficient, and in some circumstances absolutely essential, to give employees access to information that they could use to the disadvantage of their employer, trust in the confidentiality of such disclosures is desirable for all who would benefit from having the work done. Yet the original common law contexts, typically, were ones in which it was the master who had a valuable secret to safeguard. Given these historical derivations, it is perhaps not so surprising today to hear employed inventors complaining that the law of trade secrets works to the benefit of their employers and leaves their interests largely unprotected.

Trade-offs in Organizing Knowledge Production

Although the attention of industry, the legal profession, and the wider community of researchers has been focused increasingly on the property mode of organizing knowledge production, the three modes must be kept in mind in formulating effective science and technology policies. Each of the three allocative mechanisms has been found useful in some fields and at some periods in the development of modern industrial societies; but the weight of reliance has shifted among them over time, and none has been accepted as clearly superior to the others in all contexts in which useful knowledge has been sought. Even the brief economic analysis that follows readily exposes some serious drawbacks to, as well as the principal advantages of, each of the three arrangements. (For a fuller presentation of the following analysis, see, for example, Wright, 1983; Dasgupta and David, 1988.)

Theoretically, at least, prizes and research grants can be established, or procurement contracts written for amounts that would award the producers of original intellectual works commensurately with the anticipated social use-value of their creations. As a practical matter, however, the patronage and procurement solutions are burdened by the fact that public authorities generally cannot set efficient terms for prizes in advance of their results. Moreover, in markets for which the expertise about the likely costs and benefits of particular research projects is unevenly distributed, contracting will entail high transaction costs even to arrive at rather imperfect agreements. The intellectual property solution avoids these drawbacks by letting the workings of the market determine the economic rewards after the fact. Thus, the avoidance of administrative arbitrariness in awarding prizes or granting subsidies for invention has been recognized as an advantage of the patent system by economists since Adam Smith (see Jurisprudence, A.ii:31-33, cited in Smith, 1776:754, n. 69).

To secure the benefits of a rapidly accumulating stock of knowledge, it is desirable to promote speedy disclosure of new findings so that they may be disseminated, verified by replication, and put to use by others engaged in intellectual pursuits. Only in this way can the fullest scope be provided for the interactive process through which ideas proliferate and generate still more ideas. When directed toward that goal, the patronage and intellectual property systems must seek full and prompt disclosure. Each, therefore, is impelled to base the assignment of rewards in some way on the establishment of *priority* (see Dasgupta and David, 1987 and 1988). Patronage achieves this by offering prizes for discoveries or inventions of a specified kind and by awarding research grants to those who have developed reputations for research success. Among academic scientists in the modern West, such reputations as a rule are based on validated claims to priority in discoveries and inventions deemed by expert peer groups to constitute useful "contributions" to knowledge. Priority is equally central to patent awards, although whether the touchstone is priority of invention or priority of registration and disclosure varies from one national system to another. Also, statutory copyright protections are traditionally accorded to the first author to disclose (by registration with the copyright-granting authority) a particular creative "expression."

A common implication of priority-based reward systems is that they give rise to competitions—among reputation-seeking scientists and patent-seeking inventors—that are characterized by a payoff structure in which the "winner takes all," or substantially all. (This pertains to the awarding of patent rights in cases involving rivalries among inventors, but it does not imply that the original recipient of a patent automatically captures all, or even the lion's share, of the economic gains deriving from a successful invention.) As a consequence, it is likely that from the viewpoint of society there will be too many contestants in the races for priority in discovery and invention. Those entering consider only what they individually stand to gain, and they do not take into account the effect of their participation on the expected outcomes for all the other competitors. The situation resembles the inefficiencies in resource allocations that arise when there is a "common pool problem" (see, for example, Dasgupta and Stiglitz, 1980; Wright, 1983, and references therein).

Further, in addition to the crowding of the field with contestants who might be more usefully engaged in other productive pursuits, there is a tendency for private rents to be dissipated in the scramble for the prize of priority and all that it would bring. The private value of arriving at a new finding a little sooner than the second-place contestant is likely to exceed greatly the benefit that society would derive from the slight advance in the date of discovery or invention. Such allocative inefficiencies, most probably, are more serious in regard to patents than to copyrights. Problems of "racing" do not arise in regard to trade secrets, but when the law of trade secrets is used to capture the value of new discoveries and inventions, inefficiencies in resource allocation will arise from sources other than the common pool problem, and this is likely to be even more severe, for reasons explained below.

In principle, the common pool problem of excess inventive effort could be avoided under a contract research system if the procurers were as fully informed as the researchers about the likely costs and potential social value of the findings. Under such conditions a contract could be drawn up that would provide a single, successful research entity with an economic payoff that just matched the (certain) value of a patent monopoly that had been awarded for whatever duration was deemed socially optimal. As noted above, however, it is most probable that the writing of such a contract would be frustrated by the asymmetrical distribution of information between researchers and would-be customers for research results.

If the common pool problem creates a tendency toward excess investment of R&D funds in racing for patents, adjusting the terms of patent awards would seem to be a reasonably straightforward remedy. By shortening the life of the patent or narrowing the protection granted against infringements, the value of the anticipated prize could be lowered until private R&D investment was no longer socially excessive. Even with that problem fixed, however, the resulting allocation of resources would still be inefficient because the property solution, unlike patronage and procurement, inherently entails withholding access to the new knowledge itself or restricting the extent of its application by imposing license and royalty charges on the users. Three modes of use of the new knowledge may thus be curtailed: (1) it may have been the basis for the design of a new good or service that would enhance consumer satisfaction by increasing variety or offering superior quality; (2) it may have been the basis for a cost-saving production process; (3) it may have been an intermediate input in the production of further advances in knowledge.

By long-standing tradition, economists' critiques of the legal protection of intellectual property have focused primarily on the losses in economic welfare caused by the establishment of a temporary (but for copyrights, rather prolonged) monopoly of the application of information in modes 1 and 2 above (see, e.g., Plant, 1934, 1974). Exclusive possession of technological knowledge by a profit-seeking agent will restrict the extent to which that knowledge is applied for the production of commodities that embody the innovation or can be fashioned more cheaply by processes based on it. The more secure the possession is, the less the patent monopolist or copyright holder has to worry that charging a high royalty rate will induce others to seek to avoid paying it by investing in reverse engineering, "inventing around," or closely imitating his or her creation. With less risk of the entry of close substitutes to worry about, a royalty income can be extracted from the final customers by setting prices above the marginal costs of production and restricting output accordingly. The benefits of the new knowledge to society, and to consumers in particular, are thereby less than they would have been had the information been made available for exploitation on a competitive basis. Such lost benefits are referred to by economists as the "deadweight burden" of the patent monopoly.

More recently, however, growing attention to the allocation of public and private funds for R&D, and the concomitant recognition of the importance of scientific and technological advances as intermediate informational inputs into the R&D process itself, have added a new set of worries to the traditional concern with the deadweight burden of monopoly. These worries involve the adverse efficiency consequences of restrictions that intellectual property rights systems—unlike patronage and procurement arrangements—impose on access to information on research methods and results. Secrecy raises the costs to researchers and to society of the search for new knowledge. Because scientific and technological discoveries and inventions so often build on previous knowledge, many economically significant research developments (such as a commercially practical microprocessor chip) can be conceptualized as having entailed the successful solution of a large set of interrelated problems for which there exists a least-cost solution sequence. More realistically, there will be a number of such solution sequences that are substantially more efficient, in the sense of requiring less time or fewer resources, than others. Many firms with varying research capabilities can be engaged in trying to solve some or all of these problems. Because each step along the way represents a distinct "result," the economic benefits of which are likely to be worth appropriating by one intellectual property protection device or another, the payoff to priority tends to result in each firm's shrouding its efforts in every stage of the sequence under a cloak of secrecy. This is true even when the intention is ultimately to disclose a result, once it has been secured and found to be protectable under patent or copyright law.

Maintenance of secrecy by rival firms, however, makes virtually impossible the coordination of research activities required to achieve the optimal sequence. Some simulation studies by Folster (1985) suggest that the resulting losses due to wasteful duplication and delay can be very much larger than the excessive R&D expenditures attributable simply to the common pool problem. In other words, the lack of communication and coordination among the contestants in patent races would seem to be a more serious matter than the fact that too many contestants were induced to enter the race(s) to begin with. The greater the incentive is for firms to proceed with their R&D programs in complete secrecy, the more severe this source of inefficiency is likely to become. Here again, within the category of property devices for organizing the production of knowledge, patent and copyright protections possess comparative virtues (of intermediate-stage disclosure) that are lost to society when firms elect to rely on trade secrecy laws and attempt to appropriate the economic benefits of their inventions by embedding them in new goods that they can sell.

If trade secrecy per se has serious drawbacks and if, as has been pointed out, the adverse effects of the common pool problems associated with patenting and copyrighting can be mitigated by adjusting the terms of those property awards, does the foregoing analysis imply that there is not really much of a socially useful role remaining for trade secrecy law? There has not been much opportunity for the development of a consensus on this point, since economic analysis of the legal protection of trade secrets is far less developed than that of the patent and copyright systems (see, as exceptions, Kitch, 1980; Cheung, 1982; Friedman et al., 1991). This may be due, in part, to the fact that patents and copyrights sprang from statutory enactments, which often ignite political debate and public discussion, whereas the protection afforded to possessors of trade secrets by the courts is rooted in the common law.

Optimizing Intellectual Property Protection: Issues of Length and Breadth

More than 30 years ago, Fritz Machlup (1958:80) remarked that although economic analysis did not yet provide a basis for choosing between "all or nothing" where intellectual property protection is concerned, "it does provide a sufficiently firm basis for decisions about 'a little more or a little less' of various ingredients of the patent system." For some lawyers and practical policymakers, this was perhaps too self-congratulatory an appraisal (see, e.g., Priest, 1986). Nevertheless, a number of economists subsequently followed Machlup's advice and examined the question of the optimal duration and scope, or length and breadth, of patent protection, taking as given the existence of the patent system. A similar approach has been taken in examining the economics of copying and the optimal level of copyright protection.

Length of Protection

Instead of accepting the historically given length of patent protection, Nordhaus (1969) and Scherer (1972) compared the size of the incentive effect on invention with the resulting inefficiency due to the deadweight burden of monopoly. By balancing the one against the other, they showed how the optimal patent length would change with market demand and technological factors. Three principal conclusions emerged with respect to cost-reducing process inventions. First, the optimal length of patent life is probably not uniform across industries and inventions because it is sensitive to (1) the price elasticity of demand in the end-product market and (2) the responsiveness of the costs characterizing the new production process to the amount of R&D resources devoted to its invention. Second, the more elastic the demand for the new product, the shorter is the optimal length of patent protection because higher prices will cause a proportionately larger reduction of the quantity demanded, and consequently a greater "deadweight burden." Third, the optimal length of patent protection will be shorter when the technological opportunities being explored are such that greater production cost savings can be achieved with given levels of R&D expenditures. Nordhaus's (1969) work carried a fourth implication: The welfare losses resulting from setting the patent life at a suboptimal length are not very substantial except when major technological advances (cost reductions) are obtainable and when the market demand is very price elastic.

Formal analyses of the economics of copying and copyright protection have tended to follow in Nordhaus's (1969) footsteps. Hirschliefer and Riley (1979), for example, evaluate the impact of increasing copyright protection by comparing the benefits from reducing losses due to underproduction of new works to the costs incurred in the form of losses due to underutilization of copyrighted material. The conclusions derived from this approach, however, are rather more ambiguous than for the analysis of patents. The reason is that the analysis turns on one's assumptions about the substitutability of demand between, and the comparative costs of, unauthorized copies and copies produced under copyright agreement.

More stringent copyright protection would decrease the social loss due to the underproduction of intellectual works, unless copyright monopolists raised the price of their products so much that consumers increased the demand for unauthorized copies. More stringent protection might also reduce underutilization losses if obtaining an unauthorized copy cost consumers more than they would be charged by a copyright holder who had a strict, enforceable monopoly (Novos and Waldman, 1984). In the same spirit, Johnson (1985) concludes that strengthening copyright protection could enhance social welfare even without stimulating the production of new works of authorship, so long as lax restraints on copying resulted in the demand for authorized copies ("originals") being reduced greatly in relation to total consumption of the work in question. These conclusions rest crucially on the supposition that the private cost to the consumer of obtaining a close substitute by copying an authorized "original" is greater than the copyright monopolist's marginal costs. However, as Liebowitz (1985) has pointed out, the latter assumption has in many situations been invalidated by advances in copying technologies. Indeed, he suggests that the complementarity in production between authorized originals and low-cost copies could, under some conditions, mean that Johnson's (1985) appraisal was vitiated. Instead, a more permissive law regarding copying might—by allowing utilization of highly efficient copying technology—actually increase the effective demand for originals as well.

Breadth of Protection

In another branch of the literature, the policy variable of interest has been the patent's optimal breadth, or scope, of protection. Continuing in the analytic tradition of Nordhaus (1969), a number of more recent works, including some that treat the problem of optimal patent breadth and length simultaneously, focus on the balancing of incentive effects on invention against resulting inefficiencies due to underutilization.

Taking the flow rate of profit available to the patentee as a proxy for the breadth of a patent, Gilbert and Shapiro (1990) have derived conditions under which the optimal patent length is infinite: If total (consumer and producer) surplus declines at an increasing rate as patent breadth increases, it is optimal to grant an infinitely lived patent and adjust the breadth to ensure that the patentee makes at least zero net profits. As Gilbert and Shapiro would readily admit, however, the flow rate of profit available to the patentee is an unsatisfactory proxy for patent breadth because it fails to show how the given innovation is related to other innovations. They acknowledge that their static analysis ignores the fact that inventions build on each other and that a long patent grant may have "deleterious effects on the incentives of other firms to engage in related research, for fear that they will be at the mercy of the original patentee" (Gilbert and Shapiro, 1990:112). Klemperer (1990) considered the problem of optimal patent breadth in a static setting using a model of horizontal product differentiation, which ignores the possibility of vertical product improvement. The proxy for patent breadth in his model is the region of the product space covered by the patent grant. Correspondingly, two kinds of welfare losses have to be considered: (1) those caused by consumers' switching to less preferred varieties of the product that are unpatented and sold at competitive prices and (2) those caused by consumers' dropping out of the product class altogether.

Schmitz (1989) provides a more dynamic analysis of the trade-offs involved in broadening the breadth of patent protection. As a proxy variable for patent breadth, he proposes the anticipated probability of infringement. The economic importance of a particular line of research is represented by the absolute size of an increase in total surplus due to initial and secondary product development. When this parameter is increased, the relative contribution of those R&D efforts to the gain in total surplus remains unchanged, but the absolute change in the total surplus is increased. Consequently, the cost of invalidating subsequent innovations increases, and the optimal infringement probability—that is, the breadth of the patent coverage— should be reduced. The more important a particular line of research, the less subsequent developers should be constrained by the original patent claim, according to this analysis (Schmitz, 1989:4).

Kitch's (1977) "prospect theory" of patent protection and Beck's (1981, 1983) discussion of "unproductive competition" are two of the most significant early contributions to the analysis of patent breadth. An important feature of their work was the emphasis they placed on inefficiencies that might be present in the R&D process itself. They identified the source of inefficiencies in the allocation of resources to R&D not simply with what Dasgupta and Stiglitz (1980) called the common pool problem, but also with "unproductive competition for monopoly profit." Implicitly, however, the two problems are connected. Kitch and Beck's unproductive competition for patent monopoly encompassed wastage of resources on premature invention, duplicative R&D, unnecessary substitute inventions ("me-too" patents), and excessively rapid spending on research. These forms of inefficiency are clearly related to the previously discussed problem of lack of interfirm coordination or socially optimal scheduling of R&D projects, which stems from the competitive conduct of research under conditions of secrecy and from "patent racing."

As a solution to this problem, Kitch and Beck proposed to broaden the scope of patent protection to allow the rationalization of the entire development process for a given technological "prospect." Drawing an analogy to mineral resource development, Kitch and Beck have argued that by allowing the "competent" initial innovator to coordinate the subsequent development of "a technological prospect" through *efficient* bilateral monopoly contracts with other innovators, broader protection would eliminate duplicative effort, premature invention, and other forms of inefficiency in a competitive race for patent monopoly. Although they considered the possibility that the patent holder would contract with independent researchers, they implicitly assumed that the patent holder was a private firm, not a governmental procurement agency. Beck (1983:207) even proposed a competitive bidding scheme for future patents, which would be designed to "transfer the expected value of the patent owner's economic rent to the Patent Office, thus removing the economic incentive for unproductive competition."

Although this theory assumes that efficient bilateral monopoly contracts can be signed between the patent holder and independent innovators, the transaction costs in this process are likely to be nonnegligible. The entire argument projects a vision of organized and orderly development of technological prospects, the realization of which is problematic, to say the least. A major obstacle is the problem of asymmetric information and "thin markets" for specialized research capabilities, which are likely to make bargaining between the would-be "developer" and independent contractor-innovators very inefficient. In addition, although the "prospect theory" approach to patents addresses the issue of inefficiency in the development of an area of technology opened by a "breakthrough" patent, the proposed solution of broadening the scope of such patents would, except under the imagined auction scheme, further intensify the winner-takes-all nature of the payoff structure, thereby exacerbating the common pool problem.

There are circumstances, nonetheless, in which faster technological advance and consequent welfare improvement might be obtained by using a monopoly to "internalize" the process that generates innovations. David and Olsen (1989, 1991) point out that when private production of a durable good is subject to learning by doing that generates process improvements that cannot be protected as trade secrets, an industry composed of competitive suppliers may perform suboptimally from a social welfare standpoint. If the industry in question is supplying a new technology embodied in a machine or other producer good whose adoption depended on progressive reduction of its supply price, the diffusion into use and incremental improvement of the new technology would be affected adversely. The firms would move along their learning curves more slowly than is socially optimal, because they are not able to capture the benefits of future cost reductions that are a by-product of gaining more production experience. The price of the industry's product, therefore, also would fall more slowly. By correcting the externality, the grant of a monopoly franchise for production of the new good could lead to a second-best welfare optimum even when there was no prospect of future inventions being induced by the promise of patent rights.

In some respects, the above considerations resemble the concern for the efficiency of the technology development process that motivated the analyses by Kitch (1977) and Beck (1981, 1983), but this aspect of similarity should not be overstated. The monopoly franchise envisaged by David and Olsen's analysis would be designed solely to optimize the rate of incremental improvements and the resulting diffusion of the technology into use, without regard for the possibility of inducing some future breakthrough that would introduce yet another new technology. The situation is one in which patent protection would be granted not for invention but for the introduction and adaptation of a basic invention that had already been developed elsewhere. (As discussed below, these were precisely the historical circumstances in which patent grants were first used.)

To analyze further this facet of the patent system, David and Olsen (1991) develop a formal model of interdependent diffusion and learning and show that creation of a monopoly franchise may lead to an overall gain in social welfare, depending on the exact form of the learning function and on conditions governing the demand for the new product. Other things being equal, stronger learning effects at low levels of production experience tend to strengthen the case for granting a patent to the local "innovator" (not necessarily the inventor) of a technology that has yet to be brought into use. Because a learning monopolist will want to produce at a higher level than a competitively organized industry would, there will be a gain in social benefits from accelerated diffusion of the new product and reduction of its production costs. On the other hand, the learning monopolist will want to stop producing when the new good is less extensively diffused than it would be under conditions of competitive supply, which would entail some welfare loss. David and Olsen show that the point at which it is privately optimal for the monopolist to cease producing will always occur prior to the end of the time for which the exclusive patent right has been granted. Indeed, the best strategy for the monopolist may be to shut down production long before the patent/franchise expires and competitors are free to make use of the knowledge gained through production experience. Permitting this lull, in a sense, is just the social cost of using monopoly to correct the problems caused by the externalities in learning by doing.

Because the expiration date of the franchise enters into the monopolist's determination of the date on which to suspend production (and hence affects the extent of "cumulative underproduction" by the monopolist), there will exist some optimal finite length for a patent franchise that is not intended to stimulate future patent filings. Under the conditions assumed in the David and Olsen (1991) model, the optimum franchise duration is shorter when the interest rate is lower, when initial cost reductions due to learning are less drastic, and when the distribution of willingness to pay among the potential purchasers of the new commodity is more skewed toward high values. Yet even when the optimal franchise duration is brief, the associated improvement of welfare over the competitive supply alternative can be quite large.

Although temporary monopolies may provide a simple way to fix the knowledge spillover ("learning externalities") problems that cause competitive markets to generate too slow a pace of technological advance, it will not be a "first best" remedy. Indeed, monopoly cannot be looked to as the most efficient market structure from the standpoint of stimulating product innovations. Arrow (1962) pointed out long ago that an entrenched patent monopolist would have weaker incentives than a would-be entrant to engage in an R&D program that would yield substitutes, even superior substitutes, for goods that already were profit-generating items in the product line. More recently, Merges and Nelson (1990:5-6) persuasively formulated an analogous case, although in broad terms and without invoking results from any formally specified model, for restricting the breadth of patent protection. They argue that even though competitive investments in R&D can result in inefficiencies, technological development tends to proceed "much more vigorously and creatively under a regime where there are many rivalrous sources of invention, than in a setting where one or a few organizations control developments."

Although a satisfactory mathematical characterization of the process of cumulative and interactive product innovation has yet to be developed, recent models have focused on the potentially adverse impact that granting broader patents may have on the pace of technological advance. This work raises important trade-off issues that have been neglected for too long by the theoretical economics literature. The work of Scotchmer and Green

(1990) and Scotchmer (1991) examines the implications of the point that although broader protection provides stronger incentives for R&D aimed at achieving breakthrough inventions, it may seriously weaken the incentives for second-generation innovators to elaborate and improve on the work of the pioneers. Scotchmer and Green implicitly dismiss as infeasible the sort of integration and internalization of the whole line of development envisaged by the prospect theory of Kitch (1977) and Beck (1981, 1983). In their analysis, affording a broad scope of protection to the first patentee puts followers at a disadvantage in negotiating the terms of licenses for technological elements complementary to those they themselves will seek to patent or imposes on them the added costs of trying to "invent around" the blocking, first-generation patent. Many of the cumulative welfare gains attributable to breakthrough technological advances, however, derive precisely from the latter category of incremental, follow-on inventions.

Achieving socially optimal patent breadth is thus a matter of striking the best balance between the net rewards offered to inventors in the first- and the second-generation categories. One implication that would seem to follow is that as basic scientific advances reduce the costs of successfully inventing around breakthrough patents in a particular technological area, the breadth of patent protection awarded the pioneers could be increased without diminishing the net incentives that would exist for derivative, second-generation R&D projects.

Essentially the same considerations that arise from recognizing that new scientific and technological knowledge (and intellectual products more generally) spurs the further production of knowledge also arise, in principle, in regard to the protection of copyrights. This point forms a central feature of the analysis by Landes and Posner (1989:335), that is, "too much protection can raise the costs of creation for subsequent authors to the point where those authors cannot cover them even though they have complete copyright protection for their own originality [of expression]." The net effect of increased copyright protection on the supply of (equivalent) works thus depends on the balance between the encouraging incentive effects (for authors and publishers combined) and the discouraging effects of "driving up the cost of expression." Under these conditions, Landes and Posner (1989:344) find that the more the cost of expression rises with increases in the level of protection, the lower is the optimal degree of copyright protection. A social welfare rationale is thereby suggested for leniency in infringement proceedings, that is, permitting more extensive use of copyrighted material to create new derivative works and maintaining broader protection against literal copying.

INTELLECTUAL PROPERTY LAW AND SOME LEGACIES OF HISTORY

After a review like the one just completed, there is a temptation to issue an overall evaluation. In regard to intellectual property protections, however, a great array of benefits and costs cannot be quantified readily. Thus, it is difficult to do more than confirm the observation of Fritz Machlup (1958:80) in regard to the patent system:

> If one does not know whether a system "as a whole" (in contrast to certain features of it) is good or bad, the safest "policy conclusion" is to "muddle through"— either with it, if one has long lived with it, or without it, if one has lived without it. If we did not have a patent system, it would be irresponsible, on the basis of our present knowledge of its economic consequences, to recommend instituting one. But since we have had a patent system for a long time, it would be irresponsible, on the basis of our present knowledge, to recommend abolishing it.

Modern economic analysis offers little to refute the conclusion that we who would use U.S. institutions of intellectual property protection to accomplish the purposes of a modern industrial society must remain "prisoners" of their particular history. It is all too easy to miss this central message in the intricate economic analysis characteristic of the literature just reviewed. By focusing selectively on specific features of the complex structure of intellectual property protections and pointing to their putatively favorable consequences for social efficiency in resource allocation, one can convey the misleading impression that the law in this area is susceptible to easy and rapid reshaping to enhance economic welfare.

The evolution of the law in Western societies for protecting intellectual property does attest to a great adaptive capacity. Nevertheless, whereas economic efficiency would seem to call for great subtlety and differentiation in the nature and degree of intellectual property protection provided, based on differences among industries in technological and market circumstances, that evidently is not a direction in which adaptation of the law has proceeded very far. Even more evident is the fact that, today, much uncertainty and controversy surround the persistent difficulties of adapting intellectual property law to new technologies. A major source of these difficulties are the problems of achieving the semblance of consistency in the application of legal principles, of preserving the force of precedent, and thereby circumscribing the remaining areas of ambiguity and uncertainty as to the ultimate enforceability of legal claims, and the likely costs of the entailed litigation. Unfortunately, the economist's conventional approach of evaluating specific institutional arrangements and policies in isolation does not naturally accommodate consideration of these sorts of systemic concerns, with which traditional legal scholars so often are occupied. As a consequence, the literature on the economics of patents, copyrights, and trade secrets rarely takes note of the problems that arise at the interfaces between those regions of the law. Neither has it paid much attention to the interrelationships and connections between intellectual property law and the larger body of property, tort, and contract law (however, see Cheung, 1982, as an exception).

Nevertheless, it is to a large extent for reasons of the sort that conventional economic analysis has tended to overlook the fact that legal institutions evolve incrementally. Legal institutions preserve many aspects of outward continuity even when it has become apparent that the circumstances of many of the economic actors affected by the institution have changed and that a radical transformation has occurred in the inner rationale and motivation for its maintenance. Thus, although the history of intellectual property rights in the West is replete with instances of redefinition and reinterpretation in response to pressures to accommodate or advance the economic interests of those most affected by the laws, many of the structure's gross features continue to reflect the remote historical circumstances in which they originated. These legacies from the past should not be ignored, nor should their problematic aspects in contemporary contexts be minimized. In persisting, they impinge on the search for new technologies and the organization of economic activities based on the exploitation of the resulting additions to the stock of knowledge.

Patents

Patents began as instruments used by noble or republican governments in later medieval and early Renaissance Europe primarily to induce the transfer and disclosure of foreign technologies. This bit of history calls into question one causal supposition that the basic economic analysis of the patent system has fostered, that is, that the protection of intellectual property has been instituted where governments recognized there was more to be gained by stimulating indigenous inventive activity than by applying knowledge of techniques and products that could be "borrowed" freely from the rest of the world.

Patent, the English adjective, means *open*, and the noun form comes from the term *letters patent* (a literal translation of the Latin *litterae patentes*), which means simply open letters. These were the official documents by which certain privileges, rights, ranks, or titles were conferred and publicly announced. Hence, they carried the seal of the sovereign grantor on the inside, rather than being closed by a seal on the outside (see Hill, 1924:406). The "openness" involved, thus, had nothing to do with disclosure of an invention—despite misapprehensions on this point that persist today (see, for example, Bugos and Kevles, 1991). Only much later did the granting of letters patent evolve into social contrivances for stimulating original invention.

Encouraging Technology Transfer

In the fourteenth century, patents were employed to encourage the introduction of foreign technologies through the immigration of skilled artisans from abroad. Letters patent were given, for example, to the Flemish weaver John Kempe by Edward II in 1331, to two Brabant weavers to settle at York in 1336, and to three clockmakers from Delft in 1368 (see Federico, 1929a:293-295). England at this time was technologically laggard in comparison with many regions on the continent of Europe and, understandably, was endeavoring to "borrow" the more advanced industrial practices. It was hoped that the foreign master craftsmen would introduce English apprentices to the "mysterie" of their respective arts. However, because the master was not likely to remain in control of the newly skilled workers once they acquired journeyman's status, he obviously wished to be protected against the cohort of potential domestic competitors he would create.

Many of the basic features of the patent are better suited to its initial purposes and historical contexts than to the subsequent use to which patents have been put. The disclosure provisions of modern patent systems, for example, were an essential and natural aspect of the effort to induce foreign artisans to reveal a "mysterie" and train domestic craftsmen in its pursuit. Making the conduct of the trade or craft—and the consequent training of apprentices and journeymen—a condition for the privilege conveyed by the patent was quite straightforward since that was the object of the patent. Protecting instructors from the competition of their students, by giving them a monopoly of the trade, directly addressed the spillover problem because there was no way those they trained were likely to benefit except by setting themselves up in competition as soon as they learned the "mysterie." Even the duration of early English patents—14 years, with 7-year extensions possible—was not fixed arbitrarily. Seven years was the term of service of an apprentice, so the protection afforded was to last at least for two generations of trainees. Inasmuch as 7 years was the conventional term of apprenticeship irrespective of the trade or craft, there was considerable logic to making the term of the patent award uniform across all branches of industry. (As has been pointed out, however, modern economic analysis finds this aspect of the contemporary patent system difficult to rationalize.)

Granting monopolies also made sense fiscally for sovereigns whose powers of taxation and borrowing were very circumscribed. It shifted the market risks to the foreign artisan and transferred to him also the bother of collecting the excise tax in the form of the markup over his production costs. Finally, there was no need to ascertain that the grantee had originated anything, only that at the time of the grant the practice was not being carried on, and hence could be presumed to be unknown, within the sovereign's domains. The criteria of originality, novelty, and nonobviousness that have emerged as definitions of what qualifies as an invention at the U.S. Patent Office, and elsewhere, might well be seen as the makeshift results of a 200-year struggle to use the granting of patent privileges to accomplish a purpose for which it was not originally designed (see Lubar, 1990).

Most historical accounts place the origins of systematic state protection of intellectual property firmly in Renaissance Italy, from where it spread first on the continent of Europe and eventually to England. In the fourteenth and early fifteenth centuries, however, the property rights in question typically took the form of grants for the exclusive exploitation of locally unfamiliar processes or devices that had been originated elsewhere, and more likely than not by individuals other than the one seeking the privilege. Venice took the lead in these developments. As early as 1332 the Venetian Grand Council established a privilege fund for providing loans and other rewards for a foreign constructor of windmills who offered to bring knowledge of this art to the city (see Prager, 1944:713). In 1416 the council awarded Franciscus Petri, from the island of Rhodes, a patent for a superior device for the fulling (shrinking and thickening) of fabrics, which gave Petri and his heirs exclusive rights for 50 years to build, alter, and reconstruct the apparatus he would erect for that purpose (see Mandich, 1958:115-116, 149-150; Prager, 1960:379; Long, 1991:877).

In this era the practice of granting *privilegi*, which was hardly confined to Venice, sought the revelation and application of "secrets"—whether of foreign provenance or native genius. When, in 1421, the Florentine commune awarded a patent to Brunelleschi for a new design of ship he claimed could haul loads more cheaply on the Arno River (to the benefit of merchants and others), the nature of the bargain for disclosure was spelled out candidly in Brunelleschi's petition (Prager, 1946:109-110):

> He refuses to make such machine available to the public in order that the fruit of his genius and skill may not be reaped by another without his will and consent, and that, if he enjoyed some prerogative concerning this, he would open up what he is hiding and would disclose it to all.

From about this time forward, the issue of patent privileges for various devices became increasingly frequent, and by 1460, the Venetian Senate in its administrative practice was differentiating between grants of exclusive monopoly to sell products incorporating an "invention," and awards that forbade use of the device without permission while obligating the holder to grant licenses to others when "reasonable royalties" were offered (see Kaufer, 1989:4). Technology importation continued to figure as a primary objective: in 1469 a German, Johann von Speyer, received an exclusive monopoly of the trade of printing in the Venetian domain in exchange for introducing the craft.

Protecting Intellectual Property

Most modern historical accounts of the development of intellectual property protection in the West assign great significance to the Venetian Senate's passage, on March 19, 1474, of the first general patent law. This is quite understandable given the correspondence between contemporary preoccupations with stimulating invention and innovation and the language of the famous preamble (translated by Gilfillan, 1964:11):

> We have among us men of great genius, apt to invent and discover ingenious devices . . . . Now, if provisions were made for the works and devices discovered by such persons, so that others who may see them could not build them and take the inventor's honor [sic] away, more men would then apply their genius, would discover, and would build devices of great utility to our commonwealth.

Yet, most authorities view this statute as having codified prior practice rather than enunciating any novel principle (see Frumkin, 1945; Prager, 1948; Phillips, 1982; Long, 1991). The law required the registration of any "new and ingenious" device not previously made within the Venetian domain, and it prohibited all private parties except the inventor from making it for 10 years, on pain of penalties for violation of the code. Further, it appears that between 1474 and 1490, very few patents actually were issued under the Venetian code, despite the fact that right through to the middle of the sixteenth century many patent *privilegi* continued to be granted, conferring exclusive production rights for terms varying between 5 and 80 years, as well as monopolistic trade privileges (see Kaufer, 1989:6).

Despite the rising interest in invention and the spread on the continent of Europe of the use of patent grants to encourage the development of new industrial practices as an instrument of mercantilist policy in France during the mid-sixteenth century, in England the first clear provision for "patents of invention"—as distinct from technology transfer franchises sometimes referred to as "import patents"—did not emerge until the seventeenth century—and it did so then rather as an afterthought, in the course of a movement to free the economy and polity from the abuses of royal grants of monopoly privileges.

With the advent of the Tudor dynasty (1485), the use of open letters as a means of encouraging national industry gave way to the negotiation by the Crown of secret agreements designed to attract skilled foreign artisans into its service. For example, German armorers, Italian shipwrights and glassmakers, and French ironworkers were enticed to cross the English Channel in this fashion. With Elizabeth I's accession to the throne (1558), however, the previous policy of general encouragement of technology transfers was reinstituted. Between 1561 and 1571, many patents were issued by the Crown under this policy, starting with a grant to two foreigners to introduce the manufacture of hard white Spanish soap and one for the manufacture of saltpeter, an item previously imported from Antwerp (see Federico, 1929a:293-297). The royal prerogative of awarding monopolies of all sorts was exercised so extensively on behalf of Court favorites and the Crown's fiscal needs, however, that by 1601 Elizabeth was compelled to promise reforms in order to deflect a parliamentary challenge to her authority in this regard. Nevertheless, this only deferred the conflict. The abuses and retaliatory efforts to curtail the royal prerogative increased under James I, until in 1623 Parliament passed the Statute of Monopolies, which declared all Crown monopolies, charters, and patents thereafter contrary to law. An exception was allowed, however, for royal patents conferring a monopoly for 14 years or less "to the first and true inventor" of a new manufacture (see Federico, 1929a:299).

It is on the above exemption that the British patent system and its derivatives elsewhere have been erected. Even so, the modern reading of the Statute of Monopolies as "the Magna Carta of the rights of inventors" (Machlup, 1958:2-3) is somewhat anachronistic. The verb *to invent* carried far more extensive connotations at that time than it does today. For example, in a famous patent for a pump, granted by James I to Robert Crumpe in 1618, the sense of *invent* included "bringing into use, find, establish or institute manufacture" (Hill, 1924:416). In short, originality of use in England alone might be a sufficient basis, since technology transfer, commercialization, and industrial development were also seen as worthy public purposes that could be served through the award of patent monopolies.

Creating a U.S. Patent System

Patent institutions in the United States were derived from those of Britain's North American colonies, dating back to early seventeenth century grants of an ad hoc nature that resembled import franchise contracts. The first such grant, awarded in 1620 by a general court of the Virginia Company's stockholders sitting in England, went to a Mr. Somerscalls for a tobacco-curing process that was not clearly an original invention (see Bugbee, 1967:58). In 1641 the General Court of Massachusetts Bay adopted a number of provisions, including one patterned on the Statute of Monopolies and its exemption, that created a statutory basis for granting future patents individually for "such new inventions that are profitable for the Countrie" (Bugbee, 1967:61). Importation of inventions from the Old World was a natural enough proposition for New World settlers. Thus, while British courts during the eighteenth century increasingly construed the purpose of patents to be the encouragement of indigenous invention, American courts continued to consider the potential utility of providing incentives for technology transfers. Moreover, even at a later stage, in respect to the conditions of economic "openness" and competition for mobile resources, the situation of the American colonies and their successor states under the Articles of Confederation resembled those of the city-states and principalities of Renaissance and early modern Europe. In an address to a joint meeting of Congress on January 8, 1790, President George Washington, who on previous occasions had concerned himself with the subject of intellectual property, called attention to various matters requiring legislative attention, such as "the advancement of agriculture, commerce, and manufactures, by all proper means," including "giving effectual encouragement, as well to the introduction of new and useful inventions from abroad, as to the exertions of skill and genius in producing them at home."

The constitutional era ushered in a decisive shift toward preoccupation with protecting national inventive and literary activities, swept away the disparities of treatment that had arisen among the former colonies, and cemented into the structure of federal law the distinctions between patent and copyright protection that today are taken to be fundamental. Despite the considerable attention to patent-related policy issues in the American colonies during the latter seventeenth and the eighteenth centuries up to the hiatus during the Revolution itself, the first systematized patent provision in America emerged only in 1784, and then as a footnote to the copyright provisions in South Carolina's Act for the Encouragement of Arts and Sciences. This statute's purpose was to establish literary property protection for a renewable 14-year term, but it included the following interesting rider (Bugbee, 1967:93):

> The Inventors of useful machines shall have a like exclusive privilege of making or vending their machines for the like term of 14 years, under the same privileges and restrictions hereby granted to, and imposed on, the authors of books.

What makes this provision rather intriguing today is that it so closely coupled patent protection with copyright protection, assigning the former as most appropriate to "machines" and the latter to "books," but otherwise barely distinguishing the treatment of the one from the other. The language adopted by the Constitutional Convention in 1787 was influenced strongly by previous state laws and so spoke also of securing exclusive rights for "Authors and Inventors" to "Promote the Progress of Science and Useful Arts." Copyrights and patents for invention were not mentioned explicitly, nor were import franchises explicitly rejected, as the means for accomplishing this purpose.

The formal creation of a patent system during the early days of America's nationhood was thus shaped most strongly by the experience of the former British colonies and was little influenced (except in rhetoric) by the actions of their revolutionary French contemporaries. The law of 1791 that formally established a patent system in France continued the practices of *l'ancien régime*, under which inventors received royal privileges freeing them to exploit their inventions outside the confines of existing guild controls. What the French law rejected was the legal justification of the practice based on the assertion of royal prerogative. The new, revolutionary dispensation provided, instead, for the issuance of *brevets d'inventions* (commission, or equivalently, patents of invention) grounded in the "natural rights" of citizens to the fruits of their creative genius (see, for example, Hilaire-Perez, 1991; MacLeod, 1991:889-891). Americans were not so ready to accept this rationale in place of the rather different English legal theory with which they had grown up, however sympathetic in other respects they might be to the French children of the Enlightenment. Recall how disparaging even Thomas Jefferson was of the argument for natural rights in intellectual property on behalf of authors and inventors.

The U.S. Senate complied with Washington's recommendation in his address of January 8, 1790, by appointing a committee charged with considering provisions for the granting of technology importation franchises, patents for invention, and copyright protection, all within a single act. Only the latter two provisions, however, emerged from the congressional deliberations of 1790-1791 (see Bugbee, 1967:125-148). Indeed, the response of the legislators to mounting pressure for grants of copyright led to the rapid passage of the Copyright Act (1790) first, which then made it necessary to pass a separate Patent Act in the following year, thereby creating two distinct statutory bases for intellectual property protection in federal law. It is the perpetuation of this legal separation—one body of law having developed to protect inventors of "machines" and the other to protect the authors of "texts"—that causes contemporary difficulties when new technologies are found not to fit neatly into either mold. Computer software, for instance, has posed awkward problems inasmuch as this class of technology is well described as "machines which are implemented in the form of text" (for further discussion, see Samuelson, 1984, and Chapter 12 in this volume). This is not to suggest that the separation between patent rights and copyrights that developed in U.S. law sprang simply from the accidents of the legislative history of the first Federalist administration. Quite the contrary. The readiness of members of Congress to deal separately with petitions for grants of copyright, as state legislatures before them had done, reflected the long antecedent evolution of the law of copyright.

Copyrights

In its late medieval origins, as noted, the copyright privilege had nothing to do with the encouragement of intellectual creativity or originality of expression. Indeed, the very notion of claiming originality of authorship was a Renaissance departure from the scholastic tradition of seeking to cloak one's own ideas with the authority of Aristotle and the other ancients (see, for example, Long, 1991). Rights of literary property involving *published* works remained legally unprotected in Europe until the fifteenth century, when the introduction of the printing press made the rewards of publishing, or plagiarism, far greater than ever before. The new technology of printing also transformed the economics of the copying business by substantially increasing the disparity between the cost of the first (printed) copy and the unit cost of subsequent copies. Copyright law, from the beginning, has been shaped more by the economics of publication than by the economics of authorship (see Patterson, 1968; Plant, 1974:Ch.4).

Like the earliest patents of invention, the first known copyrights appeared in Renaissance Italy. By the end of the 1460s the craft of printing had been introduced in Rome and Venice, and, with the issuance during 1469-1517 of a series of privileges relating to books and printing by the Venetian Cabinet, Senate, and other governmental bodies, Venice quickly assumed the lead in Italian printing. These *privilegi* included importation franchises, the first of which (1469), as noted, awarded the German printer Johann von Speyer the exclusive privilege of conducting all printing in the city for five years in return for establishing the craft (see Prager, 1944:715). There soon followed monopolies in the form of exclusive licenses to print or sell an entire class of books for a stipulated term, prohibitions of the importation of books printed abroad, and patents for the improvement of printing and typography (see Bugbee, 1967:43-44). The question of rights of authorship was largely disregarded because much of the demand was for extant works (such as the Bible) that were in the public domain and whose authors, even when identified, were long since dead.

Toward the end of the century, however, some privileges were awarded for the protection of authors that did have the character of modern copyrights: In 1486 the historiographer of the Republic was granted exclusive control over the publication of his work. In 1493 the Venetian Cabinet gave Daniele Barbaro an exclusive 10-year grant of proprietary rights to the publication of a book authored by his deceased brother (Bugbee, 1967:45). More typical were the copyrights issued to editors and publishers for individual works written by others; these were petty monopolies prohibiting publication of the work without permission of the grantee. Publishers were soon flocking to the government to reserve well-known titles for themselves, in the hope of either publishing themselves or selling the right later to another printer. By 1517 the resulting shortage of available titles caused the Senate to restrict all such copyright *privilegi* henceforth to "new and previously unprinted works."

What was probably the first general copyright law in the world came in the form of a decree issued by the Council of Ten in Venice (1544-1545) that prohibited the printing of any work unless written permission from the author or his immediate heirs had been submitted to the Commissioners of the University of Padua. No provision was made, however, for maintaining a register of protected works (Bugbee, 1967:46). This decree was prompted by the continued unauthorized printing of works for which copyrights had been granted. A further measure directed toward more complete regulation of the printing business came in 1548-1549 with a Council decree establishing a guild into which all Venice's printers and booksellers were to be organized. An added motivation was to assist the Church in suppressing heretical literature. The same concern with censorship of a potentially dangerous new medium of communication, rather than securing the rights of authorship, prompted the royal officials of sixteenth century France to issue licenses, or privileges, for the publication of acceptable books. The French Crown, however, proved better able than the Italian city-states to resist the Church's efforts to share control of the printing business.

In the Netherlands, privileges resembling those of Venice, but without censorship provisions, were issued to publishers by state and central governments, but the primary means of regulating destructive competition involving the pirating of texts was a system of informal noninterference agreements among Dutch printers. Similar arrangements had developed among leading German publishers and were exercised through a guild and the book fairs of Frankfurt and Leipzig (see Bugbee, 1967:48). When the German book trade was interrupted during the Thirty Years' War, the Dutch quickly assumed leadership of the publishing industry in Europe. Although the flourishing printing business of the Netherlands benefited from the attraction of scholars to the comparatively free intellectual atmosphere of the Dutch towns in this era, protection of local authors' rights was not a concern, nor were the rights of foreign authors and publishers. At this time, throughout Europe, imported books, pamphlets, and pictorial material were subject to reprinting and sale without compensation for their originators. The highly successful Elzevir family of Leyden and Amsterdam was especially notorious in this regard. According to Henry Haven Putnam (quoted by Bugbee, 1967:178, n.150):

> As far as the foreign authors were concerned, the Elzevirs appear to have followed simply the dictates of their own convenience and advantage. They took what material they thought they could use, without troubling themselves to make either requests or acknowledgements. They were, in fact, the most extensive piratical publishers that the world had as yet seen, and may be said to have reduced piracy to a business system.

The Venetian printers' guild was the model for England's Stationers' Company, which was chartered by Mary Tudor in 1557. The object was to provide the Catholic sovereign with the instrumentality to control what could be printed for widespread circulation. Masters of the Company were empowered to search the premises of any printer or bookseller for works not printed in accordance with the licensing laws, and whether censorship was obnoxious or desirable in their opinion, they had a strong economic motive to enforce their monopoly by suppressing publications not licensed by the Crown. Indeed, it has been suggested that censorship in England, particularly in the mid-seventeenth century, was more a product than a cause of the Stationers' monopoly (see Patterson, 1968:101; Plant, 1974, on the Stationers' petition of 1643).

Thus, in England, copyrights began with a monopoly franchise granted for the purpose of regulating the business of printing and publishing. They had nothing to do with the encouragement of "freedom of expression," nor were they intended to promote authorship per se. Nevertheless, authors in England had personal property rights in their unpublished manuscripts, as well as contractual protections under the common law. These protections extended to a recognized interest in the integrity of the form and content of the work for which publication permission had been given, which restrained printers from making arbitrary alterations in texts once they were published and from dispensing with the need to recompense the author. In short, under these arrangements a stationer (i.e., a printer-copyist) had to obtain the author's permission to publish his manuscript even though the author did not hold the copyright (see Patterson, 1968:65-69).

The modern statutory protection of *authors'* copyrights in the United States and Britain arose in the early eighteenth century, almost as an accident. In England during the closing decades of the seventeenth century, the passing of the era of political and religious censorship made it increasingly difficult for the Stationers' Company to interest the government in the control of the new printing presses that were springing up throughout the country. When the Licensing Act that had given teeth to the Stationers' monopoly was allowed to lapse in 1694, the competition intensified as country booksellers openly flouted the doctrine of perpetual copyright that the Stationers' Company had sought to establish on the evidence of assignments registered in its record books. After 15 years of increasingly chaotic conditions of unregulated competition, the London printer-booksellers at last managed to secure new legislation, in the form of the 1709-1710 act of Queen Anne. This, the first copyright statute, did not give the publishers the perpetual rights they had sought; instead, it limited the exclusive right to printing new books registered with the Stationers' Company to a term of 14 years (following the precedent established in the case of patents under the Statute of Monopolies of 1623); and it gave the holders of copyrights on existing books the sole right to print for 21 years. Moreover, to open up the trade, the Act of Anne eliminated the guild monopoly on the holding of copyrights. Anyone could now hold the copyright for a new work—printers, bakers, cobblers, and even authors.

From the foregoing brief account of the origins of copyrights, it is evident that the signal distinction between the protection of ideas under patent law and the protection of expression under the law of copyright owes a great deal to the fact that copyrights arose in response to internal and external interests in regulating the nature of competition in the printing and publication business, an industry in which, at an early date, decreasing costs were thought to be a source of instability. Copyrights, therefore, were inherently concerned with the security of property rights in the expression of ideas—whether old ideas or new ones. Only much later did they come to be enlisted in the cause of stimulating the production of new knowledge. Is it so surprising, then, that in this new role they sometimes are found to perform rather awkwardly? Consider, as a simple case in point, the recent assignment of copyright law to the task of protecting intellectual property rights in computer software. Observers have noted that the protection afforded to original expression in copyright law offers no security for originators of novel algorithms and concepts for applications programs (such as spreadsheets and relational data bases). Yet at the same time, the opportunities that the law creates to protect original expression have had the effect of encouraging an excessive degree of variety in the "look-and-feel" of software, whereas some greater degree of standardization of the machine-user interface is widely thought to be desirable from the standpoint of economic efficiency (see, for example, Farrell, 1989; David and Greenstein, 1990).

CONCLUDING OBSERVATIONS

Historical studies reveal that although patents, copyrights, and legal protection of trade secrets have been recognizable institutions in Western societies for centuries, policies bearing on the protection accorded to intellectual property, and the juridical-institutional arrangements used to implement them, have been a mutable thing, adapted over time and across societies to the perceived needs and advantages of interested parties. The adaptations in each form of protection, moreover, have occurred within the historical context of other, related institutional arrangements affecting the costs and benefits of maintaining specific intellectual property rights. Thus, the effort to institute a uniform international regime for the protection of intellectual property rights is almost certain to cause conflict and controversy. Even though a new intellectual property regime could be Pareto improving in some situations, the need to align domestic and international laws adds further constraints that tend to render such solutions impractical. As a result, discussions of the "correct" international system for protecting intellectual property are more likely than not to degenerate into rhetorical efforts to impose institutional arrangements that may be well adapted to the national purposes and legal contexts of one country (or several similar countries) on societies that are quite different in those respects.

The supposed trade-off between promoting technological progress and technology diffusion has led to the view that strong protection of intellectual property rights must serve the former goal, at the expense of the latter. This has been a rationale for the conflicts between the technologically advanced and the developing nations over intellectual property issues: An interest in weak or minimal protection of intellectual property is imputed to the developing countries' limited capacity to innovate technologically and their comparative advantage in imitating the products and processes originated elsewhere. Yet that is not necessarily the case. Indeed, just the opposite point may be made on modern theoretical grounds and by reference to historical experience: legal protection of intellectual property rights in the form of state-sanctioned monopoly franchises can have seriously detrimental consequences for the processes of discovery and invention, whereas it may be instrumental in bringing about the successful transfer and commercial application of new scientific and technological knowledge. The arguments supporting this unorthodox contention are summarized below.

First, because invention is often a cumulative process, as scientific inquiry more generally is recognized to be, the enforcement of patent rights can interfere with further discovery. It deflects resources into "racing" for the priority prize and into inventing around the basic patent. It discourages complementary inventions, because the returns may be extracted by the patentee whose work has been built upon. Note the distinction made here between inhibiting progress rather than discouraging investment in R&D.

Second, weak and narrow patents, as in the modern Japanese system, encourage firms to cross-license and thus disseminate findings rapidly. They encourage the collective invention process—in the direction of elaboration and adaptation to particular markets, although they may discourage efforts to achieve radical, fundamental inventions (see Ordover, 1991, and references therein). This is consistent with one aspect of Kitch's (1980) "prospect" argument that broad, strong patents encourage fundamental innovations and their orderly development, but it contests the premise in the latter that a monopolist can identify and efficiently contract for the performance of cumulative, elaborative research.

Third, although it is arguable that weak patent protection regimes encourage exchanges of patent licenses among firms that are symmetrical in their technological capabilities, the opposite is more likely to be the case in regard to transfers of technical know-how from more to less capable organizations. Much of a firm's capability for absorbing and implementing patented innovations depends on its access to tacit knowledge that is complementary to the patent. If it does not possess such knowledge already because it has no experience base in the area and cannot readily hire skilled personnel from firms that do, it must contract for the required information. However, tacit information is extremely difficult to contract for, due to the problems of informational asymmetry and monitoring costs. This problem bedevils North-South technology transfers that do not involve the mediation of multinational organizations. Nevertheless, a regime of strong intellectual property protection of codified knowledge in the receiving country provides a basis—as Arora (1991) has recently demonstrated—for structuring contracts that would accomplish the transfer of uncodified and tacit knowledge that is necessary for the profitable operation of industrial processes yet remains undisclosed to the public by the patent or copyright licenser. Further, contractual arrangements to transfer tacit knowledge as part of the terms for the licensing of the use of codified and published information (e.g., in a patent) generally will require enforcement of legal protection for trade secrets, in the interests of both the licenser and the licensee.

Fourth, intellectual property rights in the form of exclusive franchise guarantees can overcome failure to exploit a patent through modifications to local market conditions due to the problem that learning of this kind will not be appropriable and, hence, there is less interest in generating learning-by-doing gains, as David and Olsen (1991) have shown. History reinforces the implications of this line of theoretical analysis by revealing extensive early use of patent privileges to encourage technology importation, both in medieval and early Renaissance Europe and in late nineteenth century Latin American countries.

A reading of the historical chronicle of the evolution of intellectual property institutions underscores several further propositions. First, the protections accorded intellectual property by nation-states have not manifested any great consistency in adhering to pure principle. Rather, they have been pragmatically altered over time in response to changing perceptions of the way the creation and dissemination of information and information products affect "national interests." They also have been tinkered with periodically to remedy unanticipated problems in the workings of institutional arrangements due to changes in the technologies employed to produce and distribute information products. Much of the late nineteenth century "reform" in national and international copyright law, for example, was provoked by developments in the technology of printing that underlay the cutthroat competition for mass markets (by the standards of the day) in cheap editions of popular novels.

Second, to be effective, statutory protections and judicial interpretations of laws defining intellectual property rights must fit within—and be compatible with—the principles of the larger framework of a society's legal institutions. Formal legal precedents and informal conventions impose historical constraints on a country's ability to fine-tune its intellectual property institutions to suit currently perceived needs. These constraints would remain even if there were widespread agreement as to the needs of the moment.

The two foregoing observations imply, in my view, a third conclusion: Proposals now being advanced to establish a uniform international regime of intellectual property protection are not practical, even though careful economic analysis would indicate that there may be considerably more points of agreement between the interests of the technologically advanced and the economically developing countries than often has been supposed.

Finally, U.S. assertion of the justice of striving to protect the "natural" ownership rights of creators of intellectual property, and its unwillingness to grant other nations any *quid pro quo* for accepting a uniformly strong international regime for protecting international property production, reflects confusions of French and British legal doctrines that are part of the American heritage concerning the subject of intellectual property. It is also quite inconsistent with some aspects of the past conduct of the United States—and that of other economically advanced countries—in the enforcement of intellectual property claims in the international arena.

ACKNOWLEDGMENTS

In preparing this paper I have benefitted from the able research assistance of Weston Headley and Phillip Lim. Editorial suggestions by the National Research Council staff contributed significantly to improving the exposition. I can claim full rights only to any errors and omissions that remain.

REFERENCES

Arora, Ashish. 1991. The transfer of technological know-how to developing countries: Technology licensing, tacit knowledge, and the acquisition of technological capability. Unpublished Ph.D. dissertation, Stanford University.

Arrow, Kenneth J. 1962. Economic welfare and the allocation of resources for invention. Pp. 609-625 in The Rate and Direction of Inventive Activity: Economic and Social Factors. Princeton, N.J.: National Bureau of Economic Research.

Beck, Roger L. 1981. Competition for patent monopolies. Research in Law and Economics 3:91-110.

Beck, Roger L. 1983. The prospect theory of the patent system and unproductive competition. Research in Law and Economics 5:193-209.

Bender, David. 1986. Protection of computer programs: The copyright/trade secret interface. University of Pittsburgh Law Review 47:906.

Benko, Robert P. 1987. Protecting Intellectual Property Rights: Issues and Controversies. Washington, D.C.: American Enterprise Institute for Public Policy Research.

Berg, Jeff. 1991. Moral rights: A legal, historical and anthropological perspective. Intellectual Property Journal 6:341-376.

Besen, Stanley M., and Leo J. Raskind. 1991. An introduction to the law and economics of intellectual property. Journal of Economic Perspectives 5(1):3-27.

Bugbee, Bruce. 1967. Genesis of American Patent and Copyright Law. Washington, D.C.: Public Affairs Press.

Bugos, Glenn E., and Daniel Kevles. 1991. Plants as intellectual, property: American practice, law, and policy in world context. Mimeo (forthcoming in Osiris, 1992).

Cheung, Steven N.S. 1982. Property rights in trade secrets. Economic Inquiry 20(1).

Chisum, D. 1989. Patents. New York: Matthew Bender.

Cooter, Robert, and Lewis Komhauser. 1980. Can litigation improve the law without the help of judges? Journal of Legal Studies 9.

Cooter, Robert, and Thomas Ulen. 1988. Law and Economics. Glenview, Ill.: Scott Foresman and Company.

Dasgupta, Partha, and Paul A. David. 1987. Information disclosure and the economics of science and technology. G. Feiwel, ed., Kenneth Arrow and the Ascent of Economic Theory, New York: MacMillan.

Dasgupta, Partha, and Paul A. David. 1988. Priority, Secrecy, Patents, and the Economic Organization of Science and Technology. Center for Economic Policy Research, Publication no. 127, Stanford University, forthcoming in Science in Context.

Dasgupta, Partha, and Paul A. David. 1990. The New Economics of Science. High Technology Impact Program Working Paper, Center for Economic Policy Research, Stanford University.

Dasgupta, Partha, and Joseph E. Stiglitz. 1980. Industrial structure and the nature of innovative activity. Economic Journal 90:266-293.

David, Paul A. 1991. Reputation and Agency in the Historical Emergence of the Institutions of "Open Science." Center for Economic Policy Research, Publication no. 261, Stanford University.

David, Paul A., and Shane Greenstein. 1990. The economics of compatibility standards: An introduction to recent research. Economics of Innovation and New Technology 1(1-2):3-43.

David, Paul A., and Trond E. Olson. 1989. Innovation Diffusion, Learning-by-Doing, and the Optimum Patent Life. Center for Economic Policy Research, Technical Paper no. 170, Stanford University.

David, Paul A., and Trond E. Olson. 1991. Technology Adoption, Teaming Spillovers, and the Optimal Duration of Patent-Based Monopolies. Working paper, Stanford University, forthcoming in International Journal of Industrial Organization.

Davis, Michael. 1989. Patents, natural rights and natural property. In Owning Scientific and Technical Information, Vivian Weil and John Snapper, eds. London: Rutgers University Press.

Dworkin, Gerald. 1981. The moral right and English copyright law. International Review of Industrial Property and Copyright Law 12(4).

Farrell, Joseph. 1989. Standardization and intellectual property. Jurimetrics Journal 30(Fall):35-50.

Federico, P.J. 1929a. Origin and early history of patents. Journal of the Patent Office Society 11:292-305.

Federico, P.J. 1929b. Colonial monopolies and patents. Journal of the Patent Office Society 11:358-365.

Folster, S. 1985. Government coordination of competitive R&D projects. D. Phil. thesis, University of Oxford.

Friedman, David D., William M. Landes, and Richard A. Posner. 1991. Some economics of trade secret law. Journal of Economic Perspectives 5(1):61-72.

Frumkin, Maximilian. 1945. The origin of patents. Journal of the Patent Office Society 27(March):143-149.

Gilbert, Richard, and Carl Shapiro. 1990. Optimal patent length and breadth. RAND Journal of Economics 21(1):106-112.

Gilfillan, S.C. 1964. Invention and the Patent System, Materials Relating to Continuing Studies of Technology, Economic Growth, and the Variability of Private Investment. Joint Economic Committee, Congress of the United States. Washington, D.C.: U.S. Government Printing Office.

Goodman, John. 1978. An economic theory of the evolution of the common law. Journal of Legal Studies 7.

Gould, Stephen Jay. 1980. The Panda's Thumb. New York: W.W. Norton.

Hilaire-Perez, Liliane. 1991. Invention and the state in 18th century France. Technology and Culture 32(4):911-931.

Hill, Thomas A. 1924. Origin and development of letters patent for invention. Journal of the Patent Office Society 6:405-422.

Hirschliefer, J., and J.G. Riley. 1979. The analytics of uncertainty and information: An expository survey. Journal of Economic Literature 17:1375-1421.

Jager, Melvin F. 1991. Trade Secrets Law, vol. 1. New York: Clark Boardman Company.

Jefferson, Thomas. 1972. The Life and Selected Writings of Thomas Jefferson, Adrienne Koch and William Peden, eds. New York: The Modem Library.

Johnson, W.R. 1985. The economics of copying. Journal of Political Economy 93:158-174.

Kaufer, Erich. 1989. The Economics of the Patent System. Chur, Switzerland: Harwood Academic Publishers, 1989.

Kitch, Edmund W. 1977. The nature and function of the patent system. Journal of Law and Economics 20:265-290.

Kitch, Edmund W. 1980. The law and economics of rights in valuable information. Journal of Legal Studies 9:683-724.

Klemperer, Paul. 1990. How broad should the scope of patent protection be? RAND Journal of Economics 21(1):113-130.

Landes, William M., and Richard A. Posner. 1989. An economic analysis of copyright law. Journal of Legal Studies 18:325-366.

Liebowitz, S.J. 1985. Copying and indirect appropriability: Photocopying of journals. Journal of Political Economy 93:945-957.

Long, Pamela. 1991. Invention, authorship, "intellectual property," and the origins of patents: Notes toward a conceptual history. Technology and Culture 32n.4:pp.846-884.

Lubar, Steven. 1990. New, useful, and nonobvious. American Heritage of Invention and Technology 6(1):8-16.

Machlup, Fritz, and Edith Penrose. 1950. The patent controversy in the nineteenth century. Journal of Economic History 10(1):1-29.

Machlup, Fritz. 1958. An Economic Review of the Patent System. Study no. 15 of the Subcommittee on Patents, Trademarks, and Copyrights of the Committee on the Judiciary, United States Senate, 85th Congress, Second Session. Washington, D.C.: U.S. Government Printing Office.

MacLeod, Christine. 1991. The paradoxes of patenting: Invention and its diffusion in 18th- and 19th-century Britain, France, and North America. Technology and Culture 32(4):885-911.

Mandich, Giulio. 1958. Principi riconoscimenti veneziani di un diritto di privative agli inventori. Rivista di diritto industriale 7:101-155.

Marx, Karl. 1970. Capital (translated into English by Moore and Aveling). London: Lawrence and Wishart.

Merges, Robert P., and Richard R. Nelson. 1990. Market Structure and Technical Advance: The Role of Patent Scope Decisions. CCC Working Paper no. 90-100, University of California at Berkeley.

Mody, Ashoka. 1990. The new international environment for intellectual property rights. Pp. 203-239 in Intellectual property rights in science, technology, and economic performance: International Comparisons, Francis W. Rushing and Carole Ganz Brown, eds. Boulder, Colo.: Westview Press.

Nelson, Richard R. 1990. What Is Public and What Is Private About Technology? CCC Working Paper no. 90-9, University of California Center for Research in Management.

Nordhaus, William D. 1969. Invention, Growth, and Welfare: A Theoretical Treatment of Technological Change. Cambridge, Mass.: MIT Press.

North, Douglass C. 1991. Institutions. Journal of Economic Perspectives 5(1):97-112.

Novos, I.E., and M. Waldman. 1984. The effects of increased copyright protection: An analytical approach. Journal of Political Economy 92:236-246.

Office of Technology Assessment. 1986. Intellectual Property Rights in an Age of Electronics and Information. Washington D.C.: U.S. Congress.

Ordover, J.A. 1991. A patent system for both diffusion and exclusion. Journal of Economic Perspectives 5(1):43-60.

Patterson, Lyman. 1968. Copyright in Historical Perspective. Nashville: Vanderbilt University Press.

Pavitt, Keith. 1987. The objectives of technology policy. Science and Public Policy 14:182-188.

Phillips, Jeremy. 1982. The English patent as a reward for invention: The importation of an idea. Journal of Legal History (May):75-76.

Plant, Arnold. 1934. The economic theory concerning patents for inventions. Economics 1(February):167-195.

Plant, Sir Arnold. 1974. Selected Economic Essays and Addresses. London: Routledge & Kegan Paul. (Chapter 4 reprints the author's essay "The Economic Aspects of Copyrights in Books," 1934.)

Polanyi, Michael. 1944. Patent reform. The Review of Economic Studies 1(1):61-76.

Prager, Frank. 1944. A history of intellectual property from 1545 to 1787. Journal of the Patent Office Society 26(November):711-760.

Prager, Frank. 1946. Brunelleschi's patent. Journal of the Patent Office Society 28(February):109-135.

Prager, Frank. 1948. Translation of Giulio Mandich, "Venetian Patents (1450-1550)." Journal of the Patent Office Society 30(March):166-224.

Prager, Frank. 1960. Translation of Giulio Mandich, "Venetian Origins of Inventors' Rights." Journal of the Patent Office Society 42(June):378-382.

Priest, George L. 1977. The common law process and the selection of efficient rules. Journal of Legal Studies 7.

Priest, George L. 1986. What economists can tell lawyers about intellectual property: Comment on Cheung. Research in Law and Economics 8:19-24.

Rosenberg, Nathan. 1990. Why do companies do basic research with their own money? Research Policy 19:165-174.

Rubin, Paul. 1977. Why is the common law efficient? Journal of Legal Studies 6.

Rushing, Francis W., and Carole Ganz Brown. 1990. Intellectual Property Rights in Science, Technology, and Economic Performance: International Comparisons. Boulder, Colo.: Westview Press.

Samuelson, Pamela. 1984. CONTU Revisited: The case against copyright protection for computer programs in machine-readable form. Duke University Law Journal.

Scherer, F.M. 1972, 1984. Nordhaus' theory of optimal patent life: A geometric reinterpretation. American Economic Review 62:422-427. Pp. 130-141 in Innovation and Growth: Schumpeterian Perspectives. Cambridge, Mass.: MIT Press.

Schmitz, James A. 1989. On the breadth of patent protection. Mimeo, SUNY-Stony Brook.

Scotchmer, Suzanne. 1991. Standing on the shoulders of giants: Cumulative research and the patent law. Journal of Economic Perspectives 5(1):29-41.

Scotchmer, Suzanne, and Jerry Green. 1990. Novelty and Disclosure in Patent Law. RAND Journal of Economics 21(1):131-146.

Siebeck, Wolfgang E. 1990. Strengthening Protection of Intellectual Property in Developing Countries: A Survey of the Literature, Robert E. Evenson, William Lesser, and Carlos A. Primo Braga, eds. Washington, D.C.: World Bank.

Smith, Adam. 1776, 1976. The Wealth of Nations, R.H. Campbell, A.S. Skinner, and W.B. Todd, eds. Oxford: Clarendon Press.

World Intellectual Property Organization. 1988. Existence, Scope and Form of Generally Internationally Accepted and Applied Standards/Norms for the Protection of Intellectual Property. WO/IN-F 129. Geneva: WIPO.

World Intellectual Property Organization. 1989. General Information. Geneva: WIPO.

Wright, Brian D. 1983. The economics of invention incentives: Patents, prizes, and research contracts. American Economic Review 73(4):707.

II

The Case For and Against a Uniform Worldwide Intellectual Property Rights System

Introduction

This section addresses the debate over movement toward a uniform, worldwide system of intellectual property rights (IPRs). The ability to protect intellectual property in countries around the world, while long recognized by the scientific, engineering, and business communities as important, did not occupy center stage as a policy issue until recently. Because of their highly technical and complex subject matter, IPR issues had been relegated to specialized administrative agencies. In the past decade, however, IPRs have become a much higher-priority issue on national and international policy agendas, particularly in regard to trade policy.

Major factors in the increasing prominence of IPR issues have been the globalization of markets and the increase in international trade in high-technology products. This environment has increased the significance of technological capabilities and the products of technology to companies and nations around the world. As multinational corporations increasingly conduct their manufacturing and marketing activities across borders, international protection of intellectual property is becoming more and more important to them.

Increased reliance on strategic alliances and joint ventures has also increased the importance of intellectual property protection, which enables firms to share their technology with partners without losing control of it. Increased competition from many nations puts a premium on the ability of firms to generate and exploit technological innovations. Considerations of international cooperation and competition also make IPR issues of greater concern to national governments, many of which have moved to better integrate this policy area into broader economic and trade policies.

A related reason for the greater visibility of IPR issues is the increased losses suffered by businesses in industrialized countries as a result of weak IPR systems, especially in developing countries. As long as the developing countries had little technological capability, firms in industrialized countries tolerated weak IPR regimes in those nations because losses due to unauthorized use of intellectual property were low. As the technological capability of developing countries grew to the point that some could quickly copy advanced technology and manufacture high-technology products efficiently, the losses to firms in industrialized countries began to grow and the governments of those countries began to pressure the developing countries to enact and enforce stronger IPR systems.

Led by the United States, the industrialized countries are now pressing for a uniform, worldwide system of IPRs. They have argued that strong intellectual property rights protection in all countries is necessary for firms to reap the economic returns from their investments in innovation, and thereby continue to invest in R&D and innovation, which will lead to future economic growth.

A move toward stronger, possibly uniform, IPR systems, however, is being resisted by many countries, particularly developing countries and newly industrialized countries, for a variety of reasons. The governments of those countries have argued, among other things, that strong IPRs would deter progress toward their national economic development objectives by increasing the cost of obtaining new foreign technology.

Negotiations on these issues are being conducted as part of the current Uruguay Round of the General Agreement on Tariffs and Trade. An effort is being made at those talks to achieve a stronger set of international standards for intellectual property protection. In addition, talks being conducted at the World Intellectual Property Organization are aimed at achieving greater harmonization among national patent laws.

The first two chapters in this section argue, respectively, for and against a uniform, worldwide intellectual property system. In Chapter 3, Robert Sherwood addresses the question of whether, on balance, strong intellectual property protection can be expected to benefit or harm developing countries. The international IPR system he posits and analyzes would have similar practical effects in different countries, although national laws would not necessarily be standardized or even harmonized. Sherwood concludes that the benefits of such a system to developing countries would outweigh the harm to those countries.

In Chapter 4, Claudio Frischtak makes the counterargument in favor of national IPR regimes that are differentiated according to level of technological and productive competence. He notes that national IPR policies are generally established to promote a nation's perceived self-interest and that, for a variety of reasons, the interests of different nations are best served by differing levels of intellectual property protection. Chapter 4 examines the question of economic effects of differentiated IPR regimes from the perspective of domestic welfare, under conditions of closed and open economies, and from the global welfare perspective.

The final chapter in this section, Chapter 5, addresses the economic effects of unauthorized use of intellectual property. Edwin Mansfield presents data on the effects of weak intellectual property protection on developing countries in terms of its influence on foreign direct investment and technology transfer. He also summarizes what is known about the economic effects of weak protection on innovating firms in terms of lost revenue and investment opportunities and the relationship between IPR protection and the rate of technological innovation.

Professor Mansfield articulates the central premise of one side of the IPR debate:

> If intellectual property rights were weakened considerably, it could have unfortunate consequences. The incentives for industrial innovation, already relatively weak in industries where patents are ineffective and entry is easy, might wither to the point where the investment in new and improved products and processes would be far below the socially optimal level. Given the central importance of industrial innovation for economic growth, such an eventuality would do considerable harm, both to the United States and to other countries.

The possibility of negative effects on innovation and economic growth must be taken very seriously. This line of thought, however, should be contrasted with the view taken by Paul David in Chapter 2—that under some conditions IPRs can have seriously detrimental consequences for the processes of discovery and invention. Unfortunately, there is little empirical evidence on the effects of IPRs on invention and innovation under varying conditions that might help resolve the difference between these two views. This adds to the difficulty of reaching international agreement on the strength and scope of intellectual property protection.

3

Why a Uniform Intellectual Property System Makes Sense for the World

ROBERT M. SHERWOOD

The central question of the debate "Does a uniform intellectual property system make sense for the world?" is whether, on balance, strong intellectual property protection can be expected to benefit or harm countries in development. The answer given in this chapter is that the benefits outweigh the harm, whatever the stage of a country's development. Thus, a uniform intellectual property system makes sense for the world.

A chapter of this length cannot go deeply into any of the points it raises. If it stimulates others to think more deeply about this subject, it will serve its purpose well.

A UNIFORM SYSTEM: WHAT IT IS AND IS NOT

Our attention is centered not on what a uniform world system of intellectual property might look like, but rather on what will happen when such a system comes into being. The consequences of such a system are more interesting than its contents. Even so, some terms of reference must be set to define a "uniform system."

The first characteristic of the uniform system being proposed is that the specific intellectual property systems of individual countries need not be *identical*. Identical national systems would require a uniformity beyond that needed to achieve the beneficial effects of a uniform system.

The diversity of jurisprudential concepts and legal systems found throughout the world implies something short of identical national intellectual property regimes in defining the proposed uniform system. Harmonization of laws, procedures, and rules in every country is not called for, although that could follow and is indeed already an objective being sought by some countries. The uniform system is not a system with a single court of appeals, or a single international intellectual property court, although that might someday prove useful. Nor does it call for a single patent or copyright office or a single worldwide patent or copyright, however cost-efficient that might be.

The second defining characteristic of the proposed uniform system is that of *congruence*. Robust similarities of outcome from one country to another, rather than identical statutory provisions, would be its hallmark.

When such congruence is achieved worldwide, those who make investment decisions, conduct research, or invent and move technology from place to place will be able to go about their business without having to think about differences between the intellectual property systems of various countries. Those differences that remain will be a proper subject only for specialist lawyers. Where congruence does not exist, people other than lawyers are troubled by system differences when making investment, research, and licensing decisions.

The third defining characteristic of the proposed uniform system is that of *stimulation*. People involved in the process of invention, technical advancement, and creative expression will be stimulated by confidence that the results of their efforts can be safeguarded from misappropriation and unauthorized copying, no matter which country becomes the location of their activity.

The knowledge that others can be prevented from unauthorized copying has been widely experienced as a powerful stimulus to invest time, resources, and effort in inventive activity. What constitutes reasonable protection can, in part, be gauged by the degree to which this stimulation is active in the technology-producing infrastructure of a country.

The proposed uniform system, then, is a robustly congruent, highly stimulative global system that embodies basic underlying concepts of protection.

A quasi-numerical approach to defining this uniform system may clarify the concept. In comparing the intellectual property systems of various countries, it has been found useful to rate them on a scale of 1 to 100. In doing this, the entire system is taken into consideration: substantive rules, administrative practice, and judicial enforcement. As examples, Germany can be rated at slightly more than 90, the United States and some of the European countries in the high 80s, and Mexico after its recent reforms at about 75, whereas Argentina and Brazil currently rank in the 30s and 40s.

Only as a system rises above a 70 will it produce positive results for that country. Those results can be measured in terms of three critical things that begin to happen: private venture capital firms become willing to invest in technology-based start-up companies, valuable technical knowledge flows more readily from university laboratories to the marketplace, and local firms become willing to devote substantial resources to internal research.

The congruence test for a uniform worldwide, high-stimulation system, then, might be defined numerically as having been met when the intellectual property systems of all countries rise above a ranking of 70.

Two examples of congruent systems help to illustrate these terms of reference. Within the European Community, the various national intellectual property systems have not been made identical or harmonized, yet people doing business across member states there pay little attention to the differences that exist between individual systems. In the United States, each state has its own approach to the protection of trade secrets and in fact there are differences, yet no one thinks much about them when planning interstate business activity. In these two examples, each state is well above the level of protection needed to provide high stimulation to research and technological activity.

Another way to look for a uniform, high-stimulation system would be to ask whether, in fact, the general reflex of the business culture of each country is to respect invention, technical knowledge, or creative expression as the property of those who create it, rather than to operate on the assumption that it is open to copying and imitation.

A uniform world intellectual property system would further mean that each national system is *comprehensive* and would have three distinct elements. First, all forms of intellectual property would be included (i.e., copyright, patents, trade secrets, trademarks, "chip" topography, and so forth.) Second, each would have efficient public administration, with transparency where discretion is involved. Third, each would have the judicial means to enforce individual rights swiftly. As a consequence, each system would be reasonably predictable.

Intellectual property systems are continuously called on to accommodate new forms of technology. Currently, for example, we are watching the world's national systems adjust to the complexities of biotechnology and software. In the proposed uniform, high-stimulation world system, countries would learn from each other as traditional forms of protection are adjusted to accommodate new technology. Is it true to say that the countries with legal systems based on common law adapt more quickly to new sets of facts than do civil code countries? Whatever the case, some countries will lead others in incorporating new forms of technology into the uniform system. This has happened in the past. In time, the desire for high stimulation, which is the underpinning of the uniform system, will bring all countries within the parameters of congruence.

This is the vision of a uniform, high-stimulation world system consisting of robustly congruent national intellectual property regimes. Having set these terms of reference, this chapter turns to an examination of what such a system can be expected to produce.

BENEFITS FOR NATIONS AND THE GLOBAL ECONOMY

This section discusses the implications of a uniform, high-stimulation system both for individual nations engaged in the development process and, briefly, for the world as a whole.

A Subject Little Studied

Without doubt, there is a fairly high burden of proof to be met in finding that a uniform, high-stimulation system makes sense for the world, particularly for the countries in development. That burden is particularly heavy because this subject has been little studied (Siebeck et al., 1990).[1]

There are reasons for the neglect. One is that Adam Smith told us to look at capital, labor, and resources in figuring out why some nations are wealthier than others. He left out innovation and knowledge. So economic theory itself has not had intellectual property on its agenda until fairly recently. Even recent work has looked mostly at the return to research investment in developed countries where high-stimulation systems are already in place.

Another reason for neglect is that since World War II, the World Bank and companion institutions have operated under the assumption that bringing more money to poor countries will help them develop. Less has happened than expected. Now, the research agenda is changing. Things such as property rights, transaction costs, and knowledge itself are beginning to be examined for their role in the development process.

Perhaps a third reason for neglect is that the subject is not easy to study. Data that would be relevant are often not kept in developing countries. For example, data on amounts spent by local firms on research are often not available or, if tax laws offer credits for such expenditures, they may be distorted. Moreover, in a sense, it is a study of things that do not happen. How many inventions would have been made by local nationals or how many local scientists would have stayed home to conduct research rather than join the "brain drain" if intellectual property had been well protected? Often we can only speculate.

Once this has been said, are there methods for careful examination of this topic or can we only conjecture? It is not a simple matter, but methods exist by which the interface between intellectual property and the develop-

[1]For one of the first papers to deal directly with the subject, see Burstein (1984).

ment process can be examined. Some are being developed now at the World Bank, and more need to be conceptualized.

One that seems promising would be to examine shifts in activity patterns after a developing country upgrades its intellectual property system. There are reports that immediately after Mexico reformed its patent law in June 1991, large numbers of patent applications were filed by Mexican nationals. Apparently Mexico's protection rose to a level at which it made a difference to local activity. It will be interesting to study this "before-and-after" opportunity in Mexico. Sharp changes that offer opportunities for research have been made in at least half a dozen developing countries in the last four to six years. Although the changes in Mexico are probably the most sweeping, reforms in Korea, Taiwan, Singapore, Brazil (copyright for software), and China present interesting before-and-after situations.

A small but striking example of a before-and-after shift comes from Colombia where copyright protection for software took effect in 1989. More than 100 Colombian nationals have since produced application software packages that have been registered with the copyright office, with hundreds more written but not registered. Many of these customized programs help run local industrial manufacturing processes. This example hints that there is a great deal of very useful technology that could be generated in developing countries by local people, given the stimulus of an intellectual property system that works.

It is commonly assumed that all technology comes from developed countries and that, by definition, developing countries cannot be expected to generate technology. As a consequence of this thinking it is presumed that weak protection for intellectual property will assist in obtaining developed country technology at little or no cost. The possibility that valuable technology could be generated from within developing countries comes almost as a shock, yet this is precisely the point of urging strong intellectual property protection in these countries, so that this possibility can be realized. It is far less likely to be realized without protection. The negative assumption gains currency particularly when attention centers on headline-grabbing technology. It may be part of the syndrome that proclaimed that every developing country ought to have its own steel mills. It fails to recognize that incremental innovation can be of great value to a developing country.

As a general comment on research methodology it may be suggested that in seeking evidence, the greatest reward will probably come from getting close to those in developing countries who are involved in the creation and transfer of technology or in free riding. Yet a methodological difficulty that has been encountered is that even local businessmen who have a good deal at stake in terms of their ability to protect innovation, have little idea of what intellectual property is or does. General ignorance about intellectual property is not surprising where the local system is so weak it plays no active role in people's planning or thinking, but it hinders discussion of the topic.

Another methodological observation is that there are, of course, many factors at work in any economy, making it hard to isolate single causes where a galaxy of factors is operating. Competing explanations are, therefore, often presented. For example, my finding that weak intellectual property protection stifles the willingness of local firms to conduct internal research was countered by the explanation that local firms do not conduct internal research because they have no money or because they are protected by closed borders, rather than because their research results would quickly be lost to competitors. The results of a survey conducted in 1988 by a Brazilian government agency shed some light on this. A high proportion of the responding companies stated that lack of legal protection reduced their willingness to conduct internal research (Sherwood, 1990:Appendix 1).

Although little studied, this subject is at a threshold of attention. As research results come in, they will throw a great deal of light, not only on intellectual property but on many aspects of the development process itself.

Research Findings From Interviews

Over the last five years, the subject has been researched by the author at a grass roots level in selected developing countries, chiefly through individual interviews. This work has been concentrated in Brazil, where more than 20 weeks were spent, but interviews were also conducted in Mexico, Argentina, Colombia, Venezuela, and in what are now Russia, Belarus, and Estonia. The effort has been to talk with people who have a direct stake in the local intellectual property system. More than 200 interviews were conducted, mainly with local businessmen, but also with university researchers, venture capital firm owners, ranchers, research park directors, state enterprise officials, and then, to help in formulating reflections, with local academic economists.[2]

Among the businessmen, there was, as noted, considerable lack of understanding about what intellectual property is and does. Still, almost all of them reported having lost valuable technology to competitors, and as a result, they were reluctant to devote significant resources to internal research. Most said that if they had the means to better protect the results of their own research, they would be willing to devote more resources to internal research.

Many businessmen reported resorting to various techniques by which

[2]For a more extensive report, see Sherwood (1990:Ch.5).

they sought to minimize loss of technology in the absence of legal protection, particularly trade secrets. One of the more frequent ways in which competitors get their hands on technology is to hire away key employees, a kind of "predatory hiring." To defend against this, businessmen segment their technology, exposing the fewest possible workers to each segment. That, and a number of other techniques, work to a limited degree to prevent technology loss, but those same techniques have a negative effect on efforts to advance technology and have a distressing effect on employee training. Human resource development suffers in silent, unnoticed ways in low-protection intellectual property environments. Most important, as noted, there seems to be a direct connection between low protection and lack of stimulation to perform in-house research.

In the universities, researchers who had come up with important inventions, sometimes to their own surprise, found they had to start their own company to commercialize their invention. When they do this, they are usually ill equipped to function as entrepreneurs, they neglect their students, and worse, they neglect their ongoing research. Some, who have studied abroad, understood that if they could effectively protect their inventions with patents and trade secrets, it would be possible to license the invention to others better prepared to commercialize the new technology. Some of the younger researchers were particularly restless on this point.

Venture capital firms find they cannot obtain useful information about the underlying technology on which start-up firms base their requests for venture capital because the typical start-up company fears losing its technology to the venture capital firm. As a consequence, venture capital firms seldom even reach the question of whether they are willing to invest in such start-up firms. Instead, they invest in firms that are not based on technology or they invest in existing companies with assets and track records.

Research park directors reported that they cannot raise private funds to support their work. This is a serious problem, particularly when government research expenditure diminishes. Moreover, the synergy expected within the parks has not materialized to the extent it does in countries with high-stimulation systems. Both deficiencies are traceable to low-protection environments. Normally, private funds will not be invested in research, other than as an act of charity, if the expected results cannot be appropriated through the application of the tools of intellectual property. Within research parks, investigators who are brought together to stimulate each other's thinking, are instead wary of sharing proprietary technical knowledge for fear it will be misappropriated by others at the center. In countries with adequate protection, this fear is overcome by enforceable confidentiality agreements and other protective mechanisms.

It appears that weak intellectual property protection inflicts very high costs on the development process. These costs are largely in the area of opportunity losses. Counting things that do not happen is frustrating, but this does not mean the costs are not great. Research methods are needed to measure these costs.

What Does Free Riding Accomplish?

Free riding in relation to intellectual property is simply shorthand for what happens when technical knowledge is treated as property in one country but not in another. The second country permits its citizens to take the technical information of citizens of the first country as a result of the design of its intellectual property system. It should be stressed that the second country thereby also permits its own citizens to take from each other.

From a short-term or static perspective, free riding may appear attractive. From a dynamic, longer-term perspective a different view emerges. It is like the village in which the local council votes one day to make bank robbery legal. The diffusion of money increases and some villagers are happy, but when the village needs a firehouse, the bank has no funds for its construction.

Even in the short term, more needs to be learned about free riding in developing countries. Theory suggests that free riding provides optimal diffusion of its object—in this case, technical information. However, free riding in relation to developing countries has been little addressed empirically and certainly not in any systematic way.

It is said that free riding has provided benefits in specific situations. For a limited range of industries, individual companies may have been able to incorporate a new product or process into their business by appropriating technology that became available through weak or absent intellectual property protection. However, to conclude that the economy of an entire country has advanced as a result is probably unwarranted. There is simply a great deal of technology that cannot be appropriated, much less advanced, without the willing cooperation of its originator.

On the other hand, the damage to a country's technological infrastructure that arises from a free-rider strategy has been little thought about, much less measured. The opportunities to conduct local research, train local technicians and researchers, attract local venture capital to the development of promising new technology, support the movement of research results from university laboratories to the marketplace, and find greater outlets for research results produced in state enterprises are among the opportunities lost to a country.

In considering methodologies for researching free riding, a distinction should be made between economic imitation and legal imitation. Some activity that is described as free riding by economists may be perfectly legal. It is often said, for example, that Japan advanced by illicit copying after World War II. It appears, instead, that Japan licensed a great deal of technology in the postwar period, paying full value, and then improved on it. There was some free riding, no doubt, but that was not the secret of Japan's success (Rahn, 1983).

From extensive interviewing in Brazil and general familiarity with Mexico's intellectual property system, it can be suggested that these countries probably have less to show from free riding than is claimed. Free-riding strategies tend to foster advances in those technologies where reverse engineering or direct copying are possible, but not in others. Where such advances are made, they tend to lag behind developments in originating countries, particularly where copying and reverse engineering do not involve the same level of knowledge that is needed to innovate.

For certain technology, the ability to free ride is remarkably easy. Software and medicine are good examples, even classic examples, of technology that is costly and risky to develop, yet quite easy to copy. In a sense this is simply a taking of products and not an appropriation of technology. Those who copy learn very little about developing software or medicine. The skills gained from copying are typically not useful in the transition to innovation.

In approaching the analysis from other perspectives, it might be asked whether free riding can be expected to foster innovation and technological growth in developing countries in the future. Has the velocity of technical change accelerated so that free riding becomes more expensive as a strategy? Does free riding condemn a country to play perpetual catch-up? Does it mean a country falls further behind as technology becomes more complex and advances with increasing speed or does free riding accelerate the catch-up process as all countries move forward? These questions deserve more attention.

It might also be asked what happens to the predicted cost savings that result from free riding. If there are cost savings for an economy, how are those savings harnessed to foster development and growth? It can be asked whether the products resulting from free riding are sold at prices that are quite high when one considers they bear no innovation costs.

Pent-up Demand

Recent press attention has highlighted the great number of foreign graduate students in U.S. universities. Behind this story is the increasing number of researchers who are trained in the United States and then return to research positions in countries with weak intellectual property systems. In the last few years, many of them have been interviewed in the course of the work reported above. They are a frustrated group.

Part of their frustration stems from poor research facilities. This may itself be a symptom of a weak intellectual property system. After all, why allocate scarce resources to research when the results can be taken by others? These returning researchers, however, have experienced the ways in which the high-stimulation system of the United States shapes the infrastructure of technology advance. They see industry conducting serious internal research and working in close association with universities—things that are unlikely to happen in the absence of such a system. They see how the system channels private financing to research.

From the interviews reported, the impression was formed that there is, indeed, what can be termed a pent-up demand for better intellectual property protection in developing countries. Several examples illustrate this.

On a recent visit to Colombia, I found that the technical team that works in the national coffee growers cooperative was frustrated by the lack of protection for biotechnology there. They have research projects in mind that could improve the stock of coffee trees, as well as some of the process technology by which coffee is prepared for market. They are unwilling to pursue this research without the benefit of better protection for the results of their research efforts. They reported that some new technology produced by a private local company had already been pirated.

On the same visit, I found that the cut-flower growers association expressed reluctance to inaugurate a research program for improved species through biotechnology without the assurance of better protection. The lack of protection already impairs the ability of local growers to import breeding stock from abroad.

In both examples, the individuals had studied abroad and were well acquainted with the influence of strong intellectual property systems on the research environment. In both cases, the individuals were perplexed when they considered approaching policymakers to discuss their frustration, assuming that nothing could be done. It was hard to see themselves as part of this pent-up demand for stronger intellectual property protection.

From Brazil, there is an interesting example of a researcher who, after study abroad, worked in a government-supported university laboratory. He has made significant inventions in medicine. Since Brazil excludes pharmaceuticals from patent protection, he has adopted the practice of flying to Europe to obtain and then license patents there. Some have been commercially successful. No economic activity has resulted in Brazil, however.

Another Brazilian example raises the question of how Brazil is helping its development by excluding certain fields from patentability. A professor from the University of Sao Paulo is the joint inventor of a bacterium that efficiently produces ethanol from sugar waste (bagasse). While a visiting professor at the University of Florida, he and two colleagues made this invention and were granted a U.S. patent. Lawyers for the university did not seek a patent in Brazil, since none would be granted under current administrative practice there. The U.S. patent has been licensed by the university, and commercial production is about to begin in the Florida sugar industry. Nothing is happening in Brazil where sugar is one of the major crops.

A recent visit to Argentina revealed still another indication of pent-up demand. The government of Argentina has long conducted research for the agriculture sector. Now, increased discipline over government spending has greatly reduced public resources for such research. This means that, in all probability, at some point the private sector will begin to pick up the slack, in efforts to apply higher levels of technology to the agriculture base. Criteria that guide private investment in research will call for higher levels of intellectual property protection. Several interviews with government and private observers point to this possibility already.

A different kind of example, again from Brazil, illustrates the damage being done to the technological infrastructure because the intellectual property system there is weak, particularly in regard to trade secrets. Again, a professor from the University of Sao Paulo is the source of this example. He stated that Brazilian companies are afraid to utilize the Japanese techniques for quality and process improvement that are so successful in other countries. These techniques involve all the employees who work with a particular process, which means they are all asked to learn about the entire process so they can suggest improvements. This flies in the face of common industrial practice in Brazil whereby process technology is segmented. Employees are exposed to as little technology as possible so that they do not become targets for predatory hiring by competitors. Predatory hiring to obtain technology is rampant in Brazil. No one has written about this yet in Brazil, but it is clearly having a silent but devastating effect on that country's industrial development.

These few examples of pent-up demand point to the negative impact that weak intellectual property systems have on the development process. As yet, this demand is poorly organized and in at least that sense is "pent up." The examples point in large part to that which does not happen. Statistics will tell us very little of the story. A good part of the argument for robust intellectual property system congruence in developing countries must, in the absence of systematic research, rest largely on anecdotes, which nonetheless suggest patterns of widespread barriers to innovative activity.

Diffusion of Benefits

The introduction of new technology into an economy has been shown not only to contribute handsomely to growth, but also to provide a high social rate of return. We have Solow, Mansfield, and others to thank for these insights (Solow, 1957; Mansfield et al., 1977). Recently Mansfield extended his study to identify intellectual property as an important ingredient in the production of new technology (Mansfield, 1988). This work centered on the U.S. economy.

What would Solow's and Mansfield's analyses show if transposed to developing countries? In private conversations, several of Brazil's well-regarded economists have offered help in answering that question.[3] They were inclined to think the analyses would be as valid for Brazil as for the United States, with an added comment by one that the introduction of new technology could have an even greater impact in Brazil because it would benefit from the store of technology already known and practiced elsewhere. They felt that strong intellectual property protection would facilitate the introduction of new technology from both internal and external sources.

It is not unwarranted to suggest that high stimulation for research and protection for incoming technology in a country like Brazil would boost economic growth and produce a high social rate of return. It does not appear that imitation and copying produce the same results.

Global Benefits

What has been said thus far has concentrated on identifying benefits that any country in development can expect to obtain for itself from installing a high-stimulation system. The benefits available to the global economy from a uniform, high-stimulation system deserve attention as well.

Intellectual property is both a stimulus to research and an aid in conducting valuable technology from place to place.

To the extent that research can be conducted through international networks, joint research programs, and shared facilities, the role of intellectual property protection is becoming more pronounced at the global level. Any country that wants its researchers to participate in internationalized research will want to be sure the protection it provides is equivalent to that afforded research participants in other countries. The greater the uniformity of protective systems, the greater is the range of potential research participants and, presumably, the greater are the research results that then flow to participating countries.

Intellectual property protection enables research to become a magnet for funds. Both public and private funds in search of worthwhile research programs are quite willing to cross national boundaries. Such willingness is dampened where the magnet effect of intellectual property protection is weak or limited. More widespread protection will provide a greater range

[3]Julian Chacel, Affonso Celso Pastore, and Annibal Villela, in private conversations.

of candidate research programs and a larger pool from which to draw research funds.

The role of intellectual property as an aid in transferring valuable technology from place to place is critical for the world economy.[4] Once research has produced results, the willingness of those who own the results to transfer them across national boundaries is boosted when their ability to maintain their rights to the results is secure. Perhaps more important, the desire of those in other countries who want to receive the results through purchase or license is increased if they in turn can feel secure about protecting what they acquire. Moreover, high-quality technology acquisition is more likely to occur if the supplier is making a willing transfer.

As a uniform intellectual property system comes into place around the world, we can expect not only higher-quality technology transfers, but also a greater willingness to conduct joint research from transnational platforms. The sharing of technical knowledge within a protective environment will have far-reaching consequences for those conducting research in all participating countries.[5]

Finally, in considering the global benefits of a uniform system, it can be projected that by fostering a faster technological pace in the economies of more countries, the proposed uniform system will accentuate the "win-win" nature of an increasingly interdependent global economy. That is to say, with more countries generating new technology, the growth that comes from its introduction can be greater as that technology becomes more widely available through the action of willing originators. Those outside the uniform system could be expected to fall further off the pace of technical advance and to experience proportionally less benefit from that technology.

REBUTTALS

This section rebuts several arguments commonly made in support of a differentiated world system in which national regimes would remain indefinitely below the rating of 70 noted at the beginning of the chapter.[6]

Special and differential treatment is a concept that comes from the General Agreement on Tariffs and Trade (GATT), which forms the basis of the world's trading system. The concept of special and differential treatment is applied by the GATT to the world's poorest countries as they participate in the world trading system. It is interesting that in the section of the Final Draft Act proposed to the Uruguay Round negotiators by Arthur Dunkel on December 20, 1991, relating to intellectual property, there are provisions

---

[4]For an assessment of the various patterns of conductivity, see Sherwood (1989).

[5]For a portrait of joint research activity at the transnational level, see Chapter 8.

[6]For a more extensive analysis, see Sherwood (1990:Ch.7).

that would permit delay in implementing the protection stipulated in that text by as much as 10 years, but there is virtually no provision for ongoing special and differential treatment. This implies that as far as the GATT negotiation is concerned, pleas for special and differential treatment with respect to intellectual property systems apparently have been ruled out.

Ethics

It is argued by some that there are ethical trade-offs to be considered when envisioning a uniform intellectual property system for the world. Very roughly, the argument is that poor countries should be granted special treatment because they are poor. This argument implies that someone wins and someone loses if intellectual property protection is strong and effective.

If intellectual property is viewed as a tool of exploitation, it is easy to follow this argument. If, on the other hand, intellectual property is viewed as a tool of development, as suggested here, then a different perspective emerges. Viewed from the perspective of opportunity gains and losses for the development process, intellectual property can be seen as part of a country's infrastructure. That is to say, it stands in the background and helps more things to happen in the country's technological base. As suggested above, it encourages innovative people to "come out of the woodwork." It serves as a magnet for local private funds, drawing them to support local research efforts, which in turn introduce growth-producing new technology into the economy. This is turn aids human resource development by providing more research job opportunities and by permitting real technical exchange within research parks and centers where synergy is expected.

The more relevant ethical consideration is how a country can continue with a weak system when a strong system holds the promise of considerable opportunities for raising the level of a country's technical base.

Dominance

It is argued that strong intellectual property protection helps multinational companies dominate a market and kill infant industries. Stated in these terms, this assertion has emotional appeal, of course, but does not describe accurately the dynamics of technological competition.

This argument assumes inter alia that only multinational companies obtain intellectual property rights and that industries in developing countries would spring to life if it were not for intellectual property protection. In fact, few developing countries have intellectual property systems that exhibit effective protection, yet infant industries do not spring up there, except perhaps where copying is relatively simple and highly profitable (e.g., pharmaceuticals, cassettes, and software). Multinationals that invest heavily in research tend to produce valuable technology, but it is not difficult to observe that a position of dominance one year can be reduced swiftly to panic the next as technological innovation roars ahead. We can see increasing instances in which companies in developing countries rise up to challenge companies in developed countries. Countries with strong intellectual property systems aid their companies in this global technological competition.

Economics

It is argued that a weak intellectual property system enables a country to gain access to foreign technology without paying for it. This approach has severe limitations as a strategy for development.

This argument carries the supposition that by disregarding intellectual property, the acquisition cost of new technology can be avoided. Although this is perhaps true in some fields, in others the cost of imitation is nearly as high as the cost of innovation. In still others, notably pharmaceuticals, the price at which the imitation is sold is often nearly as high as the original. In both cases the expected benefit from free riding is reduced.

In looking more broadly at the issue of technology acquisition cost, it should be noted that any industrial project has a technology acquisition cost, whether it is an internal or an external acquisition. If it cannot bear that cost, the project is probably not viable for other reasons.

Another frequent comment in support of weak intellectual property systems in developing countries is that these countries cannot hope to match the research expenditures of major multinationals. The IBM research budget is cited as larger than the gross domestic product of some countries.

Although that fact may be striking, it is worth looking a little deeper. Switzerland has not stopped stimulating research because it is small. A relatively small but focused research effort may have a relatively large impact on the economy of a small country. After I had visited Montevideo a half dozen times, the thought emerged that Uruguay might spring back to life and dramatically help recover its onetime role as the Switzerland of Latin America if it would stimulate local research instead of hindering it.

There is a broad range of research opportunities open to developing countries at both the international and the domestic levels. Chapter 8 in this volume makes the point that the rapid expansion of scientific knowledge today means there are more than enough research targets to go around. It notes that even the largest companies can no longer stay abreast of all that is happening in their own fields. They are turning to alliances with others, including small firms. This can include firms in developing countries. If well protected, developing country researchers will find niches where they can contribute to world-class technical advance and earn foreign exchange as well. The supposition that developing countries do not have the capability to conduct serious technological research is largely a statement of presumed lack of opportunity. Once that opportunity is perceived and then backed by a system of effective protection for intellectual assets, the means to conduct research are likely to be found.

Local researchers in developing countries can contribute to adaptation and improvement of the local technological base. Inglorious, incremental work can have a big cumulative impact. It is useful to get away from thinking predominantly of headline-grabbing technology. In the developing countries, it is a loss when the humble farm worker who might have designed a better plow is not stimulated to do so. Creativity, not the size of research budgets, is usually the critical element for results. Surely there are as much intelligence and creativity per capita in developing countries as in countries with advanced economies.

It is occasionally asserted that a country ought to have weak protection for intellectual property because it is less competitive internationally. Apparently, a weak protective system is somehow expected to improve its competitiveness over time. From what has been said already, this can be shown to be poor policy advice. Competitiveness is not likely to be improved by weak systems of protection, especially since a weak intellectual property system may be undermining local impulses to innovate.

Prowess First, Then Protection

It is asserted by some Brazilians that Brazil deserves to have a strong intellectual property system, but only once it achieves world-class stature in research. In the interim, something less is appropriate.

Does this mean that Brazil's system should be upgraded only after all fields of research have attained world-class levels of achievement? What about the interim effect on fields that are nearing, or have already reached, world-class levels? There are some in Brazil. Are they to be denied intellectual property safeguards until the lagging fields catch up? The ability of the leading fields in Brazil to attract funds to support research is already being negatively affected by the lack of intellectual property protection.

If we assume that achieving world-class stature is desirable sooner rather than later, it is not clear how achieving it sooner is boosted by a national strategy of weak protection. Strong intellectual property safeguards seem likely to speed rather than retard progress toward world-class achievement. Brazilian researchers at the verge of world-class achievements are themselves calling for a stronger intellectual property system in Brazil, whereas most of those urging continued weak protection are quite distant from research.

TRENDS TOWARD HIGH-STIMULATION SYSTEMS

A survey of national approaches to intellectual property around the world today compared with 10 years ago would show remarkable change. In the developed countries there is a high level of ferment as the basic concepts of intellectual property are tested by new forms of technology. There is also tension over how to achieve higher degrees of congruence, but that objective and those basic concepts are not seriously in question. There is also ferment in the developing countries.

Mexico, which experimented with destruction of some basic intellectual property concepts in the 1970s, reformed its system last year to bring it well above the 70 rating noted earlier. Argentina has recently put before its congress a draft patent law with a striking resemblance to the new Mexican law. Officials of the two countries conferred on the draft. Chile moved forward early last year, and the Andean Community countries are close to reform of their common intellectual property regime. India has reform under consideration.

The former Soviet Union, shortly before its collapse, made dramatic reforms in its intellectual property system in anticipation of conversion to a market-driven economy. The reforms even included a strong trade secret law, apparently designed to prevent loss of valuable technology bottled up until now in state institutions.

Starting from virtually nothing, the People's Republic of China has instituted many elements of an effective intellectual property system in the last 10 years and continues to move in this direction. Whether it plays any part in daily activity is not clear, although there are reports of enforcement litigation in some fields that point to a concrete influence.

Brazil has on the drawing board reform in several areas of its intellectual property system. In addition to the well-publicized patent law reform, this includes work on separate draft laws for protection of semiconductor chips and plant breeders' rights. This is particularly interesting because these are areas of the system that the United States has not raised with Brazil.

There may be resistance to reform from various quarters in various countries, but few are advocating less protection or no protection these days. Today, resistance often comes from a dislike of being pressured by the United States rather than from any deep sense that reform is ultimately wrong.

It appears that as countries shift from command economies to market-driven economies there is a correlative shift from weak to strong intellectual property protection. This is surely more than an accident, yet not quite an axiom. Where the state is the only actor, there is thought to be very little need for intellectual property, that is, for rules that delineate the positions of competing actors relative to new technology. (This is not true, however, where international technology exchanges are involved.) Even where the state only mandates actions by competing actors, there is reduced need for intellectual property. However, as private competition plays a bigger role in economic activity, intellectual property assumes a more prominent place in the technological infrastructure of a country.

INSTALLING A UNIFORM, HIGH-STIMULATION SYSTEM

For a national intellectual property system to work, there must first be a judicial system that works, a precondition that is often missing. Because of this, intellectual property protection has simply never been tried in many countries. Although there are laws on the books, they do not influence day-to-day activity.

Recently, the World Bank began work on a project designed to help Venezuela upgrade its judicial system. Whether any attention will be paid to intellectual property as a subset of the overall project is not clear, but the possibility of assisting in the creation of specialized courts for intellectual property would be an option worth considering.

For a national intellectual property system to work, there must also be a bureaucracy that administers the system fairly and efficiently. The fact that this precondition is also often missing is another reason that intellectual property protection has never really been tried in many countries. This is partly a problem of resources and partly a problem of training and administrative methodology.

The patent office is typically the most costly element of a national intellectual property system. Yet patent offices generate significant revenue, including foreign exchange (through fees) for the national treasury. Patent offices typically receive less from the national budget than they contribute. If patent offices were permitted to keep their fee income, greater progress could well be made in upgrading their administration.

When intellectual property systems are designed, most attention is usually given to the levels of protection to be awarded by the system. More consideration needs to be given, during system design, to administrative and judicial cost reduction opportunities.[7] For example, greater emphasis should be given to efficiently protecting trade secrets because they are very low cost and, if properly implemented, can shift the business culture of a country to greater emphasis on innovation. In the patent area, more attention can be given to utility patents as a supplement to ordinary patents, as is

[7]For a thoughtful discussion, see Estache (1990).

the practice in Germany and Japan. It is probable that many patent laws can be simplified to reduce administrative burdens.

At the international level, more can be done to aid developing countries that seek to install high-stimulation systems. Successful steps toward examination of patent applications by transnational patent offices have been taken in Africa and Europe. The Patent Cooperation Treaty gives a workable basis for linking many patent offices so that costly duplication of effort can be eliminated. The treaty could be undergirded by an international computer network using satellite connections. The resulting data base could serve collateral information purposes as well.[8]

More and better training is needed for officials who administer developing country intellectual property systems. The World Intellectual Property Organization and national patent offices in Europe, Japan, and the United States have made efforts in this regard, but the expanding need warrants more effort.

Also at the level of international cooperation, it is worth noting that by reflex, as free trade agreements are fashioned, attention is being given to greater congruence in intellectual property systems among participating countries. For example, representatives from Brazil, Argentina, Paraguay, and Uruguay met last year to consider their respective systems in view of progress toward creation of their common market. The North American Free Trade Agreement negotiations include a working group on intellectual property system congruence. Of course, the European Community has been at work on a higher level of congruence for several decades, and there have been interesting achievements in Africa.

Attention to cost reduction is not an explicit focus of any of these efforts, but centralized examination is emerging as a logical implication and this reduces costs.

The world is well on its way to a uniform, high-stimulation system. Although at a momentary impasse, activities in the GATT Uruguay Round

[8]Such a linkage system would provide for electronic filing of patent applications in the home language of each participating country. Uniform formats for applications would assist computer-aided translation to a functional language for patent examination by a central examination office (quite probably the European or U.S. patent office). The status of the application would be available on-line whenever needed. The central examination office would sustain the cost of maintaining the vital library of prior art, and this too might be available for on-line access by researchers. The linkage system could also provide access to anyone in any participating country wishing to search for the latest technology in any field of interest. Finally, patent holders could specify from time to time their willingness to license the technology represented by the patent, thereby providing an electronic brokerage facility the world does not now have. Although visionary, such a system is within the scope of current technical capabilities. Its cost is a matter for international cooperation.

that relate to intellectual property point in this direction. Action by the United States in its bilateral trade relations keeps raising the question of intellectual property. However, as more countries come to appreciate the benefits they stand to gain from installing high-stimulation systems, there will be increasing movement toward a uniform worldwide system.

CONCLUSION

Technology, not ideology, increasingly drives much of the world's activity. This means two things for intellectual property. First, ideological arguments against strong intellectual property protection will fade, permitting a fresh examination of its positive role in development. Second, the close connection between protection of intellectual property and the creation and transfer of technology will become more evident, fostering greater appreciation of this positive role.

The uniform, high-stimulation system proposed here is in effect a landscape of systems in which human creativity is encouraged. It has the ability, in conjunction with other policies, to bring into action many people who are willing to bet their energy, time, and money on being inventive and getting something for it. This can be a powerful tool of economic development, one that can release creativity within a country while linking it with activity and advances in other countries.

Returning to the Solow and Mansfield analyses, the more technology that is available to a country from both internal and external sources, the greater is the stimulus toward indigenous growth. Free riding can make technology available in some fields, but not across the spectrum. On balance, any benefits that can be ascribed to free riding are more than offset by the damage that a country's technological infrastructure sustains for lack of intellectual property protection that works well. In view of the prospective benefits to be gained by countries in development, a uniform intellectual property system makes sense for the world (see Table 3-1).

TABLE 3-1 Summary of Benefits and Harm of a Uniform System

Benefits	Harm
Boosts local private research	Diminishes "pirates"
Develops human resources	
New technology enters economy	Some technology may cost more
Increases local competition	Increases cost of some products
Innovation becomes new reflex	
Expands global research	

REFERENCES

Burstein, M.L. 1984. Diffusion of knowledge-based products: Applications to developing countries. Economic Inquiry, Western Economic Association International 22(4).

Estache, Antonio. 1990. Brazil: Issues in the design of property rights. Internal Discussion Paper, Latin America and the Caribbean Region. Washington, D.C.: The World Bank.

Mansfield, Edwin. 1977. Social and private rates of return from industrial innovations. Quarterly Journal of Economics (May).

Mansfield, Edwin. 1988. Technical change and economic growth. In Intellectual Property Rights and Capital Formation in the Next Decade, Charls E. Walker and Mark A. Bloomfield, eds. Lanham, Md.: University Press of America.

Rahn, Guntram. 1983. The role of industrial property in economic development: The Japanese experience. 14 International Review of Industrial Property and Copyright Law.

Sherwood, Robert M. 1989. New theory of conductivity in licensing. les Nouvelles: Journal of the Licensing Executives Society 24(4).

Sherwood, Robert M. 1990. Intellectual Property and Economic Development. Boulder, Colo.: Westview Press.

Siebeck, Wolfgang E., Robert E. Evenson, William Lesser, and Carlos A. Primo Braga. 1990. Strengthening Protection of Intellectual Property in Developing Countries: A Survey of the Literature. World Bank Discussion Papers No. 112. Washington, D.C.: The World Bank.

Solow, Robert. 1957. Technical change and aggregate production function. Review of Economics and Statistics.

4

Harmonization Versus Differentiation in Intellectual Property Right Regimes

CLAUDIO R. FRISCHTAK

NATIONAL INTEREST AND INTELLECTUAL PROPERTY RIGHT REGIMES

This chapter starts with a simple proposition: Intellectual property right (IPR) regimes are not established independently of what is generally perceived to be in the "national interest." That is nearly axiomatic, at least in countries that have reached a minimal level of political development. Defining the country's national interest usually falls to elected officials and policymakers. Although it is unlikely that every single government decision has the national interest as its governing criterion, the choice of major legal and developmental institutions should have (and, I suspect, normally has) this criterion at its core.

Although policymakers do not conduct complex calculations when arriving at such decisions, such processes imply an exercise in social welfare maximization. As argued in this chapter, there are few grounds in terms of either national or global welfare (other than to minimize the prospects of economic conflict or retaliation) for all countries to abide by uniform IPR regimes; there are even fewer grounds for claiming that IPR regimes should be uniformly tight. At the most general level, there is certainly no reason for countries to share social welfare functions or preference orderings that would justify uniformity in levels of protection. The fact that enforcement costs and budgetary constraints are different makes the convergence of IPR systems still less meaningful, at least from a social welfare standpoint.[1]

IPR regimes should accommodate major structural shifts in the economy, particularly the progressive maturation of a production and innovation base in the country. Yet, as suggested below, changes in IPR systems should not necessarily be unidirectional or patterned after an individual model, namely that of countries at the property rights legislative and enforcement frontier (where moving the frontier forward implies increased levels of protection). Developed countries' regimes are responding to intense technological competition by attempting to maximize the returns from their technological assets on a global scale. Even in such countries, changes in IPR systems are far from consensual, and it is sometimes argued that protection may have reached excessive levels, to the detriment of diffusion and technological innovation.

The purpose here is to discuss the relative merits of a differentiated versus a homogeneous IPR system. This chapter argues for IPR regimes that are differentiated according to the level of technological and productive competence, so as to support a country's ability to absorb, adapt, and generate technology. There is little in economic theory to support convergence of IPR systems on a cross-country basis, particularly if convergence means an increase in the level of protection in developing and industrializing countries. This is true either from an individual country standpoint or from a global welfare perspective. Furthermore, countries with large research endowments do not believe it is to their benefit to loosen their IPR standards, nor do industrializing countries see benefits (except for lesser threat of retaliation) in equating their legislation and enforcement practices with those of the country that is at the forefront of the movement toward tighter regimes, namely, the United States.

This chapter suggests that it is unlikely that an industrializing country would be serving its own interest by copying the legislation and enforcement practices of a developed country. Both may want to follow or have a set of minimum standards as guidelines,[2] but the definition of such standards should not be a mere reflection of the developed country's perception of what constitutes the minimum: rather it should be patterned after what is

[1]See the excellent discussion in Estache (1990:5).

[2]In fact, in 1979 the World Intellectual Property Organization (WIPO) introduced a model patent law that was followed by a number of developing countries and constituted a minimum standard for patent-related IPRs. The international conventions, on the other hand, do not necessarily provide for uniformity in cross-country treatment of IPRs. The most well known—the Paris Convention—signed by 99 countries (as of 1989), has as a major feature, equality of treatment between domestic and foreign patentees (article 2). The specifics of the systems are generally left for individual countries to decide.

practiced domestically. The next section discusses some of the economic reasons for differentiation in IPR regimes. It argues that differences in individual country characteristics, stage of development, and budgetary constraints make differentiated regimes superior from a domestic welfare perspective. This proposition holds as long as a country's insertion in the global economy is not affected by choice of the IPR regime. The following section examines the case when this assumption does not hold. It suggests that differentiated regimes may entail substantial costs for developing countries in terms of trade, investment, and technology transfer flows, given that most are relatively open economies. The last section concludes with a discussion of IPRs from a global welfare perspective.

CASE FOR DIFFERENTIATION OF INTELLECTUAL PROPERTY RIGHT REGIMES

There are substantial intercountry differences in IPR regimes on all key dimensions: the level and scope of protection, the mechanisms used, the strictness with which legislation is enforced, and the way violations are penalized. In the case of patents, their duration varies considerably (5 to 20 years), shorter terms being more commonly found in industrializing countries. With regard to the scope of patents, most countries have exclusionary rules, except probably for the United States and Canada (where only scientific principles, abstract theorems, and atomic weapons go unprotected).[3] Working requirements and compulsory licensing are also widespread. The latter is permitted if patents go unworked for one to four years, with shorter periods in Latin American countries, moderate (around three years) in Australia and Southeast Asia, and longer in Europe. Again, no working requirements are present in the United States or Canada.

Although there is greater agreement regarding what comes under copyright and trademark protection (the former generally comprises works of literary and artistic expression; the latter focuses on any symbol or message that serves to identify and confer reputation to a product or a firm), there still remain differences in the length of protection and, more important, in enforcement practices (including penalty levels for violators). Difficulties

[3]A WIPO review of major cross-country differences, as summarized by Estache (1990:15), reveals that for patents, "Exemptions of coverage are more frequently found in less developed countries . . . 49 countries exclude from patent protection pharmaceutical products, 45 exclude animal varieties, 44 exclude methods for treatment of human and animal body, 44 exclude plant varieties, 42 exclude biological processes for producing animal and plant varieties, 35 exclude food products, 32 exclude computer programs, 22 exclude chemical products, 14 exclude nuclear inventions, 10 exclude pharmaceutical processes, 9 exclude food processes, 9 exclude microorganisms and 7 exclude substances obtained by microbiological processes."

in enforcement are also a critical barrier for upholding the rights of firms relying on trade secrets in countries that recognize the legitimacy of this mechanism (which many do not, on the basis that the lack of disclosure disqualifies proprietary and closely held information from being protected by the IPR regime).

The differences described above are at the core of major conflicts in international and bilateral forums. Much of the discussion has centered on patents and the protectability of a group of products comprising, until recently, foodstuff, chemicals (including fertilizers and insecticides), pharmaceuticals, and their manufacturing processes. A number of developing countries have argued against protecting these product groups due to their perceived importance in fulfilling the "basic needs" of the population, and on the grounds that countries should not become hostage to firms from developed countries by granting legal rights (patents) that allow for monopolistic pricing practices. The affected firms, however, view the exclusion of certain areas from patentability as attempts to free ride and subtract from their profits. More recently, discussion on the scope of protection has been extended to new areas, such as integrated circuits, computer software, and biotechnological products and processes, including not only microorganisms but bioengineered (transgenic) plants and animals as well. In biotechnology, positions differ not only between developing and industrializing countries, but among the latter as well.[4]

How are those substantial intercountry differences explained? They are certainly inconsistent with the concept that a uniform system is somehow economically superior, unless countries were acting against their self-interest. Yet it is unlikely that most countries act most of the time against their interests. It is true that certain types of legislation, policies, and bureaucratic procedures or practices often generate groups of "special interests" that help perpetuate or shape existing institutional and policy arrangements that are not in the country's interest. This phenomenon of "capture" is well recognized in the literature and is probably present to some extent in most countries [e.g., see McConnell (1966)].[5] Yet why specifically IPR regimes that do not conform to the paradigmatic case (as defined by developed countries) are the ones resulting from bureaucratic or legislative failure is far from clear.

---

[4]Most countries, for example, with the exception of the United States and Japan, do not allow patenting of transgenic animals (except microorganisms), although the appellate committee of the European Patent Office (EPO) was reviewing an EPO decision to grant patents to such animals in late 1990. See Barreto de Castro (1991:3).

[5]The theory of capture can be considered closely related to the economic theory of regulation, the central thesis of which is that "regulation is acquired by the industry and is designed and operated primarily for its benefits." See Stigler, 1971.

What could be argued is that required changes in legal institutions or policy regimes only happen slowly, because of entrenched interests. Further, it could be argued that in the case of countries that are not so politically developed, where debate is thin and acts of government are opaque, changes occur arbitrarily and without strict correspondence to the national interest. Nonetheless, from such considerations it does not follow that such countries should pursue or pattern their legislation after a single IPR model. Many countries may in fact have IPR legislation on the books and engage in practices that are outdated, reflecting the needs and capabilities of an earlier stage of development; a case could be built that they should migrate toward other models of protection. Yet economic reasoning does not seem to suggest the superiority of a uniform model. In fact it even raises questions about the need for protection in the first place and points to its adverse welfare implications, particularly for developing countries.[6]

To Protect or Not to Protect

Although the economic rationale of patents and other property rights instruments is disputable on theoretical grounds, countries with the most productive innovation systems have fairly comprehensive IPR regimes. In some areas (such as chemicals, pharmaceuticals, or scientific instruments), where innovations are costly to generate but not as hard to imitate, a strong patent regime (or other form of protection) is probably quite necessary for firms to be rewarded for their innovative efforts. Yet the need for patents or alternative forms of protection is most clearly justified in the presence of domestic research capabilities. The presumption in this case would be that the missing link to innovation was a structure of incentives strong enough to mobilize this potential. What if the latter is missing, and the country lacks even the basic endowments to progressively build a production base in the relevant area (as in the case of very poor countries)? Then the presence of a patent or other system of protection would have little effect either way. The country would still pay for the results of research undertaken elsewhere in the form of imports. It is indeed suggestive that many poor countries abide by fairly tight IPR systems (when one considers that it

---

[6]See Primo Braga (1990:32) for an excellent and succinct review. He notes that "overall, economic theory has raised more questions about welfare implications of intellectual property than it has answered. The theory of intellectual property protection is fragmented and provides no robust answer to the questions of appropriate or optimal level of protection under various sets of real world circumstances. In particular, its relevance to developing country concerns must be considered marginal."

may help minimize frictions with major development aid donors or trading partners, while having no major adverse effects domestically).

When research capabilities are weak but the country has a substantial productive potential, capable of copying, reverse engineering, adapting, and transforming foreign innovations into marketable products domestically, then there is a strong economic incentive to produce on the basis of foreign innovations. In this case, the country would be better off by disallowing or weakening protection, on the basis that not to do so would lead to a net income (welfare) loss in the form of royalties and rents transferred abroad. Moreover, lower levels of protection would facilitate entry, drive down prices and excess profits, and maximize diffusion rates. It is not by coincidence that countries that were ranked (in 1988) as having the most inadequate IPR regimes were composed solely of those commanding significant industrial capabilities (Brazil, Taiwan, Mexico, Korea, India, China, and others; see International Trade Commission, 1988).

There are two problems with this reasoning. First, it is assumed that the country is immune to retaliation and, more generally, that trade and investment flows are not affected by a country's domestic policy choices (a discussion that is taken up below). Second, it is assumed that the absence of protection has no impact on domestic research, to the extent that research capabilities are absent. In fact, rarely is productive capacity totally dissociated from research endeavors. For research itself is an activity characterized by a continuum of subactivities in which producers engage and routinize to the extent of their needs.

This set of subactivities can be conceptualized as originating from adaptive efforts: scaling down imported processes and making them consistent with the use of local raw materials; changing product characteristics, including redesign to conform with the local environment; improving productivity through minor innovations and controlling for product quality; replicating and progressively mastering all aspects of the existing technology; and finally, engaging in formal and structured R&D, while engineering the processes to bring new products to the market.

Thus, when choosing levels of protection, countries with a productive base face the following closed-economy trade-off: how to establish a structure of incentives (in terms of levels of protection) that will stimulate domestic technological efforts, however frail and tentative, without curtailing productive activity dependent on the use of foreign innovations. It is assumed that these are innovations that can be reverse engineered or copied by other means and that, if patentable, would not be licensed to domestic producers, short of substantial additional costs.

A tentative, schematic answer is presented in Table 4-1. When competence in production and research is insignificant, whether or not to protect is immaterial. The country should follow whatever system minimizes con-

TABLE 4-1 Competence and Optimal Protection—A Schematic Representation

	Poor Countries	Late Industrializing Countries	Industrializing Countries	Industrial Countries
Production Capabilities	N	S/M	H	H
Research Capabilities	N	S	M	H
Levels of Protection	—	L	M	H

Note: N = nil; S = small; M = moderate; H = high; — = indeterminate; L = low.

flict and the probability of retaliation (although that itself would be small in any case). As production systems become more complex, but research still lags, optimal levels of protection will probably be positive but low. As production capabilities expand and technological activities intensify, protection levels should increase accordingly.

It appears that for industrializing countries, no protection is not an optimal choice; nor is uniformity the answer. Even with the decision to establish a positive level of protection on economic welfare grounds, the object of protection and the corresponding mechanism are not given ex ante. *What then should be the scope and mechanism of protection*? How should IPR regimes be tailored to individual sectors or particular technologies, consistent with the stage of development of the country or its particular circumstances?

It is important to stress that appropriate instruments of protection already exist. Defining the scope of protection and pairing technologies or product groups with the right instrument would depend on country-specific circumstances. To illustrate, take the case of mechanical technologies, the mastery of which is critical for countries to move beyond the initial stages of industrialization. A particularly appropriate system to stimulate minor inventions or improvements of a mechanical nature (where an "inventive step" is absent) would be utility models or petty patents (as used by Germany, Japan, South Korea, and Brazil, among others). These instruments protect minor innovations, generally obtained as producers attempt to improve the productivity of machinery. Protection levels are lower than they would have been if such innovations were patentable by regular means, but are nonetheless significant (as evidenced by large numbers of utility models patented in countries that allow for such a mechanism).

In the biological sphere, an interesting case is that of plant and seed varieties. Most developing countries leave them unprotected, because they are part of the food chain and thus perceived as fulfilling basic needs. Among those countries that consider plants and seed varieties legitimate objects of intellectual protection, most have significant research capabilities in plant genetics and utilize plant breeders' rights (PBRs) as the basis of protection (e.g., South Korea, Argentina, and European nations). The United States makes use of PBRs or regular patents, depending on whether the plants are reproduced sexually or asexually. The advantage of PBRs is that they do not constitute an impediment for farmers or those engaged in research to reproduce protected varieties for their own use. PBRs do require commercial seed producers to pay royalties to breeders. In countries where agricultural production is still concentrated among small and medium farms, such an arrangement might be superior to traditional patents, which suggests again that the particular protective arrangement should reflect country circumstances.

For chemicals, pharmaceuticals, and biotechnological inventions, there is the choice of protecting either products, processes, or both. For industrializing countries, it is arguable that the patentability of products might precede that of processes, being more consistent with their small to moderate research capabilities in the engineering of processes, but only marginal ability to introduce new products to the market. The continuing inability of most countries to bring out new products suggests that more or even full protection of foreign product innovations in these areas will not have a detrimental effect, except on those firms that basically copy locally unpatentable products (but have failed so far to develop research capabilities of any significance). To the contrary, greater protection might actually benefit local research institutes specializing in areas that have not attracted the attention of major international producers. Full protection might also be appropriate in the biotechnology area, where research and production are intertwined, and the lag between developing countries and those working at the research frontier is quite large and growing.

The presumption so far has been that most research activities take place outside the country and that local producers are capable of appropriating the marketable results of foreign inventions. A different perspective is offered if it is posited that certain innovations cannot be copied because they are kept as unreachable trade secrets, or because they belong to areas that do not command sufficient interest for developed country institutions to allocate significant R&D resources. In this case, protection (however strict) causes no harm, by definition. The country may in fact want to target product groups or technologies in which it is of particular interest to provide additional incentives, by establishing higher levels of protection.

An illustrative example would be of that of certain drugs for tropical diseases, for which the market in developed countries is small. Granting full and strict protection could only be to the advantage of the developing country; in fact, once again, the prevailing system in the developed country is quite immaterial from the developing country perspective. The latter may find its interest is to grant higher levels of protection, commensurate with the potential demand for the product (and the inability of local researchers to bring forth the innovation).[7] Thus, to the extent that countries have different needs or technological preferences, the free-riding motive in a developing country will compete with the incentive requirements for generating appropriate technologies, and it may be in the country's interest to target protection for such technologies at levels higher than those prevailing in the developed country.

Clearly, however, protection should be regarded as only one of the stimuli for innovation. Again, in the development of a new class of drugs for tropical diseases, potential demand may not materialize due to low income levels of those affected. Creating such demand may require a public health program with a strong drug procurement component or, alternatively, well-designed research contracts or a system of prizes. The more general point is that an IPR system should not be regarded as some *deus ex machina* of innovation; granting full and strict rights may under certain circumstances be decisive and, under others, only an accessory factor.

Specifics of Patenting

Once the methods of protection and product coverage are defined, there is still the need to establish the exact configuration of the system of protection. Patent life is a case in point. It has been shown, for example, that length of protection for a given product should be inversely related to the elasticity of demand and the social rate of discount, and positively related to R&D returns.[8] It is unlikely that markets in different countries with divergent levels of income and preferences (among other factors) would have similar elasticities or that discount rates or the productivity of the R&D system would be very close in dissimilar societies. In this sense, strict equality in the duration of patents on a cross-country (or a cross-industry basis) would not be justified.

Still uniformity might be called for in view of the difficulties of estimating with great precision the parameters that are supposed to determine

[7]This point has been elaborated, from a theoretical perspective, in Diwan and Rodrik (1989). A similar point was made, in the discussion of the Brazilian experience, by Frischtak (1989).

[8]See Nordhaus (1969:Ch.5). Nordhaus assumes a competitive world, with inventors producing small process innovations; the objective is to maximize the net welfare to society, provided the innovator's returns are sufficient to ensure that the innovation becomes available to society. The intuition behind Nordhaus's results is that the length of protection should be longer, the more insensitive demand is to price changes or the harder it is to innovate, so that it would take longer for the innovator to reap the necessary returns; similarly longer terms of protection are optimal if society can "wait" to appropriate the gains from the invention (the social rate of discount is low).

optimal patent life (demand elasticities, etc.), and the fact that in the Nordhaus model (among others), "beyond a certain number of years [of patent protection]—usually ten or less—the welfare provided by the patent system cannot be altered significantly" (Nogues, 1990:9). Thus, insofar as "the costs imposed by standardized patent terms are not very significant," patent life should be uniform among industries and technologies at least for the sake of simplifying administrative procedures (see Primo Braga, 1990:32). Yet cross-country uniformity does not necessarily follow from these propositions—much less, a clear justification for adopting a particular patent life standard.

From the point of view of the country's welfare, patent life must be made to reflect the degree of industrial and research maturity of the country, and the underlying trade-off between production/diffusion and research incentives. The shorter the time of protection, the sooner will local producers be able to copy and start up (again, on the presumption that they have the necessary imitative capabilities), the faster prices will drop, and *ceteris paribus*, the quicker the pace of diffusion will be. A similar argument can be made regarding the expiration of patents (through working requirements) or their compulsory licensing to competitors. In all cases, policymakers will be trading off production for research incentives as levels of protection decrease. As always, the probability that two countries will have similar indifference curves and "budget lines" is small; so it would be inadequate for countries to choose similar protection parameters.

Levels of Enforcement

Enforcement is a resource-intensive activity. The higher the standards, the larger the resources allocated must be to achieve a given level of compliance, or alternatively, the higher the risks of noncompliance must be. To the extent that enforcement costs are probably larger and budget constraints tighter in developing countries, laws (including those on IPRs), judicial institutions, and practices should naturally diverge from those of developed countries (if countries are attempting to allocate resources efficiently, equalizing the returns at the margin) (Estache, 1990:69 and *passim*). In sum, it can be argued that cross-country differences in protection levels are justified in view of differences in enforcement costs and available resources.

OPEN-ECONOMY CONSIDERATIONS

So far the prevailing assumption has been that trade, investment, and technology flows are invariant with respect to the choice of IPR system. Yet that is hardly a tenable assumption anymore. Except for countries with marginal production systems, all others are being forced to follow fairly high standards of protection as their basic orientation when choosing a particular IPR model. Many industrializing countries (e.g., South Korea, Taiwan, Mexico) have amended their legislation or introduced new laws in the last five years reflecting growing pressures from foreign governments and firms.

Two basic threats are driving these changes: that of retaliation, spearheaded by the United States through its own trade legislation (particularly the "super 301") and through the inclusion of IPRs in the General Agreement on Tariffs and Trade framework; and the one posed by foreign firms that refuse to deal (invest or sometimes even trade) in the absence of strong assurances that their proprietary information will not be appropriated without consent. These firms tend to concentrate in research-intensive sectors. The fundamental trade-off that policymakers face in the choice of an optimal IPR regime in the current international environment is simple to state: closed-economy considerations of potential gains from an IPR regime tailored to the country's circumstances may have to be balanced against open-economy income losses from reduced levels of trade and investment, if the choice of IPR regime is inconsistent with the trading/investment partners' interest.

Pressure from trade and investment partners is finding less resistance among developing countries with relatively mature production and innovation systems. Additionally, industrializing countries are being induced to change their IPR regimes due to changes in international economic relations (with globalization of production) and an acceleration in the pace of technical progress. Open-economy considerations are becoming more important in policymakers' decisions insofar as the welfare of countries increasingly hinges on their participation in international flows of trade and technology. As the composition of investment and trade moves away from simple commodities toward higher-value-added, more research-intensive goods, the protection of intellectual property grows in importance. The existence of an IPR regime that is similar to those in countries that have a dominant economic position signals to these countries that the industrializing country is a "trustworthy" partner insofar as protection of intellectual property is concerned.

This is in fact a critical consideration, for until quite recently, the nature of the IPR regime was either absent or quite hidden in the calculus of trade and investment decisions.[9] Even among more traditionally IPR-dependent producers, such as pharmaceuticals firms, it was hard to perceive an

---

[9]See, for example, the discussion in Frischtak (1989). It is noteworthy that throughout the 1970s and early 1980s, foreign investment flows by U.S.-based multinational firms appeared to be quite unaffected in both Brazil and Mexico by their IPR regimes, despite the countries' being rated as having the most "inadequate" regimes according to U.S. foreign investors.

association between the nature of the regime and the intensity of commercial transactions (e.g., the volume of trade, disembodied technology transferred, or investment). In case of investment decisions, in particular, the fundamental considerations were market size and growth prospects, factor supply and costs, and macroeconomic and political stability (country risk). The regulatory regime (investment licensing, foreign remission restrictions, price controls, IPRs) was of secondary importance.[10]

Yet that seems to be changing. A combination of greater competition among countries for foreign investment, and considerable shifts in perception regarding the importance of protecting proprietary information due to intensifying technological competition, suggests that IPRs are being brought to the forefront of decisions. Most concerns are still concentrated among the more IPR-sensitive producers (chemicals, pharmaceuticals, biotechnology, scientific instruments, and microelectronics products). An Organization for Economic Cooperation and Development (1987: Table 40) survey, for example, shows that inadequate IPR protection was the greatest disincentive to technology licensing, together with foreign exchange controls and government approval regulations. To the extent that licensing is a form of transaction in which firms have the least control over their technologies, and are therefore most sensitive to IPR issues, it is a "leading indicator" of firms' concerns over IPRs in direct investment and trade.

Table 4-2 summarizes how in an open economy uniform and differentiated IPR systems compare. In the table, each cell specifies the direction and intensity of impact. Thus, for example, in the case of threat of retaliation, the impact of a differentiated system is "adverse"; that is, such a system increases the threat from trade, investment, and technology transfer partner countries, and does so to a "moderate to high" degree, whereas in the case of an undifferentiated regime the impact is "positive" (i.e., the threat is removed) and "significant." The situation is similar for a country's capacity to react, or adapt to changes in technology. A differentiated system, by definition, would allow countries to change their IPR regime as they mature technologically or the technological frontier changes: thus the impact is "positive and significant." Conversely, for an undifferentiated regime: countries committed to it would be "adversely" affected in a "significant" way by their inability to modify the IPR rules of the game to suit shifts in their technological endowments (e.g., tighten those rules as they progressively accumulate such endowments) or to adapt those rules to exogenous technological changes.

How can developing countries respond sensibly to these new forces? An illustration of the complexities involved in deciding about the different

[10]According to a survey by the Council of Americas (1987).

TABLE 4-2 Impact of IPR Systems in an Open Economy

	Nature of the System	
	Differentiated	Undifferentiated
On trade flows	Adverse and small	Positive and small
On investment and technology flows	Adverse; potentially large in some areas	Positive; significant in some areas
On threat of retaliation	Adverse; moderate to high	Positive; significant
On capacity to react, to adapt to changes in technology	Positive; significant	Adverse; significant

trade-offs in an open economy comes from plant breeding in Brazil. In that country, universities and government research institutions have become quite proficient in breeding new and better varieties through classical genetics methods.[11] Most food seed varieties currently used by farmers have been produced by local institutions. Although it has a strong tradition in plant breeding through classical genetics, Brazilian research in genetic engineering is still incipient. With the advent of biotechnologically engineered plant varieties (more than 300 were being tried out in 1990), the country will have to rely to a growing extent on the research results of foreign biotechnology firms. Yet in the absence of protection, foreign investments in the area are confined to firms producing hybrids (such as corn) that are naturally (genetically) protected through the control of lineages.[12]

The inability to protect plant varieties by classical or genetically engineered methods appears to have had an overall detrimental effect on local research institutions. Products of their research can be used without compensation. Researchers and local institutions are thus being deprived of an important source of income (in the form of royalties) at a time when the government, due to a chronic fiscal imbalance, has cut public sector wages and budgetary allocations to its agencies, including those engaged in agricultural research. Even more important, domestic researchers cannot interact with their foreign counterparts engaged in biotechnology research: for one, they have nothing "to offer" that is not already freely available and

[11]Major research institutions in plant genetics have been the Instituto Agronomico de Campinas, Escola Superior de Agricultura Luiz de Queiroz, the Federal University of Vicosa, and Embrapa (Empresa Brasileira de Pesquisa Agropecuaria).

[12]Even such natural protection will not last long with the advent of new methods of genetic mapping for rapidly decoding the lineages genotypes—RFLP and RAPID—that will make it possible to "reverse engineer" seed hybrids and reproduce them accordingly (see Barreto de Castro, 1991).

despite the potential of combining the two techniques (classical genetics and biotechnology), they will not profit from it. For another, foreign firms, unable to license their biotechnology inventions, will not be willing to make them accessible to local firms and researchers, and will introduce varieties engineered in ways that cannot be copied.

In a number of instances, genetically engineered (transgenic) plants may be superior and would substitute for those obtained by classical methods.[13] In many others, however, commercial success will be attained by firms combining the two technologies. Access to the innovations of genetic engineering firms (particularly synthetic genes and their expression vectors) would be critical for applying them to genetically improved plants, thus leveraging much of classical genetics work. Granting some form of protection to local breeders—such as joining the International Convention for the Protection of New Plant Varieties (UPOV)[14]—will allow them to discuss commercial terms and exchange rights with foreign genetic engineering firms, particularly those of small and medium size that would be interested in penetrating a potentially large market for their products. For biotechnology firms, some form of protection (possibly patents) will be required, which at least opens up the possibility of licensing biotechnology products and processes. Cross-licensing and similar arrangements would be natural solutions for research units that lack each other's skills.

The absence of protection in this case clearly goes against the national interest. Large international firms in the seed production business, in association with biotechnology firms, will continue to sell in the Brazilian market either through naturally protected hybrids or by combining the two technologies. Thus, they will be introducing new genes and their vectors of expression into varieties that have been developed by local institutions through classical methods and over a considerable period of time, profiting from the results without having to share their profits with those partially responsible for the innovation.

The particular form of protection that Brazil should grant to breeders will have to be evaluated carefully; as already suggested, PBRs may have

---

[13]Take the case of delayed ripening of fruits, flowers, and vegetables through genetic manipulation. Researchers have been able to interrupt the natural production of ethylene, the gas responsible for ripening, by altering the gene responsible for releasing the key enzyme in the process—the ACC synthase. Scientists were able to rebuild the RNA of this gene, and used a bacteria to reintroduce the gene into a tomato, the object of the experiment. As a result, 99.5 percent of the production of the gas was blocked for an entire year. Needless to say, the implications of such technique are most significant: perishables can be transported over long distances without any type of refrigeration; more generally, storage losses will be reduced drastically without the need of radiation or agrotoxics for preservation purposes.

[14]This may be proposed shortly to the Brazilian Congress, possibly in the context of major reforms of the IPR system.

superior features to patents.[15] In the case of genetically engineered processes and products, the choice will focus less on the system itself (patents) than on its specifics (term, rules on compulsory licensing, penalties for violations, and disputes settlements). Biotechnology falls within the category of still thin research and production capabilities: moderately stringent rules of protection are unlikely to have adverse effects. In the fundamentally important area of plant variety improvements, strengthening IPRs in biotechnology needs to be accompanied by increased protection in the area of classical plant genetics.

CONCLUDING REMARKS: A GLOBAL WELFARE PERSPECTIVE

This chapter has argued the proposition that countries should tailor their IPR systems by taking into account their economic needs, productive and research capabilities, and institutional and budgetary constraints. In addition, they should consider how the choice of IPR regime would affect their international economic transactions. Would this approach be at cross-purposes with global welfare maximization? Before attempting to answer this question, it is useful to take a short detour and note the welfare implications of patent protection in a "North-South" context, usually modeled by having research capabilities concentrated in northern firms, while those in the South are able to appropriate research results without cost. In this class of models, free riding by the South generally improves its welfare and correspondingly diminishes that of the North whose firms, of course, always benefit from having their patents recognized in other countries.[16]

Note that this result holds as long as the South is a small part of the world market for the good subject to improvement, so that free riding by the South does not constitute a major disincentive to innovators and that cost savings associated with R&D are not substantial; if they were substantial, the gains to the South from additional R&D undertaken by northern firms in response to a strengthening of the South's IPR regime would more than compensate its income losses. Moreover, if the technological preferences of southern consumers are significantly different from those of the North (i.e., if their needs are quite specific, in terms of disease-fighting drugs, for

[15]Countries joining the UPOV system recognize the rights of breeders to a special title of protection for a specific plant variety, with the breeder required to authorize the commercial utilization of the variety. A title is awarded if the breeder can both describe genetically the new variety and show that it is characterized by homogeneity and genetic stability.

[16]See, for example, the results derived in the context of a North-South duopoly model by Chin and Grossman (1988) and the discussion in Diwan and Rodrik (1989).

example), the welfare gains from extending patent protection to the South may again outweigh income losses to northern firms (in terms of fees, royalties, etc.). When most southern needs must be satisfied by innovations specifically targeting the South's preferences, it may even be in the interest of the South to have a stronger patent protection system than the North, so as to reward R&D efforts targeting smaller and less profitable markets. In this latter case, the trade-off facing the South would be between free riding and stimulating such innovations (Diwan and Rodrik, 1989:9-14).

What are the optimal levels of protection from a *global* welfare perspective? This question can be answered rigorously only at a fairly abstract level and is not often addressed. One of the few papers that does attempt to answer it shows that if one were to value developing countries' welfare gains more highly than those of developed economies (out of an egalitarian concern), then the exact level of protection that maximizes global welfare is indeterminate. Yet numerical simulations suggest that the greater the weight attached to the welfare of developing countries, the lower their level of protection should be (Diwan and Rodrik, 1989:18-19). With a sufficiently large weight, developing countries should be allowed to free ride. Yet just as before, when the welfare of the North and the South were considered separately, these results would be overturned, depending on how specific poorer countries' needs are compared to those countries in which most innovating firms are located. The more distinct their needs or preferences are, the more would global welfare (as well as their own) be enhanced by higher levels of protection in their economies. Finally, in the case of a utilitarian social welfare function, with the welfare of all countries valued equally, one can infer from the model that patent protection should be uniform for global optimality (Diwan and Rodrik, 1989:14-19). The reason is that countries' sizes determine both their relative welfare weights and their importance to innovating firms (i.e., the potential profitability of their markets). To maximize the flow of innovation and aggregate welfare, all countries, developed and developing, should therefore "contribute" to stimulate R&D in direct proportion to their size, which calls for uniform rates of protection.

These theoretical or simulation results, derived from recent models examining the welfare economics of patent protection in North-South contexts, do not suggest that global welfare would be increased by a uniform system, except in the case where the welfare of all countries is valued equally and distribution issues are mute. In this latter case, global welfare would improve by having some countries increase their levels of protection to a common minimum. Yet how should this be effected? Clearly, if that were to go against their national interests, they would have to be either bribed (offered appropriate compensation) or pressured. Although proponents of a uniformly tighter IPR system on the basis of global economic well-being have yet to suggest compensating losers, even if they did, one would still have to find the appropriate mechanisms and instruments to effect such transfers. These could involve developed country incentives for corporations to relocate research facilities, or at least some of their activities, to developing countries; or for firms to engage in research projects that are particularly valuable to those countries and to license the results at favorable fees.

All this presumes that a move toward greater uniformity on the basis of protection parameters inspired by the more advanced countries would be Pareto superior, that is, making all countries better off and no country worse off after a suitable redistribution of income or endowments. Yet that itself is questionable. The links between IPR protection and innovation have yet to be shown to be on a scale that would justify such social engineering efforts in the name of global welfare maximization. There is no question that firms in developed countries lose potential income by having competitors in industrializing countries produce the fruits of their inventiveness. Although the scale at which this happens may be growing, it is doubtful that it justifies either a global redesign of the system or the political capital being spent on pressuring weaker countries to conform to more uniform and tighter protection standards.

The costs of changing the system of protection according to a purely ethical perspective, by taking into account individual countries' needs and relative endowments, would probably be substantial as well. If a Rawlsian criterion of justice were adopted, the objective would be to provide all countries access to innovations to fulfill their basic needs at a cost consistent with their incomes; all other innovations would be licensed on a "full-cost" basis (Rawls, 1971). Needless to say, this would require, in addition to differential rates of protection, a complex mechanism of transfers to stimulate and compensate producers of the basic needs-related innovations. Just as in the case of attempts to mold all countries to a uniform system of protection, it is likely that the cost of implementing such changes would outweigh their benefits.

REFERENCES

Bareto de Castro, L.A. 1991. Propriedade industrial e patentes industriais—implicacacoes para a agropecuaria Brasileira. Mimeo. Cenargen—Centro Nacional de Recursos Geneticos e Biotecnologias, Brasilia.

Chin, J., and G. Grossman. 1988. Intellectual Property Rights and North-South Trade. NBER Working Paper 2769, November.

Council of Americas. 1987. Coping with crisis: U.S. Investment and Latin America continuing economic crises. Mimeo.

Diwan, I., and D. Rodrik. 1989. Patents, Appropriate Technology, and North-South Trade. Washington, D.C.: World Bank PPR Working Papers 251.

Estache, A. 1990. Closed and open issues raised by Brazil's intellectual property rights: A guide for policymakers. Mimeo. Washington, D.C.: The World Bank.

Frischtak, C. 1989. Protection of property rights and industrial technology development in Brazil. World Bank Industry Series Paper no. 13, Washington, D.C.: September. Published in F.W. Rushing and C. Ganz Brown, eds. 1990. Intellectual Property Rights in Science, Technology, and Economic Performance—International Comparisons. Boulder, Colo.: Westview Press.

International Trade Commission. 1988. Foreign Protection of Intellectual Property Rights and the Effect on U.S. Industry and Trade. USITC Publication 1065. Washington, D.C.

McConnell, G. 1966. Private Power and American Democracy. New York: Knopf.

Nogues, J. 1990. Notes on Patents, Distortions and Development. World Bank PPR Working Papers 315. Washington, D.C.: The World Bank

Nordhaus, W. 1969. Invention, Growth and Welfare. Cambridge, Mass.: MIT Press.

Organization for Economic Cooperation and Development. 1987. International technology licensing: Survey results. Mimeo.

Primo Braga, C.A. 1990. Guidance from economic theory. In Strengthening Protection of Intellectual Property in Developing Countries—A Review of the Literature, W. Siebeck, ed. World Bank Discussion Papers 112. Washington, D.C.: The World Bank

Rawls, J. 1971. A Theory of Justice. Cambridge, Mass.: Harvard University Press.

Stigler, G. 1971. The theory of economic regulation. Bell Journal of Economics and Management Science (Spring).

5

Unauthorized Use of Intellectual Property: Effects on Investment, Technology Transfer, and Innovation

EDWIN MANSFIELD

I. INTRODUCTION

This chapter addresses the issue of unauthorized use of intellectual property and its effects on the international environment for innovation, technology transfer, and economic development. More specifically, it considers the economic effects on developing countries, in terms of influencing foreign direct investment and technology transfer, and examines the economic effects on innovating firms, in terms of lost revenues and investment opportunities. The chapter also indicates what information exists and is needed to evaluate the relationships among intellectual property rights protection, unauthorized use, and technological innovation. My primary focus is on the unauthorized use of the products of research and development (R&D), rather than counterfeit consumer items.

Sections II-IV summarize briefly the rationale for the patent system, the current controversies over intellectual property rights, and the changes that often have occurred during industrialization in countries' attitudes toward such rights. After a discussion in Section V of the hypothesis that intellectual property rights protection influences the transfer of technology via foreign direct investment, Sections VI-IX present the preliminary results of a study of 94 U.S. firms that attempts to measure the perceived importance of intellectual property rights protection in this regard and to compare the perceived strength or weakness of intellectual property rights protection in 16 major countries. Sections X-XIV compare our findings with those of other studies and discuss the factors behind some of our results.

In Sections XV and XVI an attempt is made to determine whether the extent of direct foreign investment is related significantly to the perceived strength or weakness of a country's intellectual property rights protection. Section XVII deals with the relationship between intellectual property rights protection, on the one hand, and the composition of direct foreign investment and the age of transferred technology, on the other. Section XVIII summarizes available evidence regarding the effects of the unauthorized use of intellectual property on the sales and profits of U.S. firms. Sections XIX-XXI discuss the results of recent studies of the effects of intellectual property rights protection on the rate of technological innovation and suggest a variety of kinds of research that might be carried out to shed new light on this very important, but inadequately explored, topic. Section XXII provides a summary and conclusions.

II. RATIONALE FOR THE PATENT SYSTEM

Intellectual property consists chiefly of patents, plant breeders' rights, copyrights, trademarks, and trade secrets. Economists have focused more attention on patents than other forms of intellectual property. Ever since the first U.S. patent laws were enacted about 200 years ago, the following arguments have been used to justify the existence of the patent system. First, these laws are viewed as an important incentive to get the inventor to put in the work required to produce an invention. Particularly for the individual inventor, patent protection is claimed to be a strong incentive. Second, patents are viewed as a major incentive for firms to carry out further work and make the necessary investment in pilot plants and other items that are needed to bring the invention to commercial use. If an invention became public property when made, a firm might be unwilling to incur the costs and risks involved in experimenting with a new process or product because another firm could watch, take no risks, and duplicate the process or product if it were successful. Third, it is said that patent laws result in inventions being disclosed earlier than otherwise, the result being that other inventions are facilitated by earlier dissemination of the information.

Despite these arguments, not all economists believe that the patent system is beneficial. Some stress the social costs arising from the fact that a patent is a monopoly right. They point out that patents have been used to establish monopoly positions in industries such as aluminum, shoe machinery, and plate glass. Also, they say that patents are not really important as incentives for innovation because long lead times ensure that most of the profits from many types of innovations can be obtained before imitators have a chance to enter the market. Further, they argue that new knowledge is not used as widely under the patent system as it should be, from the viewpoint of static efficiency. This is because the price of the information should, according to static welfare economics, be set equal to its marginal cost, which is often practically zero. However, the fly in the ointment is that this, of course, would provide no incentive for invention. In essence, a nation's patent laws must reflect a balancing of incentives for inventors and rapid diffusion of new technology.

III. CHANGING POLITICAL ECONOMY OF THE PATENT SYSTEM: EFFECTS ON ECONOMIC DEVELOPMENT

Countries differ greatly in their attitudes toward the patent system. A country like the United States that is a world leader in technology and that carries out huge amounts of research and development obviously stands to gain more from the patent system than a small, impoverished country with practically no scientific or technological capabilities. A country's attitude is likely to change as it industrializes, since the perceived gains and losses from the patent system are likely to be altered considerably in the course of the country's economic development.

Patents often seem to be of little use in a nonindustrialized developing country, for reasons advanced several decades ago by Edith Penrose (1951):

> Any country must lose if it grants monopoly privileges in the domestic market which neither improve nor cheapen the goods available, develop its own productive capacity nor obtain for its producers at least equivalent privileges in other markets. No amount of talk about the "economic unity of the world" can hide the fact that some countries with little export trade in industrial goods and few, if any, inventions for sale have nothing to gain from granting patents on inventions worked and patented abroad except the avoidance of unpleasant foreign retaliation in other directions.

However, when these countries industrialize, their views of the patent system may change, for reasons also pointed out by Penrose (1951):

> If the country is a small one, with a small internal market and fairly specialized export industries, patents in foreign markets may not only be profitable but may be an important incentive to, and protection of, invention and innovation in exporting industries . . . . [Also], to the extent that imitation can be eliminated in foreign markets through patents, design patents, trademarks, and copyrights, the products will be more easily able to retain their specialty character and thus their markets.

Whether such a country will decide to protect foreign inventions within its own borders is a somewhat different question. As Frame (1987) has pointed out, some countries seek patent protection abroad, but offer weak protection to foreign inventors at home.

IV. INTELLECTUAL PROPERTY RIGHTS: INDUSTRIALIZED VERSUS DEVELOPING COUNTRIES

There are well-known differences between the industrialized countries and the developing countries in their attitudes toward intellectual property rights. To the developing countries, such rights give inventors and innovators an undesirable monopoly on advanced technology that can be employed to raise prices and to impose unwarranted restrictions on the use of the technology. To them, the strong enforcement of intellectual property rights would do little to aid their own development; instead, it would tend to hinder their attempts to raise per capita income.

A view commonly expressed in developing countries is that knowledge should be made available at minimal cost to everyone since it is a common property of all, and that because the development of the relatively impoverished countries of the world is a goal that benefits everyone, the technology needed by these countries should be given to them at a low cost. For these and other reasons, many developing countries have relatively weak laws to protect intellectual property and less than diligent enforcement of the laws that exist. Also, they have adopted policies with regard to direct foreign investment and licensing designed to improve the terms on which they can get foreign technology.

The industrialized countries have a substantially different attitude. In their view, intellectual property rights must be respected to provide a fair return to the private investors who take the considerable risks involved in developing and commercializing a new technology. Unless such returns are available, the incentives for inventive and innovative activity will be impaired, to the detriment of all nations, rich or poor. Also, the industrialized countries sometimes assert that the establishment of stronger intellectual property rights would help to promote indigenous technological and innovative activities in the developing countries, although it is generally conceded that this is only one of many relevant factors influencing the indigenous rate of innovation.[1]

---

[1]For a discussion of recent pressures on developing countries to strengthen intellectual property rights, see Mody (1990); also, see Bale (1988), Benko (1987), Chin and Grossman (1990), Clemente (1988), Cortes (1988), Evenson and Ranis (1990), Pack (1987), Richards (1988), and other references.

V. EFFECTS OF INTELLECTUAL PROPERTY RIGHTS PROTECTION ON THE TRANSFER OF TECHNOLOGY VIA FOREIGN DIRECT INVESTMENT

Having summarized briefly some of the current controversies over intellectual property rights, and the changes that often occur during industrialization in countries' attitudes toward such rights, we turn to one of the central topics of this chapter: the relationship between intellectual property rights protection and the transfer of technology through foreign direct investment. As is well known, foreign direct investment is generally regarded as an important means of transferring technology to developing countries.[2] From both policy and analytical perspectives, it is important to obtain a better understanding of the effect, if any, that a developing country's system of intellectual property rights protection has on the transfer of technology to that country through foreign direct investment.

According to some observers, relatively weak intellectual property rights protection in a developing country may reduce the likelihood that multinational firms will invest there. Moreover, even if they do invest there, they may be willing (because of weak intellectual property rights protection) to invest only in wholly owned subsidiaries (not joint ventures with local partners) or to transfer only older technologies. For example, Robert Sherwood (1988) has argued that:

> Those who might send a leading technology to . . . a country [with weak intellectual property rights protection] would soon learn of their folly upon losing it to a competitor. There is an efficient grapevine among companies which do business internationally. If one has a bad experience in a country, all the others soon learn of it. The newer technology is not withheld to harm or abuse that country. It is kept safe at home when safeguards in a host country are defective.

Although these hypotheses may be true, there is little or no evidence to support (or deny) them. With regard to licensing, an Organization for Economic Cooperation and Development (OECD) survey indicates that exchange controls, government regulations (particularly prior approval), and weak protection of intellectual property rights were the most frequently cited disincentives to licensing in developing countries (OECD, 1987: Table 40). Very little seems to be known, however, about the effects of intellectual property rights protection on the nature and amount of technology transferred to a country via direct foreign investment. Clearly, the answer may vary, depending on the industry in question and on the characteristics of the

---

[2]For a recent study bearing on this topic, see Blomstrom and Wolff (1989).

developing country. Also, the answer may vary depending on the nature of the technology.

VI. INTELLECTUAL PROPERTY RIGHTS PROTECTION AND DIRECT FOREIGN INVESTMENT

To test the foregoing hypotheses concerning the effects of intellectual property rights protection on the transfer of technology via foreign direct investment by American firms, I chose a random sample of 100 major U.S. firms in six industries—chemicals (including drugs), transportation equipment, electrical equipment, machinery, food, and metals.[3] Information was requested from each firm concerning the importance of intellectual property rights protection to whether or not the firm would make direct foreign investments of various kinds. Complete or partial data were obtained from 94 of the firms, a very high response rate. The respondents were a mixture of patent attorneys, specialists in the firm's international operations, and top executives. The limitations of survey and interview data of this kind are well known, but with proper caution, such data can be useful.

In practically all of these industries, the proportion of firms indicating that intellectual property rights protection has a strong effect on their foreign direct investments depends heavily on the type of investments in question (Table 5-1). For investment in sales and distribution outlets, only about one-fifth of the firms reported that intellectual property rights protection was of importance. For investment in rudimentary production and assembly facilities, less than one-third said that such protection was important.[4] However, for investment in facilities to manufacture components or complete products, about half said it was important, and for investment in R&D facilities, about four-fifths said it was important.

Also, some industries, more than others, regard intellectual property rights protection as important. For all types of investments other than in sales and distribution outlets, the chemical industry (which includes pharmaceuticals) has the highest percentage of firms regarding intellectual property rights protection as important in this regard. The food and transportation equipment industries tend to have the lowest percentages, and the electrical equipment, metals, and machinery industries tend to rank in the middle. It

---

[3]The frame for this sample was the comprehensive list of major firms in *Business Week*, June 15, 1990; see Mansfield (1991) for details. Note that our results pertain only to U.S. firms. Firms from other countries may have different views concerning the role and importance of intellectual property rights.

[4]Rudimentary production and assembly facilities are ones involving basic technologies that are reasonably well known to all firms in the relevant industry.

TABLE 5-1 Percentage of Major U.S. Firms in Six Industries Where Strength or Weakness of Intellectual Property Rights Protection Reportedly Has Strong Effect on Whether Direct Investments Will Be Made

	Type of Investment					
Industry[a]	Sales and Distribution Outlets	Rudimentary Production and Assembly Facilities	Facilities to Manufacture Components	Facilities to Manufacture Complete Products	Research and Development Facilities	Mean
Chemicals[b]	19	46	71	87	100	65
Transportation equipment	17	17	33	33	80	36
Electrical equipment	15	40	57	74	80	53
Food	29	29	25	43	60	37
Metals	20	40	50	50	80	48
Machinery	23	23	50	65	77	48
Mean	20	32	48	59	80	48

SOURCE: Mansfield (1991).

[a]The number of firms in the sample in each industry is chemicals, 16; transportation equipment, 6; electrical equipment, 35; food, 8; metals, 5; machinery, 24. However, not all firms in the sample responded to all questions.

[b]The chemical industry includes pharmaceuticals.

is interesting to note that there is a very high correlation between an industry's rank in this regard and its rank in previous studies with respect to rough measures of the importance of patents in the innovation process (see section XIX below). Thus, these findings seem to conform with those of earlier studies.

Based on these results, it seems likely that, to the extent that foreign direct investment by U.S. firms is largely devoted to sales and distribution outlets and to rudimentary production and assembly facilities, a country's intellectual property rights protection will have little effect on the total amount invested by U.S. firms in that country. However, it may have a considerable effect on how much is invested in facilities to manufacture components and complete products, as well as R&D facilities.

To see whether—and, if so, how—firms regarding intellectual property rights protection as important with respect to investment in facilities to manufacture complete products, differ from those regarding it as unimportant, we compared the sales volume and percentage of sales devoted to R&D of the firms in each group.[5] The results, shown in Table 5-2, indicate that the firms regarding intellectual property rights protection as important in this respect tend to be larger (in terms of sales) and more R&D-intensive than firms that do not. However, although this is true in all industries combined and in four of the six industries, it is not true for the remaining two industries, as shown in Table 5-2.

VII. INTELLECTUAL PROPERTY RIGHTS PROTECTION AND JOINT VENTURES

Some countries press foreign firms to participate in joint ventures with local firms. These joint ventures generally require the foreign company to share technology with its local partner. Also, foreign firms manufacturing in developing countries may be asked to introduce relatively new technology and to use components produced locally. Coupled with weak patent protection, the foreign firm's technology may become available to local firms at relatively low cost.

The U.S. firms in our sample were asked to indicate whether, in their view, any of 16 countries—Argentina, Brazil, Chile, Hong Kong, India, Indonesia, Japan, Mexico, Nigeria, Philippines, Singapore, South Korea, Spain, Taiwan, Thailand, and Venezuela—had intellectual property rights protection that was too weak in 1991 to permit them to invest in joint ventures (where they contributed advanced technology) with local partners in that country. These countries were chosen because of their size and

[5]The data regarding sales and R&D expenditures pertain to 1989.

TABLE 5-2 Sales and R&D Expenditures of Firms, by Reported Effect of Intellectual Property Protection on Direct Foreign Investment in Facilities to Manufacture Complete Products

	Industry[a]						
Firms Reporting That Intellectual Property Rights Protection Has	Chemicals[b]	Transportation Equipment	Electrical Equipment	Food	Metals	Machinery	Total
Strong effect							
Mean sales[c]	656	731	349	61	10	238	150
R&D (percentage of sales)	19.1	4.6	5.8	2.9	1.6	5.5	8.2
No strong effect							
Mean sales[d]	100	100	100	100	100	100	100
R&D (percentage of sales)	2.7	4.4	9.2	0.6	1.2	6.0	5.5

SOURCE: Mansfield (1991).

[a] See note *a*, Table 5-1.

[b] The chemical industry includes pharmaceuticals.

[c] Mean sales of firms in each industry reporting that intellectual property rights protection has a strong effect on direct foreign investment are expressed as a percentage of the mean sales of those reporting that it does not have a strong effect.

[d] Mean sales of firms in each industry reporting that intellectual property rights protection does not have a strong effect on direct foreign investment is set equal to 100 (see note *c*).

importance, as well as the frequency with which they have been cited in connection with controversies over intellectual property rights protection. With two exceptions (Japan and Spain), these countries are major developing or newly industrialized countries. We include Japan and Spain to enable comparisons to be made to a developed country whose intellectual property rights protection has sometimes been a subject of controversy and to a relatively poor country in Western Europe.

More than 30 percent of the U.S. firms felt that intellectual property rights protection in India, Nigeria, Brazil, and Thailand was too weak to permit them to invest in joint ventures there (Table 5-3). On the other hand, 10 percent or less felt that this was true in Japan or Spain. As would be expected, the proportion of firms feeling that intellectual property rights protection in these countries is, on the average, too weak to permit such investments tends to be highest in the chemical industry, where patents are relatively important, and lowest in the metals and food industries.

VIII. INTELLECTUAL PROPERTY RIGHTS PROTECTION AND TECHNOLOGY TRANSFER TO SUBSIDIARIES

Many firms prefer direct investment in wholly owned subsidiaries as a channel by which to transfer their technology to other countries, particularly if they believe that licensing will give away valuable know-how to foreign producers who are likely to be competitors in the future. Also, firms prefer direct investment over licensing when the technology is sophisticated and foreigners lack the know-how to assimilate it, or when a firm is concerned about protecting quality standards. For example, if a firm licenses technology to a less-than-capable foreign firm and if the foreign firm produces defective merchandise, it may reflect adversely on the firm whose technology was used.

Each of the U.S. firms in our sample was asked whether, if it had a wholly owned subsidiary in one of the 16 countries listed, it would be willing to transfer its newest or most effective technology to such a subsidiary—or whether the weakness of the country's system of intellectual property rights protection would make such transfers very unlikely.[6] According to Table 5-4, 30 percent or more of the firms reported that they would be very unlikely to transfer such technology to India, Thailand, or Nigeria, but less than 5 percent felt this way about Japan or Spain. Singapore seems to

[6]Firms with subsidiaries (or joint ventures) in the country in question were asked this question. Firms without subsidiaries (or joint ventures) were asked whether they would be willing to transfer such technology if they had such a subsidiary. The data in Table 5-4 pertain to all firms but are highly correlated with those pertaining only to firms having such subsidiaries (or joint ventures).

TABLE 5-3 Percentage of Major U.S. Firms Reporting That Intellectual Property Protection Is Too Weak to Permit Them to Invest in Joint Ventures with Local Partners, by Industry and Country

	Industry[a]						
Country	Chemicals[b]	Transportation Equipment	Electrical Equipment	Food	Metals	Machinery	Mean
Argentina	40	0	29	12	0	27	18
Brazil	47	40	31	12	0	65	32
Chile	31	20	29	12	0	23	19
Hong Kong	21	20	38	12	0	9	17
India	80	40	39	38	20	48	44
Indonesia	50	40	29	25	0	25	28
Japan	7	40	10	0	0	0	10
Mexico	47	20	24	25	0	17	22
Nigeria	64	20	39	29	20	24	33
Philippines	43	40	31	12	0	18	24
Singapore	20	40	24	12	20	0	19
South Korea	33	20	21	12	25	26	23
Spain	0	0	10	0	0	4	2
Taiwan	27	40	41	25	20	17	28
Thailand	43	80	21	12	0	20	31
Venezuela	40	20	19	12	0	20	18
Mean	37	30	28	16	7	21	23

SOURCE: Mansfield (1991).

[a]See note *a*, Table 5-1. Some firms reported they had too little information and experience regarding particular countries to provide this information. For these countries, firms of this sort are excluded. The number of firms that had to be excluded for this reason is generally very small.

[b]The chemical industry includes pharmaceuticals.

TABLE 5-4 Percentage of Major U.S. Firms Reporting That Intellectual Property Protection is Too Weak to Permit Them to Transfer Their Newest or Most Effective Technology to Wholly Owned Subsidiaries, by Industry and Country

	Industry[a]						
Country	Chemicals[b]	Transportation Equipment	Electrical Equipment	Food	Metals	Machinery	Mean
Argentina	44	20	21	12	0	14	18
Brazil	50	40	24	12	0	39	28
Chile	47	20	21	12	0	27	21
Hong Kong	21	20	38	12	0	14	18
India	81	40	38	38	20	41	43
Indonesia	40	20	31	25	0	23	23
Japan	0	0	14	0	0	0	2
Mexico	31	20	21	25	0	22	20
Nigeria	67	20	25	25	20	23	30
Philippines	47	40	28	12	0	17	24
Singapore	12	40	21	12	0	0	14
South Korea	31	20	28	12	40	22	26
Spain	0	0	7	0	0	1 3	3
Taiwan	19	40	41	25	0	35	27
Thailand	60	80	31	12	0	18	20
Venezuela	50	20	18	12	0	18	20
Mean	38	28	25	15	5	20	22

SOURCE: Mansfield (1991).

[a]See note *a*, Table 5-3.
[b]The chemical industry includes pharmaceuticals.

be regarded reasonably well, with only about 14 percent of the firms being unwilling to transfer such technology there. The percentage of firms feeling that intellectual property rights protection in these countries is, on the average, too weak to permit such technology transfer is particularly high in the chemical industry and particularly low in the metals and food industries; the industry ranking is the same as in the previous section.

IX. INTELLECTUAL PROPERTY RIGHTS PROTECTION AND LICENSING OF TECHNOLOGY

Firms often prefer to license their technology when the foreign market is too small to warrant direct investment, when the firm with the technology lacks the resources required for direct investment, or when advantages accrue through cross-licensing. Also, as is well known, direct investment has been discouraged by the governments of some countries. Particularly in the developing countries, sometimes there has been considerable hostility toward multinational firms. Some governments feel that their sovereignty is threatened by the great power of the multinational firm over their national economies.

Each of the U.S. firms in our sample was asked to indicate whether the protection of intellectual property rights in each of the 16 countries listed was too weak to permit it to license its newest or most effective technology to unrelated firms in that country. More than 30 percent of the firms said that this was the case for India, Taiwan, Brazil, Thailand, Nigeria, and Indonesia (Table 5-5). Less than 10 percent said this was the case for Spain and Japan. The percentage of firms feeling that intellectual property rights protection in these countries was, on the average, too weak to permit licensing is particularly high in the chemical industry and relatively low in the metals and food industries. It is worth noting that more than two-thirds of the chemical firms believe that intellectual property rights protection in India, Indonesia, Nigeria, Thailand, and Brazil was too weak to permit licensing of their newest or most effective technology there.

X. COMPARISON OF VARIOUS MEASURES OF INTELLECTUAL PROPERTY RIGHTS PROTECTION

In the three previous sections, we have provided three crude measures of the perceived strength or weakness of intellectual property rights protection in 16 countries: (1) the percentage of U.S. firms in our sample feeling that protection there is too weak to permit them to invest in joint ventures with local partners; (2) the percentage feeling that protection is too weak to transfer their newest or most effective technology to a wholly owned subsidiary in that country; and (3) the percentage feeling that protection is too

TABLE 5-5 Percentage of Major U.S. Firms Reporting That Intellectual Property Protection is Too Weak to Permit Licensing Their Newest or Most Effective Technology, by Industry and Country

	Industry[a]						
Country	Chemicals[b]	Transportation Equipment	Electrical Equipment	Food	Metals	Machinery	Mean
Argentina	62	0	26	12	0	29	22
Brazil	69	40	29	25	0	73	39
Chile	47	20	22	12	0	25	21
Hong Kong	33	20	38	12	0	14	20
India	81	40	38	38	20	50	44
Indonesia	73	20	33	25	0	37	31
Japan	12	20	17	0	0	0	8
Mexico	56	20	28	25	0	36	28
Nigeria	73	20	32	38	20	25	35
Philippines	47	40	34	12	0	24	26
Singapore	25	40	24	12	20	0	20
South Korea	38	20	34	12	40	29	29
Spain	6	0	14	0	0	14	6
Taiwan	44	40	55	25	20	36	37
Thailand	73	80	36	12	0	25	38
Venezuela	62	20	21	12	0	26	24
Mean	50	28	30	17	8	28	27

SOURCE: Mansfield (1991).

[a] See note *a*, Table 5-3.
[b] The chemical industry includes pharmaceuticals.

weak to permit them to license their newest or most effective technology to unrelated firms in that country. The roughness of these measures should be emphasized. However, given the fact that other available measures also have many important limitations, these measures should be of interest.

There is a very high correlation between a country's standing based on one of these measures and its standing based on another. The coefficient of determination between the first two of the above measures averages about .73; the coefficient of determination between the first and third measures averages about .85; and the coefficient of determination between the second and third measures averages about .82. The correlation tends to be higher in the food and chemical industries than in the others. If we consider the mean of the six industries, the correlation is higher than in individual industries, the coefficient of determination being more than .90 in each of the three cases. Thus, since these three measures are so highly correlated, which one we use makes relatively little difference for many purposes. In subsequent sections, we often will use (in each industry) the mean of these three measures for a particular country as a rough index of the perceived strength or weakness of intellectual property rights protection in that country (for this industry).

Perhaps surprisingly, there is little correlation between one industry's evaluation of the strength or weakness of intellectual property rights protection in a particular country and another industry's evaluation of the strength or weakness of intellectual property rights protection in the same country. For example, consider our first measure of the strength or weakness of a country's protection system—the percentage of U.S. firms reporting that a country's protection is too weak to permit them to invest in joint ventures with local partners. Although generally there is a moderate amount of correlation (r^2 greater than or equal to .40) among the evaluations by the chemical, food, machinery, and electrical equipment industries, there is little or no correlation between these four industries and the transportation equipment industry or between these four industries and the metals industry. To some extent, the lack of correlation seems to reflect the fact that intellectual property rights protection plays a somewhat different role in each of these industries, as discussed below.[7]

---

[7]Note that the percentage of firms in the metals and transportation equipment industries with foreign subsidiaries or joint ventures in at least one of these countries is as large as this percentage in the electrical equipment and machinery industries; so differences in this regard are not responsible for the lack of correlation.

XI. COMPARISONS WITH OTHER RANKINGS OF COUNTRIES

Ours is by no means the first index of the strength or weakness of intellectual property rights protection. It is interesting to compare our results with the Pharmaceutical Manufacturers Association's (PMA) list of countries with particularly weak intellectual property protection.[8] In general, there are both a reasonable degree of correlation and significant differences. Whereas Nigeria and Taiwan tend to have relatively weak protection based on our measures, they are not on the PMA list; and although Argentina, Chile, Mexico, the Philippines, and Venezuela are on the PMA list, they are not among the weakest based on our measures. In considerable part, this is because our measures extend well beyond the pharmaceutical industry. If we look at our measures based only on the replies of the chemical industry (which includes pharmaceutical firms), our measures agree almost exactly with the PMA list, the only exception being Nigeria.[9]

It is also interesting to compare our measures with that of Rapp and Rozek (1990), who formulate an index of patent protection based upon conformity of a country's patent laws to the minimum standards proposed in the *Guidelines for Standards for the Protection and Enforcement of Patents* of the U.S. Chamber of Commerce Intellectual Property Task Force. Their procedure was based on Gadbaw and Richards (1988). Their index ranks the level of patent protection on a scale from 0 to 5, where 0 is assigned to a country with no patent protection at all and 5 is assigned to a nation whose laws are fully consistent with these minimum standards. As would be expected, there is considerable correlation between their index and ours. Some of the discrepancies may reflect the fact that their index is based solely on the laws on the books, not on the ways these laws are enforced. Also, their index is not broken down by industry. Since interindustry differences are so important, as we have seen, it is necessary for many purposes to construct a separate index for each industry, as we have done here.

XII. COMPARISONS WITH FINDINGS OF THE INTERNATIONAL TRADE COMMISSION

It is also interesting to compare our measures with estimates made by the U.S. International Trade Commission (1988:4-15), which ranked

[8]See Mogee (1989) and Rozek (1990) for this list.

[9]Frame (1987) has constructed an index based on the PMA list and the International Trade Commission data discussed in the following section.

> countries in the approximate order of negative marketplace impact . . . that resulted from inadequate intellectual property protection. In assessing negative marketplace impact, the following factors were considered—market size, share of market lost, export market losses in third countries, reduction in margins through price competition and price controls set by reference to the price of infringing material, goods, or services; use of confidential test data by others, without the respondent's authorization, in securing government approvals; lost manufacturing efficiency because of reduced volume; loss of reputation and diminished value for the company name because of counterfeiting or other infringing activity; and increased product liability costs; the added costs of intellectual property enforcement attempts; the difficulty of doing business in a straightforward, efficient manner; and opportunity losses where inadequate intellectual property protection acted as a deterrent to business activity.

The commission's rankings are based on data for 1986 obtained from 161 American firms in a variety of manufacturing and nonmanufacturing industries. When considering only the 16 countries listed in our study, the rankings (from largest to smallest losses to U.S. respondents) are (1) Taiwan, (2) Mexico, (3) South Korea, (4) Brazil, (5) India, (6) Japan, (7) Nigeria, (8) Hong Kong, (9) Indonesia, (10) Spain, (11) Singapore, (12) the Philippines, (13) Thailand, (14) Venezuela, (15) Argentina, and (16) Chile. The rank-order correlation between these rankings and our own is relatively low (about .33). In particular, Mexico, South Korea, Japan, and Spain seem to be higher on the commission's list of countries (based on negative marketplace impact) than on our list, whereas India, Nigeria, Indonesia, and Thailand seem to be lower on the commission's list than on ours. In part, this may be because the two rankings are measuring different things. The commission is measuring the reduction in profits imposed by a country's firms on U.S. firms, whereas we are looking at the willingness of U.S. firms to engage in joint ventures or to license or utilize advanced technology in a country. To see that these are two different things, note, for example, that U.S. firms may be unwilling to engage in these activities in a particular country even if the profit reductions imposed on them by that country's firms are small (perhaps because the country's firms are not very adept). On the other hand, U.S. firms may be willing to engage in these activities in another country even if the profit reductions are large, because they nonetheless find these activities profitable there. Also, the commission's rankings are influenced heavily by the entertainment industry, which is not included in this study, and by counterfeiting, an activity not taken up here.

In Appendix G of its report, the commission shows the number of times that the firms in its sample reported inadequacies in a country's patent protection regime and inadequacies in remedies and enforcement in 1986. If this number is compared with our measures, one finds that the correlation between them is very low. This may be because the commission's questionnaire asked firms to list countries "in approximate order of importance to you, which you would most like to see adopt fully adequate and effective intellectual property protection" (U.S. International Trade Commission, 1988:D-22). Thus, the countries that are cited most often are not necessarily those in which U.S. firms would be least likely to license or utilize advanced technology. Instead, they are those where U.S. firms felt that their reductions in profit due to weak intellectual property rights protection were greatest. As noted above, these may not be the same thing. Also, countries such as South Korea have acted to strengthen their system of intellectual property rights protection after 1986. This too may help to explain the low correlation.

XIII. REASONS INTELLECTUAL PROPERTY RIGHTS PROTECTION IN SOME COUNTRIES IS REGARDED AS INADEQUATE

As we have seen, a substantial percentage of U.S. firms in some industries regard the protection of intellectual property rights in many of the 16 listed countries as being too weak to enable them to make certain kinds of investments in those countries or to transfer particular types of technology there. Based on interviews with officials of many of these companies, it appears that in deciding whether a particular country's system of protection is too weak, they are especially interested in the answers to three broad questions. First, are the country's laws of sufficient scope to protect their technology? For example, some countries do not permit chemical or pharmaceutical inventions to be patented. Second, does an adequate legal infrastructure exist in the country? In some countries, there are few patent attorneys or other specialists dealing in this area of expertise. Third, are the relevant government agencies in the country able and willing to enforce the laws and to provide prompt and equitable treatment to foreign firms? In some countries, there are reports of corruption and of local firms winning in court with uncanny regularity.

It is not very difficult to see why many U.S. firms would feel that some of the 16 countries have inadequate systems of intellectual property rights protection. In India, no product patents are granted for drugs, chemicals, alloys, optical glass, semiconductors, and intermetallic compounds. In Thailand, firms have complained about the lack of patent protection for chemicals, pharmaceuticals, food and beverages, and agricultural equipment, as well as the weak protection of trademarks and copyrights.[10] In Brazil, no patent

[10]See Sen (1990) on India and Schumann (1990) on Thailand. In the May 25, 1989, report of the U.S. Trade Representative on "Special 301," Thailand and India were leading on its "priority watch list," followed by South Korea, Taiwan, and the People's Republic of China.

protection for chemicals, pharmaceuticals, and foodstuffs exists, and the protection of trade secrets is weak.[11]

In Taiwan, foreign firms have claimed that patent protection for chemicals and pharmaceuticals has been inadequate and that there has been no unfair competition law dealing with false advertising, imitative product packaging, and inaccurate marks of origin. In 1986, a revised patent law was passed that extends full patent protection to chemical and pharmaceutical products. Also, firms unregistered in Taiwan can pursue trademark infringement cases in local courts, and copyright protection has been extended to computer software. Nonetheless, many problems remain; for example, violators can file "invalidation claims" to delay court cases, making plaintiffs defend the legality of their patents or trademarks.[12]

Even where patent protection exists, the term of the patent may be relatively short. In India, the patent term is seven years for food, medicines, and drugs. In some countries, the patent holder must work the invention within one to three years after the patent is granted; otherwise, the patent is subject to compulsory licensing or may lapse. Firms object that they cannot manufacture their products in every nation where they expect patent protection. When compulsory licenses are granted, the royalty rate is often set at 0.5 percent or less of sales, which producers regard as very low. In India and the Philippines, pharmaceutical patents are subject to compulsory license on demand, even if the patent holder does work the invention there.[13]

XIV. FACTORS RESPONSIBLE FOR INTERINDUSTRY VARIATION IN THE EVALUATION OF PROTECTION IN PARTICULAR COUNTRIES

As emphasized above, industries differ considerably in their evaluation of intellectual property rights protection in particular countries. Based on

---

[11]See Frischtak (1990) and Sherwood (1988). Also, metallic admixtures and alloys are not patentable unless they have "specific intrinsic qualities precisely characterized by the nature and proportions of their ingredients or by special treatment." Nonetheless, as shown in Tables 5-3 to 5-5, this seems to have had little or no effect in discouraging U.S. metals firms, illustrating once again the importance of an industry-by-industry analysis. According to some U.S. metals firms, they often can incorporate sensitive technologies in "black boxes" that can be protected.

In April 1991, a new industrial code was forwarded to Brazil's Congress for consideration. This new code would cover pharmaceuticals and other products and processes not currently protected. See Suzigan (1991).

[12]See Schumann (1990) and references in note 13.

[13]See Goans (1986), Hill (1985), Matthews (1988), and the President's Commission on International Competitiveness (1985).

interviews with firms, it is clear that intellectual property rights protection plays a different role for each industry. In some industries such as metals and transportation equipment, it is relatively difficult for competitors to make effective use of a firm's technology without many expensive and complex complementary inputs. In other industries such as chemicals, it is relatively easy for local firms to imitate an innovator's new products. Differences of this sort help to explain why there is little or no correlation between the chemical industry's evaluations of particular countries and the metals or transportation equipment industry's evaluations of these same countries. Because these industries face different problems, they tend to see a particular country in a different light.

Further, a particular country's laws often affect different industries in quite different ways. As pointed out in the previous section, some countries that have adopted a patent system do not extend patent protection to pharmaceuticals and chemicals (or sometimes to food). Clearly, such countries are likely to receive very low marks from the chemical industry, even though other industries—often ones in which patent protection is of less importance in any event—do not regard these countries very negatively. (For example, many food firms do not seem to respond so negatively to countries with weak or nonexistent patents on foods.)

Two cases in point are Argentina and Venezuela. Both of these countries receive blistering evaluations from the chemical industry, due in part to Argentina's laws denying patent protection to pharmaceutical products and Venezuela's lack of patent protection for pharmaceutical products or chemical preparations, reactions, or compounds. (Almost two-thirds of U.S. chemical firms said protection in both of these countries was too weak to permit them to license their newest or most effective technology there.) Yet outside the chemical industry, U.S. firms give both of these countries relatively good marks. (Only about 15 percent of U.S. nonchemical firms said protection in these countries was too weak to permit such licensing.)

Still another factor that may account for interindustry variation in the evaluation of protection in any particular country is the fact that local firms in one industry in this country may be more aggressive in exploiting weak laws and enforcement than local firms in another industry. Thus, even though the intellectual property rights protection really does not vary between these two industries, U.S. firms perceive it to be weaker in the former industry than in the latter. Some observers believe that this helps to explain our findings regarding Argentina; in their view, Argentina's drug firms are much more aggressive in this regard than are other segments of Argentinean industry.

XV. A SIMPLE STATISTICAL ANALYSIS BASED ON COUNTRY DATA

Clearly, the strength of intellectual property rights protection is only one of many factors influencing foreign direct investment. Studies by Dunning (1980), Root and Ahmed (1979), and others have identified a number of variables that may affect the amount of direct investment in a country. For example, Root and Ahmed conclude that six variables are particularly important: (1) per capita gross domestic product (GDP); (2) level of corporate taxation; (3) ratio of exports to imports; (4) extent of urbanization; (5) percentage of GDP attributable to wholesale and retail trade, transport, and communication; and (6) frequency of change of the national executive. Also, they indicate that differences among countries in population should be taken into account (Root and Ahmed, 1979).

To see whether the weakness or strength of a country's intellectual property rights protection seems to be related to the amount of U.S. foreign direct investment in that country when the above variables are held constant, we assume that

$$I_j = B_0 + B_1X_{1j} + B_2X_{2j} + B_3X_{3j} + B_4X_{4j} + B_5X_{5j} + B_6X_{6j} + B_7X_{7j} + B_8X_{8j} + e_j,$$

where I_j is the extent of U.S. foreign direct investment in the jth country in a given year, X_{1j} is the population of the jth country (in 1986); X_{2j} is per capita GDP in the jth country (in 1986); X_{3j} is the level of corporate taxation in the jth country (as estimated for a wholly owned subsidiary with specified and comparable characteristics in each country); X_{4j} is the ratio of the jth country's exports to its imports (during 1983-1986); X_{5j} is the extent of urbanization in the jth country (as measured by the percentage of people in cities of 100,000 or more); X_{6j} is the percentage of the jth country's GDP attributable to wholesale and retail trade, transport, and communications (in 1983-1986); X_{7j} is the frequency of change of the national executive of the jth country (during 1963-1977); X_{8j} is the average over our six industries of the mean of the three measures of the weakness of the jth country's intellectual property rights protection in Tables 5-3 through 5-5; and e_j is a random error term.

By using least squares and omitting Japan because it is a highly developed country, estimates of the B's were obtained. Two kinds of dependent variables were used: the change in the U.S. direct investment position in the jth country, and the U.S. capital outflow to the jth country.[14] For each kind

[14]According to the U.S. Department of Commerce, the direct investment position is the book value of U.S. investors' equity in, and net outstanding loans to, their foreign affiliates. (A foreign affiliate is a foreign business enterprise in which a single U.S. investor owns at least 10 percent of the voting securities or the equivalent.) The change in direct investment position equals capital outflows plus the valuation adjustment. Capital outflows equal reinvested earnings plus intercompany debt outflows plus equity outflows. See Scholl (1990).

of dependent variable, data were used for four periods: 1988, 1989, 1990, and the mean of the three years. Thus, eight regressions were run, each pertaining to a different kind of dependent variable and period. Regardless of which of these dependent variables is used, the relationship between the strength or weakness of a country's intellectual property rights protection and the amount of U.S. direct investment in the country is never close to being statistically significant.[15] That is, the estimated value of B_8 is always far from statistically significant. Moreover, when other nonsignificant independent variables are dropped from the regressions, the estimated value of B_8 remains statistically nonsignificant. Further, if Spain (as well as Japan) is omitted because, unlike other countries in the sample, it is outside Africa, Asia, or Latin America, the results remain the same. Moreover, if independent variables (e.g., measures of human capital formation) suggested by other studies are included in the analysis, the results are unaffected. Whether the estimated value of B_8 is negative or positive depends on which of these many regressions one picks, but in no case is it close to being statistically significant.

XVI. A MORE DISAGGREGATED STATISTICAL ANALYSIS

The simple analysis in the previous section is based on only 15 observations, since it lumps together all industries. A richer analysis can be conducted by recognizing that there are differences among industries as well as countries in the strength or weakness of intellectual property rights protection and by seeing how these differences, as well as those among countries, seem to be related to U.S. foreign direct investment in particular countries in specific industries. Given that our data in Tables 5-3 through 5-5 are broken down by both industry and country, a more disaggregated analysis of this sort is feasible. In this section, it is carried out.

Let I_{ij} be the change in U.S. direct investment position in the ith industry (i = 1, 2,...,6) in the jth country (j = 1,2,...,16), and let P_{ij} be the mean of the three measures (in Tables 5-3 through 5-5) of the weakness of intellectual property rights protection in the ith industry in the jth country. That is, P_{ij} is the mean of the percentage of firms in the ith industry that feel it would not be advisable to invest in joint ventures, transfer new technology to a subsidiary, or license new technology to firms in the jth country. We assume that

$$I_{ij} = A + \theta_i + \phi_j + \gamma P_{ij} + z_{ij},$$

[15]This remains true if Japan is included. These regressions were run by Jeong Lee, as part of the work on his doctoral dissertation.

where θ_i is an industry effect reflecting relevant differences in industrial characteristics, ϕ_j is a country effect reflecting relevant differences in national characteristics (such as the total effect of the seven factors in the previous section), and z_{ij} is a random error term.[16]

Using least squares, I estimated A, γ, and the θ's and ϕ's. The dependent variable takes three forms: the change in U.S. investment position in the ith industry in the jth country in 1990, the change in 1989, and the sum of the changes in both years. Regardless of whether Japan—or both Japan and Spain—are excluded or included and regardless of which period is used, the results provide no evidence that U.S. direct investment tends to be higher in industries and countries where intellectual property rights are relatively strong. What evidence there is seems to be in the opposite direction, but this evidence is never statistically significant. That is, the estimated value of γ never differs significantly from zero.

It may also be of interest to present the results when other dependent variables are used. If investment position abroad and capital outflows are the dependent variables[17] and if Japan and Spain are included in the analysis, the estimates of γ are almost always negative, but far from being statistically significant. If Japan is excluded, the estimates of γ are always positive, but never significant, when investment position abroad is the dependent variable; and they are always negative, but never significant, when capital outflow is the dependent variable. If both Japan and Spain are excluded, the estimates of γ are always positive, but never significant, when investment position abroad is the dependent variable; and they are generally negative, but never significant, when capital outflow is the dependent variable.

To sum up, the analysis of this section (like that in the previous section) provides no statistically significant evidence that the strength or weakness of intellectual property rights protection is related in a major or consistent way to the extent of U.S. foreign direct investment in a given country. However, the crudeness of this analysis should be recognized. In particular, the nature of intellectual property rights protection has been changing in many of these countries, and investment decisions in 1990 (and earlier years) may have been influenced by previous, as well as more recent, levels of intellectual property rights protection. (When the 1991 investment data become available, this problem can be avoided to a greater extent than at

---

[16]Of course, this model is crude in many respects. For one thing, the value of γ, assumed constant in the equation, may vary from industry to industry and from country to country. Given the limited amount of data and of previous quantitative analysis in this area, this model seems to be a reasonable beginning, but its crudeness should be stressed.

[17]Of course, investment position abroad is a stock, not a flow. Consequently, it is heavily dependent on earlier decisions.

present.) Moreover, other limitations cited in previous sections should be recalled. (We are continuing to extend this analysis.)

Even if these results are taken at face value, they do not mean that the strength or weakness of intellectual property rights protection has no effect in this regard. What they do suggest is that intellectual property rights protection is only one of a large number of factors influencing whether U.S. firms increase or reduce their direct investments in a particular country. Thus, the effect of this factor is often swamped by the effects of other factors such as the size and growth of the country's domestic market, the extent of factor supply and rate of increase of factor prices, and the degree of stability of the macroeconomic environment.

This is entirely in accord with the results of our survey and interviews discussed earlier in this chapter. As shown in Table 5-1, the bulk of the firms in our sample felt that for many types of investments, such as sales and distribution outlets, and rudimentary production and assembly facilities, the strength or weakness of intellectual property rights protection is not important. Given that these types of investments are quantitatively large, the results of this section are entirely consistent with our earlier findings in Table 5-1. Moreover, even for those types of investments in which intellectual property rights are important, our interviews indicated that many other factors are important too. This seems to be quite consistent with previous case studies.[18]

XVII. INTELLECTUAL PROPERTY RIGHTS PROTECTION AND COMPOSITION OF DIRECT FOREIGN INVESTMENT

Developing countries are interested in the composition of direct foreign investment, as well as its total volume. Governments realize that the amount of technology transfer to their citizens and firms depends on the kinds of investments made by foreign firms, not just on the dollar volume of such investments. In particular, investments in facilities to manufacture components or complete products are likely to raise the country's technological level to a greater extent than investments in sales and distribution outlets or in rudimentary production and assembly facilities.

According to section II of this chapter, firms tend to be much more likely to regard intellectual property rights protection as important for the former than for the latter types of investment. Thus, a country's system of intellectual property rights protection may influence the composition of direct foreign investment. Whereas U.S. firms may be quite willing to invest

[18]See Frischtak (1990). This finding seems to be in accord with case studies of Nigeria and Turkey; see Adikibi (1988) and Kirim (1985).

considerable amounts in sales and distribution outlets and in rudimentary production and assembly facilities in countries with weak protection, they may be much less inclined to invest in R&D facilities or in facilities to manufacture components or complete products. Such facilities may be more likely to go to countries with stronger protection systems.[19]

However, preliminary analyses based on detailed data collected from 11 major U.S. chemical firms suggest that there is little or no relationship between the strength or weakness of intellectual property protection in a country and the composition of a firm's investment there. For each firm, Jeong Lee determined the percent of its total investment in each of these 16 countries (where it had substantial investments) devoted to sales outlets and rudimentary facilities. For none of these firms was there a significant correlation between this percentage and my index of protection. Although a country's system of intellectual property protection may influence the composition of U.S. firms' investments there, its effects (except in wholly owned subsidiaries) seem to be overwhelmed by those of other factors.

For a few chemical firms, it has been possible to estimate the age of a small sample of technologies transferred via foreign investment to these countries. For present purposes, the age of a technology is defined as the difference between the year the technology was transferred and the year the technology was first used by this firm. The results suggest that U.S. firms tend to transfer somewhat newer technology to countries with relatively strong intellectual property rights protection than to countries with weak protection. However, the sample size is so small that the results should be regarded only as suggestive. Fragmentary data in the machinery industry suggest the same thing, but they, like the chemical data, are too limited to be more than suggestive.[20]

XVIII. EFFECTS OF UNAUTHORIZED USE OF INTELLECTUAL PROPERTY ON SALES AND PROFITS OF U.S. FIRMS

We turn now from the effects of unauthorized use of intellectual property on technology transfer and direct investment in developing countries to its effects on the sales and profits of U.S. firms. This, of course, is a topic that has attracted a great deal of attention from policymakers here and abroad. In 1988, the International Trade Commission (ITC) published a study focusing on the economic effects of weak intellectual property rights

---

[19]Of course, when the firm can defend the new technology through incorporation in "black boxes" or other means, such technology may be sent to countries with weak protection, but such defensive mechanisms are often unavailable or ineffective.

[20]These data are old and pertain to only a few firms. Unfortunately, data of this sort are extremely scarce, which explains why these fragments seem to be worth presenting at all.

TABLE 5-6 U.S. Industry Estimates of Worldwide Sales Losses Due to Weak Intellectual Property Protection, 1986

Industry	Aggregate Estimated Loss by Responding Firms ($ millions)	Number of Firms Reporting		
		No Loss	Loss	Total
Aerospace	120	2	5	7
Building materials	739	0	6	6
Chemicals	1,334	2	18	21
Computers and software	4,130	6	25	31
Electronics	2,288	6	11	17
Entertainment	2,060	0	12	12
Food and beverages	86	2	8	10
Forest products	665	0	7	7
Industrial and farm equipment	622	1	9	10
Metals and metal products	29	1	6	7
Motor vehicles and parts	2,194	0	4	4
Petroleum refining	1,295	3	6	9
Pharmaceuticals	1,909	0	10	10
Publishing and printing	128	0	11	11
Rubber products	511	1	4	5
Scientific and photographic instruments	5,090	1	6	7
Textiles and apparel	251	0	11	11
Other	151	0	8	8
Total	23,845	26	167	193

SOURCE: International Trade Commission (1988).

protection. A questionnaire was sent to a nonrandom sample of about 700 firms, most of whom were members of the Fortune 500. Of the 431 firms responding to the questionnaire, 269 reported that intellectual property (patents, copyrights, trademarks, trade secrets, mask works, proprietary technical data) was of more than nominal importance to their business in 1986, and 167 firms reported that their aggregate losses in sales in 1986 were in excess of $23 billion. Non-R&D-based industries such as entertainment were included in the study. The biggest sales losses were reported by the scientific and photographic instruments, computer, and electronics industries (Table 5-6).

According to the respondents, about $2 billion of sales were lost in the United States because of U.S. imports of infringing goods, about $6 billion of U.S. exports were lost because of inadequate intellectual property protection, and about $3 billion in royalties and fees were lost. Because these figures pertain to only part of the sample, they would obviously seem to be underestimates of the total impact on the sales of U.S. firms in 1986. For

45 firms, the ITC staff made crude estimates of the loss in profits on this account. The estimated aggregate loss was about $750 million, which was approximately 0.7 percent of sales. According to the ITC staff, this may have amounted to about a 10 percent profit reduction for these firms.

A number of industry trade associations and related groups also have estimated the sales and profit losses in their own industries. The U.S. National Agricultural Chemicals Association issued a 1985 report estimating that the U.S. agricultural chemical industry lost about $200 million in 1983. The U.S. Pharmaceutical Manufacturers Association has estimated that U.S. firms lost sales of about $200 million in five countries (Argentina, Brazil, South Korea, Mexico, and Taiwan) in 1984.[21]

The sources of these estimates generally issue warnings that their results may be in considerable error. For example, the International Trade Commission (1988:4-1), says the following:

> The Commission could identify no better means of developing estimates than asking a broad range of firms in the industries most probably affected for the core evidence on U.S. losses from inadequate intellectual property protection—estimates that could admittedly be biased and self-serving. The study, however, built in some cross-checks: data, while estimates, are submitted under oath; data requested on costs of identification and enforcement provided an opportunity for followup inquiries on any discrepancies between losses and enforcement efforts; and estimates were obtained by industry and by country from trade associations and American Chambers of Commerce abroad as a cross-check of the cumulative results of responses by firms . . . . Whereas none of these cross-checks assures high definition or conclusiveness of results, the study found the results of the submissions of firms to be logically consistent internally.

Without detailed information as to the ways in which the firms made the estimates on which these figures are based, it is impossible to evaluate the accuracy of the findings of these studies. This is not to say that data such as those in Table 5-6 are not of interest, but it is very difficult to estimate the sampling errors or biases they contain.

XIX. EFFECTS OF INTELLECTUAL PROPERTY RIGHTS PROTECTION ON THE RATE OF TECHNOLOGICAL INNOVATION

In trying to determine whether it is in a particular country's interest to afford strong protection of intellectual property rights, one of the central questions is: How much effect does such strong protection have on the rate

[21]For a summary, see Mogee (1989).

of technological innovation? Little empirical research was carried out to help answer this question until a decade or so ago, when several studies—by Taylor and Silberston (1973), Mansfield et al. (1981), Mansfield (1986), and Levin et al. (1987)—were carried out. Although these studies focus only on the patent system (rather than other forms of intellectual property) in industrialized (rather than all) countries, and although they do not distinguish between domestic and foreign inventions, their findings are relevant.

All of these studies found that patents are much more important in some industries than in others. Among a random sample of 100 firms from 12 industries (excluding very small firms) in the United States, patent protection was judged to be essential for the development or introduction of 30 percent or more of the inventions in only two industries—pharmaceuticals and chemicals. In another three industries (petroleum, machinery, and fabricated metal products), patent protection was estimated to be essential for the development and introduction of about 10-20 percent of their inventions. In the remaining seven industries (electrical equipment, office equipment, motor vehicles, instruments, primary metals, rubber, and textiles), patent protection was estimated to be of much more limited importance in this regard (Mansfield, 1986). According to another study, product patents were regarded as much more important by the drug and organic chemical industries than by most others (and process patents were regarded as most important by the drug and chemical industries; Levin et al., 1987).

Without question, the patent system enables innovators to appropriate a larger portion of the social benefits from their innovations than would be the case without it, but this does not mean that patents are very effective in this regard. Contrary to popular opinion, patent protection does not make entry impossible, or even unlikely. Within four years of their introduction, 60 percent of the patented successful innovations included in one study had been imitated. Nonetheless, patent protection generally increases the cost (to the imitator) of imitation. According to Mansfield et al. (1981), the median estimated increase in imitation cost was 11 percent. In the ethical drug industry, patents had a bigger impact on imitation costs than in other industries, which helps to account for the fact that patents are regarded as more important in ethical drugs than elsewhere. (The median increase in imitation cost was about 30 percent in ethical drugs in contrast to about 10 percent in chemicals and about 7 percent in electronics and machinery.)

According to some observers, the unauthorized use of intellectual property may grow as more and more players in various parts of the world enter high-technology industries. If intellectual property rights were weakened considerably, it could have unfortunate consequences. The incentives for industrial innovation, already relatively weak in industries where patents are ineffective and entry is easy, might wither to the point where the investment in new and improved products and processes would be far below the socially optimal level. Given the central importance of industrial innovation for economic growth, such an eventuality would do considerable harm, both to the United States and to other countries.[22]

Obviously, changes in the protection of intellectual property rights are likely to have different effects in some countries than in others, and there is no simple way to determine what is in some sense best for the world as a whole. Even in the United States, we lack reliable estimates of how much the volume of inventive and innovative activity would change in response to a weakening or strengthening of intellectual property rights protection. Various kinds of research are needed, some of which are discussed in the following two sections.

XX. NEEDED RESEARCH ON THE EFFECTS OF STRONGER INTELLECTUAL PROPERTY PROTECTION ON TECHNOLOGICAL CHANGE IN DEVELOPING COUNTRIES

It is frequently argued that stronger protection of intellectual property rights would help to promote indigenous technological and innovative activities in the developing countries.[23] This may be true, particularly in those countries that already have reached a minimal level of industrialization and have a reasonable amount of scientific and technological resources. However, there is very little information on which one can base an estimate of how large this effect may be. In this section, I sketch out three types of studies that might be helpful in this regard.

First, a study might be conducted to determine the effects of stronger patent protection on the size and composition of the R&D expenditures of firms located or headquartered in selected developing countries (and the rate of commercialization of new products and processes). Although surveys of business firms have well-known limitations, it would be interesting and useful to find out what the leading executives of a sample of firms in these countries believe would be the effects of stronger patent protection on the size and composition of their firm's R&D expenditures. Findings of this sort would be rough, but nonetheless of use.

---

[22]Of course, although a minimum degree of protection of intellectual property rights seems to be required to foster innovation in particular areas, this does not mean that increases in protection are always socially desirable. For example, see Levin et al. (1987). According to a simple model constructed by Chin and Grossman (1990), developing countries gain by protecting intellectual property if their share of the relevant market is large or if prospects for productivity gains through R&D are sufficiently bright in the industry. For substantial innovations, they find that global welfare is likely to increase with intellectual property rights protection.

[23]For example, see Clemente (1988) and Haagsma (1988).

According to interviews carried out by Robert Sherwood, many companies in Brazil are reluctant to undertake R&D because they know that their rivals can acquire the new technology simply by hiring away their key personnel (Sherwood, 1990). Besides having relatively weak patent protection, firms in Brazil seem to have little recourse to stop loss of trade secrets to competitors in this way. His results suggest that in addition to influencing the amount spent on R&D, the relatively weak protection of intellectual property rights has reduced the productivity of the research and development that is carried out. For example, there is less cooperation among firms in research parks in Brazil and Mexico than in countries such as the United States, and foreigners are less likely to send world-class technology to Brazil (also see Tocker, 1988). The quantitative importance and frequency of occurrence of effects of this sort might be probed in a systematic survey of firms in selected developing countries.

To complement, extend, and check on the results of such a survey, an econometric study might also be made of the effects of strengthened patent protection on firms' R&D expenditures. In a number of countries, patent protection has been strengthened in recent years. For example, in Japan, new chemical (and drug) products could be patented in 1975 and later years, but not before. Using standard econometric techniques, one may be able to estimate the effects of such changes on industrial R&D expenditures. Indeed, Kawaura (1988) has already taken some steps in this direction. In developing countries such as South Korea and Taiwan, it would be interesting to estimate the effects to date on industrial R&D expenditures of the recent strengthening of patent protection regarding drugs and chemicals. Although subject to obvious limitations, the results would be useful.

Second, a study might be carried out to explore the costs and benefits to developing countries of modifying their patent systems. Thus, Robert Evenson (1984) has pointed out:

> In developing countries a relatively high proportion of time is devoted to adaptive invention, much of which is not patentable. Many of these countries have vented frustration over the terms on which technology is purchased in international forums. Few have shown imagination in designing legal systems suited to their competitive position in international invention. Most invention from these countries is adaptive. Yet they have generally not modified their patent systems to encourage adaptive invention. They have instead opted to weaken the scope of patent coverage in an attempt to discourage foreign patenting. In this the slow-growth industrialized economies and the developing economies have been successful. Unfortunately, they have also discouraged national invention in the process.

A study could be carried out to determine the sorts of modifications that developing countries might consider, the potential costs and benefits of each such modification, experiences in other countries with such modifications, and the practical problems in getting these modifications enacted.

Third, a study might be carried out to estimate the effects of stronger intellectual property rights protection on the size and composition of R&D expenditures by multinational firms in developing countries. During the early 1980s, approximately 8 percent of the company-financed R&D of American firms was performed outside the United States. About 60 percent of this R&D was done in Germany, Britain, and Canada, but some was carried out in developing countries. Among the reasons for carrying out R&D outside the United States was the presence of environmental conditions abroad that cannot easily be matched at home, the desirability of doing R&D aimed at the special design needs of overseas markets, the availability and lower cost of skills and talents that are less readily available or more expensive at home, and the greater opportunity to monitor what is going on in relevant scientific and technical fields abroad (Mansfield et al., 1982).

Some observers have suggested that if the protection of intellectual property rights were strengthened, a large amount of the overseas R&D carried out by multinational firms might be performed in developing countries. Because of external economies, this might promote technological change in these countries. Due to the limited scientific and technological resources in most developing countries, as well as other factors, it seems unlikely that a sizable increase in such R&D will occur in many parts of the Third World. Nonetheless, it would be useful to obtain information from various multinational firms as to the conditions under which they would seriously consider establishing or expanding R&D facilities in developing countries and the importance of strong intellectual property rights protection relative to other factors in making the R&D location decision. A considerable amount of research has been carried out concerning the factors influencing the location of R&D facilities. By building on that work, it may be feasible to obtain information of this sort.[24]

XXI. NEEDED RESEARCH ON EFFECTS OF STRONGER INTELLECTUAL PROPERTY PROTECTION IN DEVELOPING COUNTRIES ON INNOVATION IN DEVELOPED COUNTRIES

Besides affecting the rate of technological change inside their own borders, the developing countries, by providing weak intellectual property rights

---

[24]Richards (1988) has suggested that U.S. firms and government agencies might be willing to increase R&D expenditures in those developing countries that strengthened the protection of intellectual property rights. Bale (1988) has stated that Hewlett Packard's investment in R&D in Singapore and Taiwan would "probably" increase, given the general strengthening of intellectual property rights in these countries.

protection, influence the rate of innovation in developed countries. Firms in the drug, chemical, and other industries in Europe, Japan, and the United States can expect to receive less profit from a particular new product or process than would otherwise be the case. Thus, it seems reasonable to believe that some R&D projects that might otherwise be profitable are not carried out, and some innovations that might otherwise be commercialized are judged not to be worthwhile. Given the fact that the private returns from industrial innovation tend to be considerably less than the social returns (Mansfield et al., 1977), this depresses innovative activity in the developed countries, which in any event may be below the socially optimal level, the result being that the world economy grows less rapidly than otherwise would have occurred.

Research is badly needed to shed light on how large or small these effects are. It may be possible to estimate for a sample of firms the extent of the loss in profit from selected innovations that has been experienced due to weak intellectual property rights protection in developing countries. Using these estimates, one may be able to determine the percentage decrease in discounted profit that would be expected on this account for innovations of various kinds, and the proportion of various kinds of innovations that no longer would be profitable on this account. Rough estimates might also be made of the social losses (to developed and developing countries) resulting from the fact that these innovations are not carried out.

By using data obtained from market research firms specializing in the drug, chemical, and other industries, as well as data published by various Third World countries and information from members of these industries, losses in sales due to weak patent protection might be approximated. By applying the results of various studies of cash flows from innovations in these industries, the effects on the net present value criterion of various proposed innovations could be estimated. Based on the firms' internal records, estimates might be made of the number of proposed innovations that were turned down but would have been accepted if patent protection had been stronger. Also, rough estimates might be made of how frequently new products that would be profitable with patent protection are not proposed in the first place because of the lack of patent protection in developing countries. If the relationship between estimated and actual net present value would have been the same for innovations turned down or not proposed for this reason as for those actually carried out, one might be able to estimate the private returns that were forgone. If the relationship between private and social returns would have been the same for these innovations as for those actually carried out, the social returns that were forgone might also be estimated. Of course, this analysis would be very rough, but at least it would be a beginning.

To illustrate the factors involved, consider drugs to treat tropical diseases. According to drug companies, weak patent protection in developing countries has discouraged research on such diseases. For example, Richard Furland (1988), chairman of the Squibb Corporation, has stated that "most developing nations in South America, Africa, parts of Asia, do not accept patents . . . . And there's very little research being done on tropical diseases because people know that if they develop a drug, the market will be immediately taken over by other people." It may be possible, based on intensive interviews with leading scientists, technologists, market researchers, and others inside and outside the relevant industries, along with statistical analysis of the available data, to shed light on the extent to which weak patent protection has discouraged R&D of various kinds, including that directed at drugs for tropical diseases. Obviously, the results would be rough, but rough results (if based on careful study and interpreted with proper caution) are better than none at all.

To extend these results, it may also be possible to obtain econometric estimates of the effect of stronger patent protection in recent years in countries such as Japan, Korea, and Taiwan on R&D expenditures by U.S. firms, particularly in the pharmaceutical and chemical industries. Using standard econometric techniques, one may be able to estimate the sensitivity of industrial R&D expenditures in the United States to changes in patent protection in selected foreign countries, including some from the Third World. The available data would probably permit the disaggregation of R&D expenditures in some industries such as pharmaceuticals, which would be highly desirable. The findings might be an important check on the results of the studies suggested earlier in this section and would complement them.

XXII. CONCLUSIONS

At least five conclusions seem to follow from the foregoing analysis and discussion. First, the great majority of the U.S. firms in our sample report that the strength or weakness of intellectual property rights protection has an important effect on some, but not all, types of foreign direct investment decisions. Whereas about 80 percent of the firms in our sample maintained that this factor was important with regard to investments in R&D facilities, only about 20 percent said that it was important with regard to sales and distribution outlets. Also, some industries—notably, the chemical (including drugs) industry—regard intellectual property rights as much more important than others, such as the food and transportation equipment industries. In most industries, large and relatively R&D-intensive firms are more likely than other firms to regard intellectual property rights protection as important.

Second, based on the views of these firms concerning whether or not intellectual property rights protection in 16 major countries allows them to invest in joint ventures, transfer new technology to a subsidiary, or license new technology to each of these countries, it is possible to construct a crude index of the perceived strength or weakness of intellectual property rights in each country. In general, the countries in this sample perceived to have the weakest protection are India, Thailand, Brazil, and Nigeria; those perceived to have the strongest protection are Spain, Japan, Hong Kong, and Singapore. However, there is often little correlation between one industry's evaluation of the strength or weakness of intellectual property rights protection in a particular country and another industry's evaluation of the same country. For example, there is little agreement between the chemical industry and the transportation equipment industry.

Third, there seems to be no statistically significant relationship between the perceived strength or weakness of a country's intellectual property rights protection, as measured by the above index, and the extent of U.S. direct investment in that country in the late 1980s and early 1990s. Based on our interviews with company executives, this is not surprising since they stressed the fact that intellectual property rights protection was only one of a great many relevant variables—and frequently not the most important one. Based on their responses, one might expect that the composition of U.S. direct investment would be affected by a country's perceived strength or weakness of protection, but data for 11 chemical firms show little such correlation. Preliminary results suggest that U.S. firms tend to transfer somewhat newer technology to countries with relatively strong intellectual property rights protection than to countries with weak protection.

Fourth, according to estimates collected by the International Trade Commission from 167 U.S. firms, their aggregate losses in sales in 1986 due to weak intellectual property rights protection were more than $23 billion. For 45 firms, the ITC staff made crude estimates of the loss in profits on this account. The estimated aggregate loss was about $750 million, which was about 0.7 percent of sales. According to the ITC staff, this may have amounted to about a 10 percent profit reduction for these firms. Estimates of sales losses have also been made by industry trade associations, such as the National Agricultural Chemicals Association and the Pharmaceutical Manufacturers Association.

Fifth, based on recent studies, it seems to be generally agreed that patents are regarded as much more important in some industries (pharmaceuticals and chemicals, in particular) than in others. Although it is frequently argued that stronger protection of intellectual property rights would help to promote indigenous technological and innovative activities in the developing countries, there is little or no information on which one can base an estimate of how large or small this effect may be. Also, whereas weak intellectual property rights protection in developing countries seems likely to depress the incentives for technological innovation in the developed countries, no estimates have been made of the magnitude of this effect. Some directions in which research might be carried out to shed light on these very difficult—and centrally important—questions have been suggested.

ACKNOWLEDGMENTS

This chapter draws freely on preliminary findings of research I have carried out for the World Bank. Thanks go to Jacques Gorlin, Zvi Griliches, Robert Miller, Ashoka Mody, Mary Ellen Mogee, and Robert Sherwood for helpful comments on an earlier draft.

REFERENCES

Adikibi, A. 1988. The multinational corporation and monopoly of patents in Nigeria. World Development 16:511-526.

Baldwin, R. 1979. Determinants of trade and foreign investment. Review of Economics and Statistics.

Bale, Harvey. 1988. A computer and electronics industry perspective. In Intellectual Property Rights and Capital Formation in the Next Decade, C. Walker and M. Bloomfield, eds. Lanham, Md.: University Press of America.

Beaumont, William. 1986-1987. The new patent law of the People's Republic of China. IDEA 27(1).

Benko, Robert. 1987. Protecting Intellectual Property Rights. Washington, D.C.: American Enterprise Institute.

Blomstrom, M., and E. Wolff. 1989. Multinational Corporations and Productivity Convergence in Mexico. C.V. Starr Center for Applied Economics, New York University.

Chin, J., and G. Grossman. 1990. Intellectual property rights and North-South trade. In The Political Economy of International Trade, R. Jones and A. Krueger, eds. Oxford: Basil Blackwell.

Chudnovsky, D. 1985. The entry into the design and production of complex capital goods: The experiences of Brazil, India, and South Korea. In Capital Goods in Economic Development, M. Fransman, ed. London: Macmillan.

Clemente, C. 1988. A pharmaceutical industry perspective. In Intellectual Property Rights and Capital Formation in the Next Decade, C. Walker and M. Bloomfield, eds. Lanham, Md.: University Press of America.

Cortes, Costa, M. 1988. A view from Brazil. In Intellectual Property Rights and Capital Formation in the Next Decade, C. Walker and M. Bloomfield, eds. Lanham, Md.: University Press of America.

Dahlman, C., and F. Sercovich. 1984. Exports of technology from semi-industrial economies and local technological development. Journal of Development Economics 16:63-69.

Dahlman, C., and F. Valadares Fonseca. 1987. From technological dependence to technological development. In Technology Generation in Latin American Manufacturing Industries, J. Katz, ed. New York: St. Martin's Press.

David, Paul. 1986. Technology diffusion, public policy, and industrial competitiveness. In The Positive Sum Strategy, R. Landau and M. Rosenberg, eds. Washington, D.C.: National Academy of Sciences.

Deolalikar, A., and R. Evenson. 1990. Private inventive activity in Indian manufacturing: Its extent and determinants. In Science and Technology: Lessons for Development Policy, R. Evenson and G. Ranis, eds. Boulder, Colo.: Westview Press.

Dunning, Johni. 1980. Toward an eclectic theory of international production: Some empirical tests. Journal of International Business Studies (Spring/Summer):9-13.

Elias, Victor. 1978. Sources of economic growth in Latin American countries. Review of Economics and Statistics (August).

Evenson, Robert. 1984. International invention: implications for technology market analysis. In R&D, Patents, and Productivity, Z. Griliches, ed., Chicago: University of Chicago Press.

Evenson, Robert and G. Ranis, eds. 1990. Science and Technology: Lessons for Development Policy. Boulder, Colo.: Westview Press.

Frame, J.D. 1987. National commitment to intellectual property protection. Journal of Law and Technology 2 (2):209-227.

Fransman, Martin. 1985. Conceptualizing technical change in the Third World in the 1980s: An interpretive survey. Journal of Development Studies 21:572-652.

Frischtak, Claudio. 1990. The protection of intellectual property rights and industrial technology development in Brazil. In Intellectual Property Rights in Science, Technology, and Economic Performance, F. Rushing and C. Brown, eds. Boulder, Colo.: Westview Press.

Furland, Richard. 1988. Betting the farm on research. New York Times (September 27).

Gadbaw, R., and T. Richards, eds. 1988. Intellectual Property Rights: Global Consensus, Global Conflict? Boulder, Colo.: Westview Press.

Goans, Judy. 1986. Intellectual property abroad. Business America (October 27):2-7.

Grubaugh, S. 1987. Determinants of direct foreign investment. Review of Economics and Statistics 69:149-152.

Haagsma, A. 1988. A View from the European Community. In Intellectual Property Rights and Capital Formation in the Next Decade, C. Walker and M. Bloomfield, eds. Lanham, Md.: University Press of America.

Hill, Eileen. 1985. Commerce Department program seeks greater protection for U.S. intellectual rights. Business America 8:3-9.

Hsia, R. 1973. Technological change in the industrial growth of Hong Kong. In Science and Technology in Economic Growth, B.R. Williams, ed., New York: Macmillan.

Katz, Jorge. 1984. Technological innovation, industrial organization, and comparative advantages of Latin American metal work industries. In Technological Capability in the Third World, M. Fransman and K. King, eds. London: Macmillan.

Katz, Jorge et al. 1987. Productivity and domestic technological search efforts. In Technology Generation in Latin American Manufacturing Industries, J. Katz, ed. New York: St. Martin's Press.

Kawaura, A. 1988. The Impact of Stronger Patent Protections: A Case Study of the Japanese Pharmaceutical Industry. East-West Center, Honolulu, Hawaii.

Kim, Hyunku. 1986. The impact of indigenous R and D and technology transfer on productivity growth in a newly industrialized economy: The case of Korea. Ph.D. dissertation, University of Pittsburgh.

Kirim, A. 1985. Reconsidering patents and economic development: A case study of the Turkish pharmaceutical industry. World Development 14:503-521.

Kravis, I., and R. Lipsey. 1982. The location of overseas production and production for export by U.S. multinational firms. Journal of International Economics.

Lall Sanjaya. 1981. Developing Countries in the International Economy. London: Macmillan.

Lall, Sanjaya. 1984. Exports of technology by newly industrializing countries: An overview. World Development.

Levin, Richard, A. Klevorick, R. Nelson, and S. Winter. 1987. Appropriating the returns from industrial research and development. Brookings Papers on Economic Activity (3):783-820.

Lim, Y. and P. Fong. 1991. Foreign Direct Investment and Industrialization in Malaysia,

Singapore, Taiwan, and Thailand. Paris: Organization for Economic Cooperation and Development.

Machlup, Fritz. 1957. An Economic Review of the Patent System. Study no. 15, U.S. Senate Judiciary Committee, Subcommittee on Patents, Trademarks, and Copyrights, Washington, D.C.

Mansfield, Edwin. 1982. Patents, innovation, and U.S. technology policy. American Patent Law Association Quarterly Journal.

Mansfield, Edwin. 1985. How rapidly does new industrial technology leak out? Journal of Industrial Economics (December):217-223.

Mansfield, Edwin. 1986. Patents and innovation: An empirical study. Management Science (February):173-181.

Mansfield, Edwin. 1988a. Intellectual property rights, technological change and economic growth. In Intellectual Property Rights and Capital Formation in the Next Decade, C. Walker and M. Bloomfield, eds. Lanham, Md.: University Press of America.

Mansfield, Edwin. 1988b. Industrial innovation in Japan and the United States. Science 241:1769-1774.

Mansfield, Edwin. 1988c. The speed and cost of industrial innovation in Japan and the United States: External vs. internal technology. Management Science (October).

Mansfield, Edwin. 1988d. Industrial R and D in Japan and the United States: A comparative study. American Economic Review 78:223-228.

Mansfield, Edwin. 1991. Intellectual Property Rights, Foreign Direct Investment, and Technology Transfer. Center for Economics and Technology, University of Pennsylvania.

Mansfield, Edwin, and Anthony Romeo. 1980. Technology transfer to overseas subsidiaries by U.S.-based firms. Quarterly Journal of Economics 95:737-750.

Mansfield, Edwin, Mark Schwartz, and Sam Wagner. 1981. Imitation costs and patents: An empirical study. Economic Journal (December):907-918.

Mansfield, Edwin, et al. 1977. Social and private rates of return from industrial innovations. Quarterly Journal of Economics (May):221-240.

Mansfield, Edwin, et al. 1979. Foreign trade and U.S. research and development. Review of Economics and Statistics 61:49-57.

Mansfield, Edwin, et al. 1982. Technology Transfer, Productivity, and Economic Policy. New York: W.W. Norton.

Matthews, Chris. 1988. NIC's make progress but do not make grade on intellectual property. Business International (May 23):155-157.

Maxwell, Philip. 1987. Adequate technological strategy in an imperfect economic context. In Technology Generation In Latin American Manufacturing Industries, J. Katz, ed. New York: St. Martin's Press.

Mody, Ashoka. 1990. New international environment for intellectual property rights. In Intellectual Property Rights In Science, Technology, and Economic Performance, F. Rushing and C. Brown, eds. Boulder, Colo.: Westview Press.

Mogee, Mary Ellen. 1989. International trade in chemicals: Intellectual property problems and issues. Great Falls, VA: Paper prepared for Organization for Economic Cooperation and Development, May 18.

Organization for Economic Cooperation and Development. 1987. International Technology Licensing: Survey Results. August.

Pack, Howard. 1987. Productivity, Development, and Industrial Development. New York: Oxford University Press.

Penrose, Edith. 1951. The Economics of the International Patent System. Baltimore: Johns Hopkins University Press.

President's Commission on Industrial Competitiveness. 1985. Global Competition: The New Reality. Washington, D.C.: January.

Primo Braga, C. 1991. The North-South debate on intellectual property rights. In Global Rivalry and Intellectual Property. Ottawa: The Institute of Research on Public Policy.

Rapp, R., and R. Rozek. 1990. Benefits and costs of intellectual property protection in developing countries. Journal of World Trade 24 (2):75-102.

Richards, T. 1988. Intellectual property rights: Reconciling divergent views. In Intellectual Property Rights and Capital Formation in the Next Decade, C. Walker and M. Bloomfield, eds. Lanham, Md.: University Press of America.

Root, F., and A. Ahmed. 1979. Empirical determinants of manufacturing direct foreign investment in developing countries. Economic Development and Cultural Change:751-767.

Rosenberg, N., and C. Frischtak, eds. 1985. International Technology Transfer. New York: Praeger.

Rozek, Richard. 1990. Protection of intellectual property rights: research and development decisions and economic growth. In Intellectual Property Rights In Science, Technology, and Economic Performance, F. Rushing and C. Brown, eds. Boulder, Colo.: Westview Press.

Scherer, F.M., et al. 1959. Patents and the corporation. Unpublished.

Scholl, Russell. 1990. International investment position: Component detail for 1989. Survey of Current Business (June):54-65.

Schumann, Gunda. 1990. Economic development and intellectual property protection in Southeast Asia: Korea, Taiwan, Singapore, and Thailand. In Intellectual Property Rights In Science, Technology, and Economic Performance, F. Rushing and C. Brown, eds. Boulder, Colo.: Westview Press.

Sen, Falguni. 1990. Intellectual property rights and the management of R and D in India. In Intellectual Property Rights In Science, Technology, and Economic Performance, F. Rushing and C. Brown, eds. Boulder, Colo.: Westview Press.

Sherwood, Robert. 1988. The benefits developing countries gain from safeguarding intellectual property (and related papers). Unpublished.

Sherwood, Robert. 1990. Intellectual Property and Economic Development. Boulder, Colo.: Westview Press.

Siebeck, W., with R. Evenson, W. Lesser, and C. Primo Braga. 1990. Strengthening Protection of Intellectual Property in Developing Countries. Washington, D.C.: World Bank Discussion Paper 112.

Stobaugh, Robert. 1969. Where in the world should we put that plant? Harvard Business Review (January-February):129-136.

Suzigan, W. 1991. Brazilian industrial policy and the challenge of competitiveness. Seminar at Woodrow Wilson Center, Washington, D.C., October 18.

Taylor, C., and Z. Silberston. 1973. The Economic Impact of the Patent System. Cambridge: Cambridge University Press.

Teitel, Simon. 1988. Science and technology. In Economic and Social Progress in Latin America. Washington, D.C.: Inter-American Development Bank.

Tocker, E. 1988. A chemical industry perspective. In Intellectual Property Rights and Capital Formation in the Next Decade, C. Walker and M. Bloomfield, eds. Lanham, Md.: University Press of America.

United Nations Conference on Trade and Development. 1981. Planning the Technological Transformation of Developing Countries. New York: United Nations.

United Nations Educational, Scientific and Cultural Organization. 1983. Statistical Yearbook. New York: United Nations.

U.S. International Trade Commission. 1984. The Effects of Foreign Product Counterfeiting on U.S. Industry. Washington, D.C.: USITC Publication 1479.

U.S. International Trade Commission. 1988. Foreign Protection of Intellectual Property Rights and the Effect on U.S. Industry and Trade. Washington, D.C.: USITC Publication 2065.

Westphal, Larry, Y. Rhee, and G. Pursell. 1981. Korean Industrial Competence: Where It Came From. Staff Working Paper No. 469. Washington, D.C.: World Bank.

Westphal, Larry, L. Kim, and C. Dahlman. 1985. Reflections on the Republic of Korea's acquisition of technological capability. In International Technology Transfer. N. Rosenberg and C. Frischtak, eds. New York: Praeger.

Yeutter, C. 1988. Negotiating intellectual property rights protection. In Intellectual Property Rights and Capital Formation in the Next Decade, C. Walker and M. Bloomfield, eds. Lanham, Md.: University Press of America.

Discussion

Discussion of the foregoing topics at the conference focused on what conclusions, if any, can be drawn about the desirability of a worldwide, uniform system of intellectual property right (IPR) protection, versus a system of national IPR regimes that are differentiated according to the country's stage of economic and technological development. No clear consensus arose from the discussion on the superiority of one system over the other. In fact, one of the most prominent themes that was repeated again and again was the lack of an adequate evidentiary base for determining the optimal global IPR system or for making policy prescriptions.

The question was raised of whether stronger IPRs are in the interests of developing countries, or under what conditions they might be in their interest. This question is central to the debate, because if IPRs are not in the interests of developing countries, those countries may naturally resist strengthening their IPR regimes. Moreover, if IPRs are not in the interests of developing countries and if the United States is successful in imposing a uniform, strong IPR system worldwide, the system may result in harm to the economies of those nations. On the other hand, if stronger IPRs can be shown to be in the interests of developing countries, it would simplify the movement toward a uniform worldwide IPR system and would result in benefits to the developing countries and to the world.

As shown in the chapters in this section, many factors must be considered in weighing the benefits and costs of stronger IPRs to any particular country or group of countries. It has even been suggested that under certain conditions (e.g., if a developing country has a strong demand for a product not produced elsewhere), it may be in the interest of a developing country to provide stronger protection for inventions in that product technology than is provided in industrialized countries. However, although conference participants recognized as critical the question of benefits to developing countries, researchers pointed out that there are practically no data to supply an answer.

Some preliminary data useful in the analysis of the costs and benefits of strong IPRs are supplied by Mansfield's research on the effect of weak IPR systems on foreign direct investment (FDI) by U.S.-based companies. If it could be shown that the size and composition of FDI in developing countries are negatively affected by weak IPR regimes, those countries might be influenced to strengthen intellectual property rights. During the discussion, however, the question was raised of whether the same factors motivate FDI by firms based in Japan, Europe, and the newly industrialized countries that motivate U.S. firms. As firms from other countries become more significant players in FDI, developing countries may become less concerned about foreign direct investment by the United States. Moreover, if firms from other countries do not have the same incentives and concerns with respect to FDI and IPRs as U.S.-based firms, the United States can expect less cooperation from their governments in international IPR negotiations. Again, there are no data from research studies to address these questions, although anecdotes based on personal experiences suggest that Japanese companies, for example, view FDI very differently than U.S. firms do.

Another question raised during the discussion was one that is seldom raised in IPR debates—that is, are stronger IPRs in the interest of industrialized countries? It is often taken for granted that this is true. However, the current debate in the United States about patenting of software and genes has raised the possibility that intellectual property protection can be too strong and can squelch vital incremental-improvement inventions, perhaps strengthening the position of firms that have made basic inventions. This issue is related to one discussed by Paul David in Chapter 2—that enforcement of patent rights under some conditions can inhibit technological progress.

An alternative to the current (differentiated) international IPR regime, namely, a uniform, worldwide system of strong IPR protection, was explored in this section. The simple proposition that uniformity is a good thing needs to be examined carefully, however. The point was made during the conference that although uniformity in itself has certain theoretical benefits, the details of the uniform system are important. As in the case of technological standards, there may be broad agreement on the need for standards but intense disagreement on what the standards should be, based on the differing interests of the negotiating parties.

Even if all countries could be persuaded to accept the same IPR laws, those laws could lead to very different effects in different cultures. One discussant argued, for example, that although the Japanese and U.S. patenting institutions are quite similar (see Chapter 6), they have led to quite different effects. In the United States, the IPR system has reinforced the tendency toward the not-invented-here syndrome, whereas in Japan it is said to support a more cooperative approach to technology. Similarly, the same IPR laws may have quite different effects in nations that are at different stages of economic and technological development or productive capability.

The absence of good data and other information on the benefits and costs of strong IPRs to developing countries will likely affect the outcome of the current General Agreement on Tariffs and Trade (GATT) negotiations on trade-related aspects of intellectual property rights. Discussants questioned how far the United States can expect to push the developing countries to strengthen their IPR systems when it cannot be shown that the current level of protection is too low or that stronger protection would be in their interest. If the developing countries cannot be convinced of the benefits to them of strong IPRs, it may be necessary for the United States to make concessions in other areas of the talks to achieve the international IPR regime it desires. How willing is the United States to make the necessary trade-offs? Judging from U.S. intransigence on such issues as agriculture in the GATT talks, one discussant surmised that the United States is unlikely to make such a trade-off for stronger IPRs and that a uniform, worldwide system of strong IPRs is therefore unlikely to emerge from the current round of talks.

III

National and International Approaches to Intellectual Property Rights

Introduction

The prospect of creating a global, uniform system of intellectual property right (IPR) protection faces a formidable challenge given the differences in national IPR regimes worldwide. Although the fundamental premise for granting property rights to inventors and artists—that is, to promote creativity and innovation ultimately to the public's benefit—is a common theme in many, if not most, national systems, the legal regimes of each country have evolved to reflect the culture, philosophy, and commercial history of its people. As Paul David illustrates in Chapter 2, the historical development of intellectual property law has been influenced by perceived national needs, such as to increase technology transfer from abroad, to encourage indigenous innovation, to sustain and regulate individual industries, and to enforce an author's natural rights in his or her creation. As a result, IPRs must be seen as unique policy tools engineered to satisfy national, not international, needs and capacities.

The essays in Chapter 6 highlight some of the major issues concerning the IPR systems in Japan, the European Community (EC), India, and the newly industrializing economies. They describe the specific factors motivating change in each of the national or regional legal structures, and they address the likely reaction of those systems to increased pressure for harmonization.

Although disparities between the developing and the industrialized worlds often receive the most public attention, nations that hold common assumptions about the benefits of strong IPRs also differ markedly in the implementation of those rights. The United States, for example, stands alone in the industrialized world with its practice of granting patents to the "first to invent," rather than the "first to file." Japan, whose level of IPR protection is comparable to that of the United States, uses a narrow interpretation of patent claims, which encourages inventors to cross-license. The Japanese system may be better suited to protecting incremental innovation rather than major, sweeping inventions. Detractors claim that the Japanese system favors Japanese inventors, whose industrial power has been based on seizing the commercial value of an invention by refining and building on breakthroughs that already exist.

James Armstrong, who discusses the IPR system of Japan in Chapter 6, concludes that it does not present an insurmountable barrier to eventual harmonization with U.S. practices. He points out that the law in both nations is dynamic and flexible, and he argues that, however different the two systems, the overriding determinant of eventual harmonization will be the fact that both view strong IPRs as essential to a modern industrial economy. Both Japan and the United States will have to adjust as technology and the world economy change.

In his discussion of the European Community in Chapter 6, Bryan Harris points out that even when harmonization is a collectively established objective, it may be constrained by other factors. The EC's explicit objective of achieving harmonization has been thwarted by politics, industrial opposition, the question of whether harmonization will truly maximize the collective economic interests of the EC, and the sovereignty concerns of the EC's member countries. Suggesting that the EC represents a small scale version of the eventual global debate on harmonization, Harris submits that harmonization for its own sake cannot be justified without a greater understanding of basic issues, such as the relationship between the economic interests of intellectual property owners and of intellectual property users and the question of whether IPRs continue to be a consistent and appropriate legal and economic concept in the face of technological change and the development of international industrial relationships.

Describing India's IPR system in Chapter 6, Deepak Nayyar notes that like all countries, India strives to strike a balance between the interests of producers of scientific and technological knowledge and those who use it. That particular balance point is determined, in India's view, by a nation's level of economic development. Acknowledging the importance of technology for development, Nayyar argues that India faces a resource availability problem that can be solved only within a framework of IPR policies that favor the dissemination of technology. Nayyar claims that the "Dunkel draft," a proposition put forward in the Uruguay Round of the General Agreement on Tariffs and Trade (GATT) discussions on Trade Related Aspects of Intellectual Property Rights (TRIPS), ignores the essential philosophy of the systems of India and other LDCs, and calls for standardization at a much higher level of protection than now exists in India. He does, however, emphasize the importance of considering the TRIPS discussions in the larger context of the multilateral trade negotiations, with their potential for cross-sectoral trade-offs.

In his section of chapter 6, Carlos Primo Braga focuses on the newly industrialized economies (NIEs). He postulates that the recent strengthening of IPR laws among NIEs is a product of the historical correlation of stronger IPR laws with rising levels of economic development and the external trade retaliation pressures of the United States and Europe. He suggests that the NIEs have found it in their economic self-interest to become more closely aligned with the practices of their major trading partners, even beyond what might be typical for their stage of economic development. In Primo Braga's view, this trend is likely to continue.

In Chapter 7, Jacques Gorlin presents an overview of the IPR provisions currently under discussion in the Uruguay Round of the GATT and the World Intellectual Property Organization (WIPO). The push for increased IPR protection through these forums has been strongly opposed by the less developed countries (LDCs), whose IPR systems diverge the most from what has been proposed. In many LDCs, IPR laws are designed to move innovations quickly into the public sector, and they give the inventor only brief and restricted rights to the invention. In general, patent terms are shorter, and compulsory licensing may be used to force a transfer of technology by the inventor, particularly if the patent is not being "worked" in the country. Patents may not be granted for innovations in the fields of food and medicine, because food and medicine are considered to be common rights of all peoples. The rationale of developing countries is that an innovator should not be allowed to maximize individual profit when the nation as a whole is poor and needy. Yet critics observe that a weak level of protection eliminates altogether the incentive to apply for patents, bring products to market, and ultimately innovate, which is the basis for economic growth.

Chapter 7 also focuses on the provisions of the TRIPS Dunkel draft. Included among the key issues of the proposal are the establishment of a 20-year term for patents, limits on the use of compulsory licensing, and the creation of an enforcement mechanism. The provisions would allow for an adjustment period for LDCs. Although the proposal was tabled in December 1991, no agreement has yet been forthcoming. Gorlin suggests that the fate of the IPR package is tied to the success of the overall GATT discussions, which are to be concluded in 1993.

The chapters in this section do not discuss in detail the specific provisions that separate one national system from another. Rather, they provide a representative picture of the source of national interests and rationales that resist the idea of international harmonization, as well as the forces that are creating pressure for movement in that direction. All nations, including the United States, recognize the growing importance of technology and innovation as elements of economic success. The challenge is to find common ground among national IPR regimes that can form the basis of an international system that can offer the benefits of technology and innovation to all.

6

Comparative National Approaches to Intellectual Property Rights

Japan

JAMES E. ARMSTRONG III

Where do we start? We say the word "Japan," and it is a polarized word. The polarized charge means a lot of things. We saw a vigorous week of Japan bashing in December 1991 during the fiftieth anniversary of Pearl Harbor. The Japanese patent system is also a source of controversy, considered by its outspoken American critics as a nontariff trade barrier.

As I try to compare our patent system with that of the Japanese, I would like to raise some questions in your mind. I am not going to try to give you any answers because for some of the problems that we face today, there are no clear answers. Reflecting on Dr. David's comments in Chapter 2, I suppose I am in an opposite camp in defining what a patent is. To me a patent is not a monopoly. A patent is an industrial property right that gives someone the right to exclude others for a limited varying period of time, depending on the country in which the patent is held. If I make an invention covered by one of those "improvement" patents there may be a patent that dominates me, and I in turn may dominate someone else. I may not have the right to use commercially what I invented. So, unless I have a complete right to do everything without limitation, the patent cannot be, in my view, a monopoly.

Let us, in considering comparative national approaches, begin with a country as different from ours as Japan. I first visited Japan about 40 years ago, and it seemed very, very different to me as a young U.S. Air Force officer. After years of seeing many different things in Japan, I would now like to try to find out what is the same. First, where are the similarities? Then we can analyze the differences.

Two similarities come to mind quickly when comparing the U.S. patent system or intellectual property system and the Japanese system. First, the benefits of a patent system per se to a modern industrial economy are well recognized by both countries.

What are those benefits? By disclosing knowledge, rather than keeping it secret, an intellectual property system serves as a stimulus to further technological development. Now, when your computer doesn't work, you may wonder whether technological development is good or bad. I often do, but an intellectual property system does stimulate dissemination of knowledge rather than hiding it. It also gives you a market edge. It gives you, for a limited time perhaps, the right to exclude others if you are an economic unit trying to compete. It also serves as a legal component for technology transfer. Dr. David writes about the know-how component which is usually, in my experience at least, 80 percent of technology transfer, but patents are an additional further legal component that serves to tidy things up. These are certain benefits. If you think about it, there may be many more.

A second similarity is that both systems, the United States and the Japanese, are dynamic; they change. I have been in this business—that is, the patent business—for 37 years now, and I cannot recognize U.S. patent law today as it was when I first knew it and began its practice in 1955. It is so different. Why is it so different? As Dr. David suggests, it had to be responsive to the society in which it exists and our society has changed dramatically since those placid days of the 1950s. The "fifties mind-set" is something that flashed into my head this morning. To me the years were a kind of nice, soft, hazy interlude, those Eisenhower years. It seemed things moved so quietly. Now nothing moves quietly. Why? I suppose Dr. Melvin Calvin, Nobel Prize winner from Berkeley, whom I heard speak about 13 or 14 years ago, was the ultimate prophet when he said that "the microprocessor will change our lives in ways that none of us will ever believe." Give it 10 years. Well, we have seen it in 10 years. Give it 20 years. The whole pace of transmission and dissemination of information is different. International communism collapsed as a result.

The Japanese are presently attempting to file patent applications in the Japanese Patent Office electronically. I have had many opportunities to talk with Japanese colleagues on the subject of electronic filing. I have spent at least, on average, three months in Japan every year for the last 21 years, am fluent in Japanese, and have many chances to talk and interact. I view this new Japanese paperless system as an American inspiration. The former

Commissioner of Patents, Gerry Mossinghoff, went to Japan and first made the "paperless" pronouncement. It was his first visit to Japan. In fact, before going, he called me in for a friendly chat and told me that what he was going to say was that in the future there will be all electronic filing for patents. The Japanese were petrified. I must have had about 50 Japanese friends come up to me and say, "What are we going to do?" Well, the Japanese started electronic filing, and we are still talking about it. That fact might tell us something.

Now, what the Japanese have created I am not sure. They have created a situation in which Japanese patent attorneys have had to purchase—much to the profit of Toshiba and others—some very expensive equipment. They must send a floppy to the Japanese Patent Office, together with the electronic transmission, so the system cannot be called truly "paperless." On the other hand, because of this electronic feed, the Japanese Patent Office is building one of the most marvelous technical data bases that one can imagine. The ultimate result is hard to evaluate now, but of course, this is an age in which the ultimate result, in any case, is very difficult to evaluate.

A quick point: the life of the law is not logic but experience, and we are in a challenging age in which we must continue to learn. We have to learn to deal with these electronic beasts. We are in a period in which all nations are striving for intellectual property harmonization, at least among the industrialized countries. The European Community, Japan, and the United States are trying to harmonize their respective patent systems to benefit inventors and commercial enterprises throughout the world.

What are the differences in the various systems? First to file versus first to invent? Is the utility model good or bad? Is an opposition system good or bad? Is the limitation from the filing date of a patent application on the ultimate life of the patent good or bad? These are things we can debate. Back in those times of the fifties mind-set, I was a young man and a young attorney. I thought that the U.S. system was absolutely the best thing in the world. After many, many years I am open and receptive to other ideas, and I think that the modern world, to use the modern parlance "global village," which I first heard from Bruce Merrifield, is such that we are going to have to adopt and exchange ideas and try to harmonize.

I will leave you with a story on harmonization. On October 28, 1991, I visited one of my Japanese corporate clients, and after we disposed of our business at hand, my host said, "We would like to ask you some questions." There is an organization that many of you know called the Pacific Intellectual Property Association, or PIPA. PIPA has study groups working in cooperation with WIPO (World Intellectual Property Organization) as subcommittees on very specific points concerning harmonization. One point of difference is concerned with the doctrine of equivalents which, without going into technical patent jargon, simply is founded on the question, "Should we take a patent claim literally and thus put the inventor in a straitjacket, or should we be expansive in interpretation while remembering we have to be fair to the public?" The public must know what a patent claim means in order to avoid infringement.

This is the social, economic thrust of equivalents. The United States, as we do in our pendulum society, has swung toward a liberal interpretation of patents. Twenty years ago, the Japanese also interpreted claims nuclearly, or as they said, according to the German system. Under the German system a claim is given a broad interpretation. The Japanese system of interpretation was gradually narrowed, and that is the center of the present debate. Let me continue the story. My Japanese hosts posed to me two problems that were under study in committee on the American side and on the Japanese side. They gave me a complete outline of the Japanese position and the American position and they said, "What do you think?" This conversation was all in Japanese. I got up with my chalk in hand, as I like to do. I said, "Let me take the challenge of taking my own position so that I won't be influenced by either the American group or your group."

In conclusion, I found the Japanese position to be methodical and logical in a Japanese sense, but very incompatible with American tastes. I found the American presentation to be superficial, winging it without going into any real depth on the problem. However, I found that there was a certain existing harmony. If the Japanese could only have looked at it with an understanding of the history, philosophy, and precedent of the U.S. law, they would have seen that American objectives were not off the mark. They could not do so because one's native language is a mind set for everyone. There was a sound basis under American law for achieving what they wanted, but not by Japanese rationale. I leave you with that thought. Harmonization is not easy.

The European Community

BRYAN HARRIS

In many respects the subject of this report concerns, on a worldwide scale, some of the central problems with which, on the scale of half a continent, the European Community is at present preoccupied. The problems are

- whether there is an intrinsic merit in harmonizing intellectual property rights or whether, on the contrary, individual states should be left to pursue what they see as the most advantageous ways of protecting those rights;

- whether the emphasis should be on strict protection of intellectual property rights as a reward to inventors and authors and as an incentive to investment, or whether, on the contrary, the emphasis should be on the widest possible dissemination of the technology and literature normally protected by intellectual property laws;
- whether intellectual property, as a legal and economic concept, is consistent, appropriate, and up to date or whether, on the contrary, it has to be bent out of all recognition when new social or technological requirements demand; and
- whether all can be illustrated from current trends in the European Community.

At first sight, the European Community (EC) appears to favor the maximum degree of harmonization. There is a provision for harmonizing national laws, under Articles 100 and 100A of the European Economic Community (EEC) Treaty. There is already on the EC statute books a first directive harmonizing national trademark laws, and there are various proposals for harmonizing national copyright laws. However, there are some constraints on harmonization. One is that the harmonization provisions are subject to the overriding need to show that they directly affect the establishment and functioning of the common market. According to the Green Paper on Copyright, which the Commission of the European Communities issued three years ago, the legal powers do not extend to law reform for its own sake. Another constraint is that the member states are reluctant to cede all their powers to a community legislature, and the Maastricht Treaty, which is expected to come into force on January 1, 1993, has a specific provision in favor of what Brussels calls "subsidiarity"—that is, the retention of legislative power at the national level unless it is manifestly more effective at the EC level.

Since European Community intellectual property measures are subject to the economic objectives of the common market, they have to meet the requirements of the EEC treaty on at least two points: the provisions on the free movement of goods and services, and the provisions on competition or antitrust. The judgments of the Court of Justice of the European Communities on intellectual property rights are almost entirely concerned with the problems of reconciling these rights with the provisions of the EEC treaty. The results are not always entirely clear or consistent. In *Allen and Hanbury's* v. *Generics UK* (case 434/85), the provisions of the EEC treaty on the free movement of goods took precedence over patent rights; in *Warner Brothers et al.* v. *Christiansen* (case 158/86), the protection of copyright took precedence over the provisions on the free movement of goods. In *Volvo* v. *Veng* (case 238/87), national laws on the protection of models and designs were upheld in the face of EEC rules on competition; but in the recent judgment of the Court of First Instance in *BBC* v. *The Commission of the European*

Communities (case T-70/89), the protection of copyright was clearly subordinated to EEC rules directed against the abuse of a dominant position.

Even European Community legislators have had to keep their eyes firmly on the reconciliation of the strictly defined economic aims and the legal rights appurtenant to intellectual property. In the patent field, the EC has been trying to bring into being a community patent, which will be valid throughout the 12 member states and will thus avoid all the problems of cross-border disputes. However, patentees are not overenthusiastic because relatively few wish to take out patents in more than three or four Member States and the cost of the community-wide patent may be hard to justify. There is a moral here for global protection: Uniformity has obvious merits, but must not be marketed at too high a price.

Although trademarks are an important form of intellectual property, they do not have the same bearing on science and technology as patents and copyright, but two aspects of the European Community's experience in this field are relevant and worth a brief mention. The first is that although the economic pressure to "globalize" the use of trademarks is strong and has benefited some firms trading in Europe, such as the Mars Corporation, there is still a cultural and linguistic resistance to the process. Thus, there is not quite the degree of support for a pan-European trademark system that the community authorities had expected. The second is that while the EC is nevertheless going ahead with its proposals for a community trademark, it is hamstrung by a purely political dispute over where the trademark office should be located. This is a salutary reminder that the concerns of intellectual property experts are in the last event always subordinate to the political process and that legislation on intellectual property is ultimately determined by political considerations.

To some extent this is illustrated by developments in copyright legislation in the European Community. Because the various proposals now under discussion in Brussels are beyond the scope of this report, one specific proposal will suffice to make the point. New technology has greatly facilitated the ability to copy protected works, which is particularly true in the field of audio and visual recording. There is a tendency in some of the member states of the European Community, though not in the United Kingdom or Ireland, to respond to the new technology by imposing a levy on blank recording tapes, the proceeds of which are supposed to go to copyright owners. The struggle among the blank tape industry, the record industry, and the representatives of artists and consumers is not yet resolved. However, a compromise does appear to have been reached in the United States and a similar compromise may well be reached in Japan, based on a small levy shared between the manufacturers of both hardware and software in respect of digital but not analogue, and audio but not video, recording tapes and equipment. The question is whether the European Community will take the same route. If the United States, Japan, and Europe all adopted the same approach, it would be a striking example, from the field of intellectual property, of harmonization in an important sector of world trade. Unfortunately, there is a real risk that this salutary outcome may be frustrated by some of the less attractive features of European Community politics: in particular, a wish to demonstrate that the EC can do better than the United States and Japan and should, in any case, be different.

From a theoretical point of view, there are many loose ends in the community's approach to intellectual property matters. It is far from certain that either know-how or trade secrets are likely to be treated as forms of intellectual property, though for the purposes of EEC antitrust law, the licensing of know-how and the licensing of patents are treated rather similarly. One of the reasons for the uncertainty is the difference between the common law approach and the continental laws on unfair competition: the gulf between them is still quite wide. At the heart of the problem, however, is the question of the taxonomy of intellectual property rights: whether, for example, some of the categories of rights are validly described as a form of property at all; whether the property criterion inhibits the creation of new categories of legal relationships to meet new technological or social circumstances; whether there is really enough in common between patent rights and copyright, for example, to justify their being in the same general legal classification. Patents and trademarks are infringed as a rule only "in the course of trade," whereas copyright protection is far more extensive. This is a fundamental difference. Patents and copyrights can be protected even if their economic value is nil; fundamentally different rules apply to trademarks. (In passing, a conference held in Paris in November 1991 discussed modern methods of valuing intellectual property rights and cast a new light on many of the assumptions about their economic worth.) In both legal and economic terms it is difficult, at any rate on the basis of developments in the European Community, to arrive at a definition of intellectual property that applies to the whole range of existing intellectual property rights, let alone future candidates for recognition.

As for the extrinsic importance of intellectual property rights, more work needs to be done within the European Community to define not only the relationship between the monopoly inherent in those rights and the rules against monopoly abuse, but also the relationship between the economic interests of intellectual property owners and those of intellectual property users. From time to time the Commission of the European Communities and the European Parliament conduct hearings into intellectual property matters, and a letter published recently in the London *Times* commented on the fact that invitations were not being extended to consumers or to licensees and other commercial users of intellectual property. The complaint reflects a feeling, at any rate in Europe, that unless intellectual property interests are prepared to widen the scope of their discussions, they will find themselves overtaken by events and that, ultimately, the really important issues will be dealt with at the international level, less by the World Intellectual Property Organization and more by successive rounds of the General Agreement on Tariffs and Trade. However, whether or not that is a desirable trend, and how far it is being encouraged by the European Community, are beyond the terms of reference of this chapter.

India

DEEPAK NAYYAR

This section attempts to explore the contours of the debate on national and international systems for the protection of intellectual property rights. It outlines the salient features of the system in India, compares it with the systems in industrialized countries, sets out the underlying economic rationale, and examines its relevance for the developing world. Concerns about the international system for intellectual property rights, now proposed in the Uruguay Round, emerge from the analysis. The object is to highlight the strategic issues rather than to provide a systematic or complete discussion of the complex problems.

First, the important characteristics of the patent system in India are described and contrasted with the system in the United States and other industrialized countries. Second, the economic rationale of the system for protecting intellectual property rights in India is outlined. Third, the relevance of the Indian view from the perspective of developing countries is considered. In conclusion, I would like to situate the discussion in the context of the Uruguay Round of multilateral trade negotiations, where India and the United States have been major participants in the debate on a possible international regime for the Trade Related Aspects of Intellectual Property Rights (TRIPS).

CHARACTERISTICS

There is an elaborate system for the protection of intellectual property rights in India embodied, inter alia, in the Patents Act of 1970, the Copyright Act of 1957, the Trade and Merchandise Marks Act of 1958, and the Design Act of 1911. The law of patents is, of course, at the heart of the system. It is neither necessary nor possible to provide an exhaustive description. The salient features of the patent system in India, as elsewhere, are incorporated in the scope, nature, use, and term of patents.

Exclusions from Patentability

Apart from the universal exceptions of public order, law, morality, and injury to human, animal, or plant life and health, the Patents Act in India excludes (1) methods of agriculture or horticulture; (2) any process for the treatment of human beings, animals, or plants; (3) substances intended for use as food or medicine or drugs; and (4) substances produced by chemical processes. The exclusion of microorganisms, plants, and animal varieties is implicit. In sharp contrast, the system in the United States provides for no exclusions except, perhaps, for human beings. Consider, for example, the seven areas of biotechnology: methods for treatment of humans and animals; animals and animal varieties; plants and plant varieties; microorganisms; substances derived from biotechnology; biological processes for the production of animals or plants; methods of horticulture and agriculture. None of them is patentable in India. All of them are patentable in the United States. However, in most other countries, including countries in the European Community where the first three are possibly nonpatentable, there is a fair amount of diversity in exclusions from patentability.

Rights Conferred

It is necessary to make a distinction between product patents and process patents. For chemicals, pharmaceuticals, and food products, the patent law in India permits patentability of processes alone, not of products. The rights conferred by a patent in India are very similar, though not identical, to those of the industrialized countries.

1. For product patents, the rights conferred apply to making, using, and selling the product, which extend to importing and offering it for sale in some industrialized countries of Europe.
2. For process patents, the rights conferred apply only to the use of the process. In the United States, a process patent also confers all rights for products obtained directly from that process.

Burden of Proof

When a patent has been granted and an infringement is claimed, the current general law applicable in India casts the burden of proof on the party that is claiming infringement. In the United States and several other industrialized countries, the burden of proof is reversed.

Compulsory Licensing

To curb monopolistic or restrictive practices and enable governments to use patents in the public interest, compulsory licensing is provided for in most intellectual property rights systems. In India, such provisions exist to meet situations in which reasonable requirements of the public interest with respect to the patented invention are not satisfied (e.g., the patented invention, whether product or process, is not worked on a commercial scale) or where the patented product is not available to the public at reasonable prices. The Patents Act, thus provides for

1. compulsory licensing, on application, in such situations as those described above; and
2. an automatic licenses-of-right system in the case of food, pharmaceuticals, and chemicals, where patents are deemed to be endorsed with the term "licenses of right," on completion of three years from the date of sealing the patent.

The provisions are, of course, subject to the payment of a royalty to the patent holder. In the United States and other industrialized countries, provisions for compulsory licensing and government use are limited to established violations of antitrust laws and public noncommercial purposes, although the practice in some sectors (e.g., in space research), is less restricted.

Term of Protection

The law of patents in India provides for a term of 14 years from the date of filing complete specifications; with respect to process patents for food, drugs, and medicines, however, the term is limited to 7 years. In contrast, in the United States as in most industrialized countries, the term of patents is between 15 and 20 years from the date of filing complete specifications.

RATIONALE

The implicit rationale for, or philosophical foundation of, the intellectual property rights system in India is embodied in three underlying objectives.

First, it seeks to strike a balance between the interests of producers on the one hand and consumers on the other, that is, between those who develop the scientific knowledge or innovation and those who use the goods or services derived from it. Needless to say, every country attempts the same, but the point at which the balance is reached depends on a country's level of development. The level of income in the economy and the stage of development in the society are thus particularly important in this context.

The logic of exclusion from patentability follows from this objective. Methods of horticulture and agriculture, as well as food, are excluded because such a large proportion of the population is dependent on agriculture for its livelihood, and the purchasing power of the poor, even for food, is limited. Drugs and medicines are excluded because millions do not have access to basic health care.

Second, it attempts to ensure rewards for the owners of knowledge or for the innovators but, at the same time, places a limit on the monopoly profits or quasi rents that may be appropriated by the entity that commercializes the technology or transforms the scientific knowledge into a marketable product. This is the logic of compulsory licensing. There are two underlying principles set out in the Patents Act: (1) patents are granted to encourage inventions and to secure that the inventions are worked in India; and (2) patents are not granted merely to enable patentees to enjoy a monopoly for the importation of the patented article.

Third, it attempts to create an environment that is conducive to the diffusion of existing technologies and the development of new technologies, insofar as technology is a basic determinant of development in a society that is a latecomer to industrialization. The patentability of processes but not products in some sectors, and the reduced term of protection for patents, derive from this objective.

RELEVANCE

It would be reasonable to ask whether the Indian approach to intellectual property rights is relevant for developing countries in general or those at similar levels of income and technological development. In my judgment, the answer must be in the affirmative, although there may be differences in degree, emphasis, or nuance.

First, technology is strategic in the process of industrialization. The direction and speed of technological development influence not only the pace but also the quality of economic growth. Thus, an economy that industrializes should be able to move from importation through absorption and adaptation of technology through to the stage of innovation, at least in some sectors, on the path to sustained industrialization. In the pursuit of this objective, late industrializers in Europe, Asia, and Latin America have sought to facilitate their technological transformation through intellectual property rights systems that are, or were, conducive to catching up with the industrialized countries. It is important to recognize that unlike comparative advantage based on natural resource endowments, comparative advantage derived from knowledge or skills can be acquired only through a framework of policies that foster rather than hinder the learning process. Economic history is replete with examples of technological leapfrog. Clearly, at this juncture in the world economy when absolute poverty is an important international concern, developing countries need to capture rather than forgo such opportunities.

Secondly, intellectual property rights systems must recognize differences in levels of development between economies. There are two dimensions of this proposition: (1) What purpose does a good serve if it is available only at a price that is beyond the reach of the majority of people in a society? For instance, medicine or computer software at international prices is simply not affordable in a country with the average income levels of India. (2) There are sectors in which the benefits of knowledge need to be socialized, rather than privatized, for human development. For example, the increasing commercialization of plant-breeding research in the developed countries, supported by patent systems, has far-reaching implications for food and agriculture in developing countries. In a world where a very significant proportion of humankind does not have enough to eat, scientific research on plant genetics or plant varieties should be a public resource rather than private property.

From the perspective of developing countries, therefore, it is both necessary and desirable to create a *differential*, rather than a *uniform*, international regime for the protection of intellectual property rights. Quite apart from the wider considerations set out above, the proposed uniform regime across countries raises two specific issues that need to be highlighted. First, the real constraint for several late industrializers in the developing world is that they do not possess the critical minimum in terms of resources for research and development; hence, technological leads and lags may be determined not so much by scientific ability as by resource availability. Second, there is a basic contradiction between the protection of intellectual property rights through a patent system that does not allow late industrializers to develop such technologies on their own and systems of restrictions on, or licensing of, exports of technologies that are closely held (or captive) so that late industrializers cannot import such technologies.

THE URUGUAY ROUND

In recent years, there has been a sharp acceleration in the pace of technical progress, particularly in sectors such as information, communications, and biotechnology. This has led countries that are technology leaders and technology exporters to seek a major change in the international regime for patent protection to include new products and processes particularly in the sphere of biotechnology, for copyrights to include computer software and informatics, and for strengthening related aspects of the system to protect intellectual property rights extending as far as trade secrets. The underlying logic is that technical progress in many of these sectors is more susceptible to replication, which may erode the rewards for innovators.

Thus far, the international system for the protection of intellectual property rights has been embodied in the legal and institutional framework provided by the World Intellectual Property Organization. However, the industrialized countries have launched a strong initiative in the Uruguay Round of multilateral trade negotiations to create an extended and tighter international system for the protection of intellectual property rights, with provisions for dispute settlement and enforcement as part of the multilateral trading system. The text of the draft agreement, circulated by the Director General of the General Agreement for Tariffs and Trade (GATT) in December 1991, seeks to expand the scope of the intellectual property rights system, increase the life of privileges granted or rights conferred, extend the geographical spread where the privileges or rights can be exercised, reduce the restrictions on the use of rights conferred, and above all, create an enforcement mechanism with retaliation across sectors.

This important departure from the system of intellectual property rights, or patent law, of a country such as India must be recognized rather than ignored. Exclusions from patentability would be confined simply to animals and animal varieties, and plant and plant varieties. It would no longer be possible to limit patentability to processes alone, which would statutorily extend to products. The burden of proof would be reversed. Importation would be deemed as the equivalent of working a patent. Compulsory licensing would be possible only under a very restrictive set of conditions, while automatic licenses of right would disappear. The term of protection for patents would be extended to 20 years. Needless to say, the acceptance of these changes would necessitate amendments across the board in the patent law of India and several developing countries.

The implications of this proposed regime for the absorption, diffusion, and adaptation of technologies, let alone for innovation, in developing countries are far reaching. Much needed technologies may no longer be available at affordable costs. The emergence of a domestic technological capacity may be preempted. Transfer of technology may slow down. The incidence of restrictive business practices by transnational corporations may increase. These are just some of the important implications and consequences which suggest that the emerging international system for the protection of intellectual property rights is bound to be inequitable and inimical from the perspective of developing countries.

The need for a more balanced and equitable system is obvious. The interest of technology followers and technology importers is just as important as the interest of technology leaders and technology exporters. It is essential to ensure rewards for innovators, but surely the protection of monopoly profits or quasi rents for transnational corporations should not take precedence over the interests of consumers in a world characterized by uneven development. It would seem that the proposed agreement on Trade

Related Aspects of Intellectual Property Rights about to be concluded as part of the Uruguay Round, does not have such a balance. The interests of the industrialized countries are the focus of attention, while the interests of the developing countries are the object of neglect.

In conclusion, let me stress that it would be a mistake to consider the debate on TRIPS in isolation. It must be situated in the context of the political economy of multilateral trade negotiations in the Uruguay Round, with linkages across sectors and issues, that seek to change the rules of the game for the international trading system. What is more, it needs to be recognized that the contentious and controversial negotiations in GATT only skim the surface. The phenomenon is not simply about the rules of the game for international trade. It is far more complex than that and must, therefore, be placed in its wider context. The rise of transnational corporations, combined with prodigious technical progress, has brought about a fundamental change in the organization of production, marketing, and distribution in the world economy. It has pushed the frontiers of international trade far beyond goods, into services, technology, information, and knowledge, dismantling the traditional divide between them. This process has just begun. Technical progress has always been labor saving. What is new about recent developments is that informatics and robotics are displacing not only the muscles but also the brains embodied in labor. This is likely to have a profound impact on output, employment, and trade in the world economy. Most of these developments are concentrated in a few industrialized countries and, within these countries, in a few corporate entities. The degree to which the national interests of industrialized countries coincide with the corporate interests of transnational firms is uncertain. The national interests of developing countries, however, are very different, in view of the far-reaching implications for the development process.

The Newly Industrializing Economies

CARLOS ALBERTO PRIMO BRAGA

The objective of this section is to provide a brief survey of the current status of intellectual property right (IPR) systems in newly industrializing economies (NIEs).[1] This analysis focuses on the NIEs for two reasons: (1)

[1]An IPR system (or regime) has many dimensions. First of all, it encompasses many different instruments (patents, copyrights, trademarks, trade secrets, plant breeders rights, protection for mask works, etc.). The strength of a system, in turn, reflects not only the level (extent of limitations on titleholders' rights), duration, and scope of protection provided, but also the capacity of the system to enforce the conferred rights.

these are, according to my perception, the economies best positioned to benefit from IPR protection in the developing world; and (2) in the mid-1980s, producers in these economies were frequently accused of "piracy" of intellectual property. NIEs are defined here as those developing economies that by 1989 had an income per capita of at least U.S. $2,000, a share of manufacturing in gross national product of at least 30 percent, and exports of manufactured products accounting for more than 40 percent of total export revenues. Brazil, Hong Kong, Malaysia, Mexico, Singapore, the Republic of Korea, and Taiwan qualify as NIEs according to these criteria. It is also worth mentioning that all of them were exporting more than U.S. $3 billion per year of knowledge-intensive products by the end of the 1980s.[2] These are economies that either have recently "graduated" as developed countries (surpassing the income per capita threshold of U.S. $6,000) or that are among the top middle-income economies, according to World Bank criteria (World Bank, 1991). For the objectives of this section, however, what makes the NIEs particularly interesting is the fact that a few years ago, with the exception of Hong Kong, all of them were listed among the so-called problem countries (i.e, they were perceived to have defective IPR systems from the perspective of industrialized nations).[3]

The points that I would like to explore are the following: (1) since the mid-1980s, NIEs have strengthened IPR protection in their territories. Actually, many other problem countries (e.g., Argentina, Chile, Indonesia) outside the NIEs category have also enacted reforms of their IPR systems over this period; and (2) there is a close relationship between this cycle of reforms in developing countries and external pressure exerted by developed countries. Some NIEs, however, have approached IPR reform in the context of broader economic reforms. In other words, external pressures are not the only forces shaping these reforms.

---

[2]Knowledge-intensive (or high-tech) products are usually defined as those products for which investments in the creation of knowledge are responsible for a substantial share of their production costs. Industries engaged in the production of knowledge-intensive products typically have a strong interest in IPR protection. For further details on high-tech trade statistics see Primo Braga and Yeats (1992).

[3]Organizations such as the International Intellectual Property Alliance and the Pharmaceutical Manufacturers Association, as well as U.S. governmental institutions (e.g., the Office of the United States Trade Representative and the Patent and Trademark Office), organized several of these lists. It is worth mentioning that a country could, for instance, have a "good" patent law and yet be considered a "problem country" either because of inadequate enforcement of its patent law or because some other aspect of the IPR system (e.g., copyright law or trade secrets) was considered inadequate. For a summary inventory of the so-called problem countries, see Rozek (1990).

RECENT IPR REFORMS IN NEWLY INDUSTRIALIZING ECONOMIES

The attitudes of developing countries toward IPRs changed significantly over the last decade. It is enough to remember, for instance, that many of these countries favored a revision of the Paris Convention in the early 1980s. The objective of this revision was "to weaken the international standards of industrial property protection" (Kunz-Hallstein, 1989:269). Unable to accomplish this objective, developing countries reacted strongly to attempts to introduce Trade Related Aspects of Intellectual Property Rights in the GATT negotiations.[4] Yet, an analysis of the evolution of IPR systems over the last five years shows a clear trend toward higher levels of protection in the developing world.

The case of the NIEs is quite illustrative in this context. Recent developments in trade relations between the United States and the NIEs highlight the above-mentioned trend. Over the last three years, Mexico, South Korea, and Taiwan were removed from the "priority watch list" that identifies the main problem countries according to the Super 301 provision of the 1988 Omnibus Trade and Competitiveness Act. In July 1990 the retaliatory action against Brazil, introduced in October 1988 as a result of a Section 301 investigation focusing on Brazil's lack of patent protection for pharmaceutical products, was discontinued.

Those in favor of high standards of protection may still find many flaws in the current IPR systems of the NIEs, particularly with respect to enforcement. The debate, however, has lost most of its moral overtones, which tended to divide the world in a Manichaean fashion between the forces of light (the supporters of IPR) and the forces of darkness (the "pirates"). In short, as the public debate on IPR is now conducted, there are no more "bad guys" among the NIEs.

At the intellectual level, this may simply reflect a recognition of the inadequacy of attempts to frame the debate in terms of natural law concepts.[5] More fundamentally, however, it reflects the broad scope of the reforms being implemented by these economies.

Probably the most dramatic example in this context is provided by Mexico (Villarreal, 1991). On June 27, 1991, Mexico introduced its new Law for the Development and Protection of Industrial Property. This new law expanded the scope of protection to technological fields, which were until then excluded from patentability (or for which issuance of patents

[4]For further details, see Primo Braga (1989).

[5]For a brief discussion of the role of fairness and morality in shaping IPR laws, see Primo Braga (1990a).

would be allowed only after 1997), and which include biotechnology (including genetic procedures to obtain animal and plant species or their varieties), plant varieties, microorganisms, chemicals, and alloys. The duration of patent protection, which used to be 14 years from the granting of the patent, was extended to 20 years from the date of filing the application. Limitations on patentees' rights (e.g., the granting of compulsory licenses) were restricted to exceptional circumstances. Protection for trademarks and industrial designs was also enhanced, and the law introduced more explicit protection for trade secrets.

The new law is also intended to improve the conditions for enforcement of IPRs by creating a new institution to help the Mexican Patent and Trademark Office: the Industrial Property Institute. It is also worth mentioning that Mexico amended its copyright law in 1991 in an attempt to correct some of its perceived weaknesses. The main change in this context was the adoption of tougher penalties for copyright violations.

Among the Asian NIEs—which, in general, already had higher levels of protection than the Latin American NIEs by the mid-1980s—the last few years have also been characterized by additional reforms designed to strengthen IPR protection: (1) Singapore enacted a new Copyright Act in 1987, expanding its scope and significantly increasing penalties for infringement. (2) Taiwan's Patent Law was amended in 1986, reversing the burden of proof to the alleged infringer and increasing penalties for IPR infringements. In 1991, a new Fair Trade Law was enacted, which provides for protection of trade secrets. Copyright law is also being revised with the goal of strengthening protection to a level similar to the one prevailing among Berne Convention signatories. (3) Korea amended its Patent Act in 1986, extending the term of patent protection (from 12 to 15 years), reverting the burden of proof, and increasing the requirements for compulsory licensing. Enforcement efforts have significantly increased since then, and a trade secrets law is being drafted. (4) Malaysia in turn revised its copyright law and acceded to the Berne Convention in 1990.[6]

Finally, let us take a look at Brazil, which was one of the main opponents of the movement toward higher levels of IPR protection in the early 1980s. The Brazilian Software Law of 1987 extended copyright protection to computer programs. Enforcement efforts to protect IPR have increased significantly over the last few years, particularly with respect to software and home-video cassettes. In 1991, the Collor administration submitted to the Brazilian Congress a draft law reviewing Brazil's system of "industrial protection." Among the main changes proposed in the new law are the

[6]For further details on the IPR systems of the Asian NIEs, see Schumann (1990) and U.S. Trade Representative (several years).

reduction of technological fields excluded from patentability, for example, patent protection for pharmaceutical products, the extension of the duration of a patent to 20 years from filing, and a more explicit provision for the protection of trade secrets. This new law is still being debated in the Brazilian Congress, but it is quite clear that Brazil is also moving—even though at a slower pace—in the same direction as the other NIEs.

The recognition that all NIEs have strengthened IPR protection since 1986 leads us to the next question: What forces have fostered these changes?

THE FORCES BEHIND IPR REFORMS IN THE NEWLY INDUSTRIALIZING ECONOMIES

Some analysts, including myself (Primo Braga, 1989, 1990b), believe that historically the level of IPR protection has been positively correlated with the level of economic development. Such a proposition could, in principle, be used to rationalize the recent "wave" of reforms as follows: The "trade-off between encouraging the diffusion of existing technology through unlicensed imitation and stimulating the creation of new technology becomes steeper over time" (Frischtak, 1989:1), as a country develops, accumulating human capital. Accordingly, there is a "development threshold" after which the protection of IPR generates net welfare gains and the political economy of the process would tend to favor innovators against imitators. If one assumes that the NIEs have reached this threshold, the ongoing cycle of reforms could be understood as the natural outcome of domestic pressures, with external forces acting as catalysts in the process.

An alternative characterization would stress the role of external pressures, particularly those exerted by the United States in paving the way for reforms. Actually, there is an obvious match between U.S. actions, either via unilateral initiatives (e.g., Section 301 investigations) or via bilateral negotiations, and the pace of reform in NIEs. It is also worth mentioning that the multilateral negotiations on TRIPS—a U.S. initiative—have likewise contributed to putting IPR protection on the agendas of policymakers around the world. From this perspective, the reforms reflect mainly the economic weight of the threat of trade retaliations.[7]

It seems to me that the truth is somewhere between these two alternative models. There is no doubt that external pressures have played a major role in the process. Yet, there is no simple relationship between the magnitude of the external pressure applied and the dimensions/characteristics of the reforms implemented. In the case of Brazil, for instance, despite sig-

[7]For a formal analysis of the relevant welfare function in this context, see Primo Braga (1989).

nificant external pressure, the pace and scope of the ongoing reform cannot be characterized as dramatic. On the other hand, the sweeping Mexican reform—Mexico now presents levels of protection similar to ones prevailing in the industrialized world—took most observers by surprise.

FINAL REMARKS

It seems that the greater the degree of openness (a concept encompassing both trade orientation and the treatment accorded to foreign capital) of an economy, the higher is the probability that it will pursue "systemic convergence" with its major economic partners. The recent evolution of IPR systems in the NIEs provide an illustration of this proposition. Countries pursuing "systemic convergence" (e.g., Mexico) are willing to upgrade their systems beyond the levels that would be typical for their stage of development.[8]

The net welfare impact of these reforms, however, remains an empirical question. Past experience shows that a strong IPR system is not a necessary condition for technological development. It is worth mentioning that some of the NIEs had already achieved comparative advantage in knowledge-intensive products in the 1980s and were able to attract significant flows of foreign direct investment, despite the flaws of their IPR system.[9] It can be argued, however, that given the increasing globalization of economic activities, systemic convergence has become a necessary condition for countries to pursue an outward oriented development strategy.

There is a presumption that these reforms will foster domestic R&D and foreign direct investment flows, contributing to an expansion of the innovative capacity of these economies. The potential anticompetitive implications of the reforms, however, should not be forgotten.[10] For those involved in research focusing on the economic role of science and technology in the developing world, the NIEs provide fertile ground for empirical analyses of the net welfare effects of strengthening IPR protection.

Finally, it is important to acknowledge the significant progress achieved in the TRIPS negotiations, as reflected in the draft final act of the Uruguay Round, presented by the GATT Secretariat in December 1991. The multilateral solution provides our best hope to avoid the proliferation of IPR-related

---

[8]For an analysis of predicted patent protection levels according to stage of economic development, see Rapp and Rozek (1990).

[9]See Primo Braga and Yeats (1991) for data on revealed comparative advantage in high-tech products.

[10]For conflicting evaluations of the welfare impact of introducing patents for pharmaceutical products in a developing country, see Rapp and Rozek (1990) and Challu (1991).

trade frictions. It is quite clear that the present text does not please all parties involved—for example the lack of retroactive "pipeline" patents is considered a major shortcoming by certain segments of the pharmaceutical and chemical industries; the transition periods for the implementation of IPR reforms are considered excessive by some. Future negotiations will determine to what extent the TRIPS agreement will become an effective force in the promotion of systemic convergence. In its absence, however, the IPR debate at the international level can easily become a discussion about rent shifting. This would be unfortunate because it would preclude a more balanced evaluation of the role that IPR may play in the developing world.

REFERENCES

Challu, P., et al. 1991. Patenteamento de Productos Farmaceuticos: Consecuencias. Buenos Aires: Sociedad Impresora Americana.

Frischtak, C.R. 1989. The Protection of Intellectual Property Rights and Industrial Technology Development in Brazil, World Bank Industry Series Paper No. 13. Washington, D.C.: World Bank.

Kunz-Hallstein, H.P. 1989. The United States proposal for a GATT agreement on intellectual property and the Paris Convention for the Protection of Industrial Property. Vanderbilt Journal of Transnational Law 22(2):265-284.

Primo Braga, C.A. 1989. The Economics of Intellectual Property Rights and the GATT: A View from the South. Vanderbilt Journal of Transnational Law 22(2):243-264.

Primo Braga, C.A. 1990a. Guidance from economic theory. In Strengthening Protection of Intellectual Property in Developing Countries: A Survey of the Literature, W. Siebeck, ed. World Bank Discussion Paper No. 112. Washington, D.C.: World Bank.

Primo Braga, C.A. 1990b. The developing country case for and against intellectual property protection. In Strengthening Protection of Intellectual Property in Developing Countries: A Survey of the Literature, W. Siebeck, ed. World Bank Discussion Paper No. 112. Washington, D.C.: World Bank.

Primo Braga, C.A., and A. Yeats. 1992. The simple arithmetic of existing multilateral trading arrangements and implications for a post-Uruguay Round world. International Trade Division, World Bank, processed.

Rapp, R.T., and R.P. Rozek. 1990. Benefits and costs of intellectual property protection in developing countries. Journal of World Trade 24(5):76-102.

Rozek, R.P. 1990. Protection of intellectual property rights: research and development decisions and economic growth. In Intellectual Property Rights in Science, Technology, and Economic Performance: International Comparisons, F.W. Rushing and C.G. Brown, eds. Boulder, Colo.: Westview Press.

Schumann, G. 1990. Economic development and intellectual property protection in Southeast Asia: Korea, Taiwan, Singapore and Thailand. In Intellectual Property Rights in Science, Technology, and Economic Performance: International Comparisons, F.W. Rushing and C.G. Brown, eds. Boulder, Colo.: Westview Press.

U.S. Trade Representative. Several years. Foreign Trade Barriers. Washington, D.C.: U.S. Government Printing Office.

Villarreal Gonda, R. 1991. The New Mexican Industrial Property Law. Processed.

World Bank. 1991. World Development Report 1991: The Challenge of Development. New York: Oxford University Press.

7

Update on International Negotiations on Intellectual Property Rights

JACQUES J. GORLIN

GENERAL COMMENTS

Currently, the two principal international organizations that deal with intellectual property matters are the General Agreement on Tariffs and Trade (GATT) and the World Intellectual Property Organization (WIPO). This chapter reviews the current state of play in the GATT negotiations on intellectual property—the Trade Related Aspects of Intellectual Property Rights (TRIPS) negotiations—and in the WIPO negotiations on a patent law harmonization treaty, a trademark law harmonization treaty, a possible protocol to the Berne Copyright Convention, and a dispute settlement treaty. Although other negotiations are currently underway in WIPO and the United States continues to pursue bilateral efforts under the Special 301 provisions of the Omnibus Trade and Competitiveness Act of 1988, the negotiations discussed below best typify the new "competitive" situation in the field of international intellectual property negotiations.

The GATT negotiations on TRIPS and discussions in WIPO share similar objectives: the strengthened protection and improved enforcement of intellectual property rights via multilateral instruments. However, the overwhelming interest of the principal developed countries in a trade-based multilateral regime for intellectual property as part of the current GATT Uruguay Round, and the fact that the TRIPS negotiations are in their final stage have pushed the TRIPS negotiations to the fore. While attention is currently focused on the GATT TRIPS negotiations, the WIPO negotiations are, to a large extent, on hold.

NEGOTIATIONS ON TRADE RELATED ASPECTS OF INTERNATIONAL PROPERTY RIGHTS[1]

On December 20, 1991, GATT Director General Arthur Dunkel tabled a draft final act, which offered "a concrete and comprehensive representation of the final global package of the results of the Uruguay Round." Included in this 450-page text was an agreement on TRIPS. The TRIPS text reflected the combined efforts of the individual country negotiators and, where consensus could not be reached, the views of Ambassador Lars Anell of Sweden, the chairman of the TRIPS negotiating group, and the GATT Secretariat. Although the document was presented on an almost "take-it-or-leave-it" basis by Mr. Dunkel as a final package, the reaction of U.S. Trade Representative Carla Hills best characterizes the current status of the document: "It is important to emphasize that the Director General's document is only a draft; it is not a finished legal text." Both developed and developing countries have proposed changes in TRIPS as well as in other elements of the Uruguay Round package.

The TRIPS agreement covers copyright and related rights, trademarks, geographical indications, industrial designs, patents, layout designs (topographies) of integrated circuits, and protection of undisclosed information. In addition, it contains sections on basic principles, such as national and most favored nation treatment; internal and border measures that countries will have to implement to enforce the intellectual property rights covered in the agreement; and transitional and institutional arrangements.

The following is a brief summary of the key provisions of the draft agreement as they affect technology-related intellectual property[2]:

Copyright and Related Rights

1. Parties to the agreement (i.e., countries) are required to provide Berne Convention protection. The moral rights provisions of Berne, however, are excluded from coverage under TRIPS.
2. Computer programs are protected as literary works under the Berne Convention.

---

[1]In providing this brief status report and summary, the author has not assessed the TRIPS provisions and their relative effectiveness in meeting the intellectual property-related objectives of the various parties to the negotiations.

[2]Readers are urged to consult the TRIPS text that is contained in GATT document MTN.TNC/W/FA.

3. Compilations of data or other material are protected as such.

4. Authors of computer programs and their successors in title are provided with an exclusive rental right.

5. For the purposes of dispute settlement under a TRIPS agreement, nothing in the agreement can be used to address the issue of the exhaustion of copyright rights.

Patents

1. "Patents shall be available and patent rights enjoyable without discrimination as to the place of invention, the field of technology and whether products are imported or locally produced." This provision would require the United States to do away with Section 104 of the Patent Act, which prohibits reference to acts of invention that take place outside the United States in determining the right to a patent, and would require Canada to end its compulsory licensing system for pharmaceutical products. In addition, countries would have to recognize the importation of patented products as meeting working requirements for purposes of compulsory licenses.

2. Plant and animal inventions and the biotechnological processes for their production are excluded from coverage under TRIPS.

3. Patents are protected for 20 years from the filing of a patent application.

4. Although compulsory licenses are not prohibited, their use is subject to certain conditions on the circumstances and manner in which they may be granted. Among the restrictions are adequate notification and remuneration and judicial review. The use of dependent patent compulsory licensing is also circumscribed.

5. For the purposes of dispute settlement under a TRIPS agreement, nothing in the agreement can be used to address the issue of the exhaustion of patent rights.

Semiconductor Layout Designs

The TRIPS agreement deals with the major weaknesses of the Washington Chip Treaty: the term of protection has been extended to 10 years; the innocent infringer provisions have been strengthened; and compulsory licensing of chips is subject to the same conditions as patent compulsory licensing.

Trade Secrets

For the first time, trade secrets or "undisclosed information" are protected in an international instrument from third-party acquisition "in a manner contrary to honest commercial practices." In addition, governments are required to protect proprietary data that they require for marketing approval of new chemical entities "against unfair commercial use."

Enforcement of Intellectual Property Rights

1. The TRIPS text requires countries to have available enforcement procedures so as to "permit effective action against any act of infringement of intellectual property rights covered by this Agreement, including expeditious remedies to prevent infringements and remedies which constitute a deterrent to further infringements." Procedures covered in the agreement are both civil and administrative in nature and include provisional measures, with proper safeguards, where expeditious action is necessary.

2. In addition, the TRIPS text requires countries to have special border measures that would permit the suspension of the release of suspected infringing imports by the custom authorities. These special border provisions, which are mandatory for counterfeit trademark and pirated copyrighted goods, may be extended to goods involving industrial designs, patents, integrated circuits or undisclosed information.

Transitional Periods Before Parties Must Adhere to the Entire TRIPS Agreement

The transitional arrangements in the TRIPS text are prospective and do not provide any retroactive "pipeline" protection for pharmaceutical and agrochemical products.

The following transition periods apply before a Party (i.e., a country) must adhere to the provisions of the TRIPS agreement:

1. All parties have one year following the date of entry into force of the TRIPS accord.

2. Developing countries and countries that are in the "process of transformation from a centrally-planned into a market, free enterprise economy" have an additional four years for a total of five years.

3. The 46 countries on the United Nations' list of least developed countries have an additional six years (eleven years from entry into force). If a least developed country requests an extension, the Council on Trade Related Aspects of Intellectual Property Rights, established by the TRIPS agreement to monitor the operation of and compliance under the agreement, "shall . . . accord extensions of this period."

4. In addition, developing countries do not have to provide patent protection for pharmaceutical and agrochemical products for an additional five years (ten years from entry into force).

WORLD INTELLECTUAL PROPERTY ORGANIZATION DISCUSSIONS

Because WIPO discussions have not advanced to the point of the GATT TRIPS negotiations, where a single unbracketed document has been produced, this section discussion on the four WIPO negotiations is focused more on their procedural than on their substantive status.

Patent Law Harmonization Treaty

The first part of the bifurcated diplomatic conference on patent law harmonization was held in the Hague from June 3-21, 1991. Formal decisions on the text were put off until the second session of the diplomatic conference, which is not expected until July 1993 at the earliest. The decision to bifurcate the diplomatic conference was linked to the one-year delay in GATT negotiations caused by the breakdown of the Brussels ministerial meeting in December 1990.

The WIPO patent law harmonization exercise seeks the development of a treaty that will simplify and expedite the obtainment of patent protection around the world and will strengthen that protection once granted. As opposed to the GATT TRIPS agreement, which contains minimum standards of protection and enforcement, the WIPO patent law harmonization treaty sets forth a number of concrete provisions that will have the effect of harmonizing certain administrative and substantive laws and rules for obtaining and enforcing patents in adherent countries.

Under the current draft treaty, adhering countries would be required to

- grant patents to the inventor first filing an application (Article 9);
- grant patent protection for products and processes in all fields of technology (Article 10);
- provide a minimum patent term of 20 years from filing of the patent application (Article 22);
- provide a grace period of one year for disclosures of inventors (Article 12);
- accept patent applications satisfying certain minimum standards regarding content and format (Articles 3 and 4);
- accept and give dates to applications in the English language (Article 8); accept and process related inventions in a single application (Article 5); require publication of applications a fixed time period after filing (Article 15); require courts to give a fair breadth of interpretation to patent claims (Article 21); and provide a reversal of the burden of proof for process patents (Article 24).

The treaty would mandate certain changes in U.S. patent law. The most controversial change would require the United States to amend its patent law from a "first-to-invent" to a "first-to-file" system for determining the right to a patent. This would make U.S. law consonant with the almost universal approach found outside the United States that grants the right to a patent to the first inventor to file for the patent, not necessarily to the first to make the invention. U.S. negotiators have indicated their willingness to make this major change in U.S. law provided the final treaty represents a "balanced package" that includes concessions from other countries that would provide quicker and more certain patent protection for U.S. inventors abroad. Among these beneficial provisions that must be included in the final text are

- acceptance of English language specifications in patent applications; a one-year grace period for the filing of patent applications after an inventor has disclosed his invention;
- a speedier examination process in which the patent search must be completed within 18 months from filing, and examination completed within five years from filing; acceptance of patent applications that satisfy a minimum standard format; the filing of related inventions in a single application; the requirement for a reasonable breadth of interpretation to patent claims; and elimination of pre-grant opposition procedures.

The patent law harmonization exercise is essentially viewed as a forum to resolve differences among the varying but relatively adequate systems of patent protection found in the developed countries. Nevertheless, the treaty would require changes in the laws of all countries, including the developing countries; thus a number of issues on the negotiating table involve critical North-South differences, such as patent term and coverage. Because these issues are also the subject of the TRIPS negotiations in the GATT and—more importantly—are considered more amenable to resolution in the GATT than in WIPO owing to cross-sectoral leverage, the WIPO negotiators delayed their patent harmonization discussions until the GATT could first resolve the overlapping North-South negotiations. Although the diplomatic conference was not actually postponed, the net effect of the bifurcation was the same.

There is optimism in some circles that, once the GATT TRIPS negotiations are successfully concluded, a patent harmonization treaty containing improvements in European, Japanese, and U.S. laws will also be concluded.

Trademark Law Harmonization Treaty

Discussions on a trademark law harmonization treaty are not as advanced as those on patent law harmonization. While the first two meetings of the Committee of Experts have focused on the substantive provisions of a draft trademark treaty put forward by the WIPO Secretariat, a consensus has not yet developed among the participating countries on the desirability of having a WIPO trademark law harmonization treaty or whether the treaty should contain minimum or maximum standards.

WIPO had put discussion of a trademark harmonization treaty on the back burner pending the conclusion of the GATT TRIPS negotiations. With the uncertain future of the Uruguay Round, WIPO has once again stepped up its consideration of trademark law harmonization. At the third session of the Committee of Experts, which was held in June 1992, the experts reviewed a draft treaty that sought to address administrative procedures instead of substantive trademark law, which had been the focus of the previous drafts. A diplomatic conference is not expected for at least another year.

Possible Protocol to the Berne Convention

The two-part meeting of the Committee on Experts, which took place on November 4-8, 1991, and on February 10-11, 1992, represented the first step in the long WIPO process for development of a possible protocol to the Berne Convention. The meetings discussed, among other issues, the inclusion of computer programs in the proposed protocol.

Widely different opinions on the inclusion of computer programs were presented at the meeting, and as a result, the issue of protection of computer software was postponed for later consideration by the Committee of Experts. Similarly, it was agreed that it would be premature to deal with "computer-produced" works and that artificial intelligence systems should not be included in the proposed protocol. However, it was agreed that the proposed protocol should deal with the question of protection of data bases and that a future working document should include a study of the possibility of also protecting data bases that contain large amounts of data or information but do not meet the originality criterion, such as catalogues of goods offered for sale. Currently, no future meetings of the Committee of Experts have been scheduled.

Settlement of Intellectual Property Disputes Between States

In response to the criticism that the weakness of WIPO-administered treaties is in their lack of dispute resolution mechanisms, a Committee of Experts has held three meetings to discuss the outlines of a WIPO dispute settlement treaty. The most recent meeting took place July 1992. The next session is not expected before the fall of 1993. A diplomatic conference will not take place until the experts have met one or two more times.

A draft treaty has been prepared by the WIPO Secretariat. The treaty calls for the use of consultation, good offices, conciliation, and mediation for the resolution of disputes. It establishes procedures for the creation of panels but does not give any authority to the WIPO Assembly to adopt the panel's reports. There was also a wide divergence of opinion on whether the dispute settlement treaty should be limited to the enforcement of WIPO-administered treaties.

Discussion

Given a diversity of national interests, what is the probability that a global intellectual property right (IPR) system such as that being proposed in the negotiations of the Uruguay Round of the General Agreement on Tariffs and Trade (GATT) can gain widespread acceptance? This question, and the forces moving nations both closer and further apart on IPR issues, were addressed by the conference audience and panel members in a discussion of the perspectives presented in Chapter 6.

During the conference, it was pointed out that many of the newly industrialized economies (NIEs) have recently implemented higher levels of intellectual property protection. One audience member questioned whether this development stemmed from the NIEs' perception that the benefits of rewarding innovation in their countries now outweigh the costs of having to pay for intellectual property, or whether the motivating factor was the belief that the loss of income due to U.S. trade retaliation would offset the opportunities and benefits of "free riding" on foreign technology.

As his essay in Chapter 6 indicates, Carlos Primo Braga believes that both factors played a role. At the conference, he added that although bilateral pressure from the United States has been credited with success in bringing the NIEs closer to the standards of industrialized countries, this success may have been helped by an internal push for increased IPR protection, and vice versa. In the case of Brazil, the efforts of domestic parties to acquire greater protection for biotechnology products was accelerated by the threat of retaliatory U.S. trade action in response to alleged infringement of protection for pharmaceutical products.

During the conference discussion, Deepak Nayyar suggested that transnational corporations (TNCs) operating in developing countries are also using their influence to encourage stronger IPR laws. The incentive for stronger laws may be attributed to technological progress in telecommunications and computers that have allowed TNCs greater flexibility in the international marketplace. The ability to easily transfer and access the data of subsidiaries around the globe gives a parent company the power of centralization. The more consistent IPR regimes worldwide, the easier it becomes to treat the products of subsidiaries as if they were all in one place. The global dispersion of production, distribution, and marketing has given transnational companies an opportunity to take advantage of the world economy and its resources. Standardization of IPR laws is a further step toward the "globalization" of markets, the removal of uncertainty in transactions, and the elimination of requirements for special strategies for each country or market.

One discussant at the conference wondered if yesterday's NIEs are today's industrialized countries because they had a lax IPR regime in the past. In an environment in which IPR protection is weak, a nation could theoretically move along the technology learning curve without paying for foreign technology, while attracting capital investment in less costly reproduction capabilities (because of the absence of R&D costs). The accumulation of capital and expertise can push a country into the next stage of industrialization. It has also been noted that the NIEs have instituted stronger levels of protection only since their level of industrial development has risen, a fact that supports the argument that, until there is innovation and development, there is no need for IPR protection. Carlos Primo Braga responded that despite the appearance, there is no evidence that the *absence* of strong IPRs has somehow contributed to or propelled economic growth and development. Bryan Harris added that conclusions about the effects of IPR laws are difficult to draw because some countries do not have intellectual property laws, but others have laws yet do not enforce them. It is not clear whether this difference has any differential effect on a nation's growth.

James Armstrong was asked at the conference if there was a similarity between Paul David's thesis that national IPR regimes have developed in response to the particular development of industries and the recent reciprocal criticisms of the United States and Japan, each of which has claimed that the patent system of the other is geared to protecting and promoting local technology over foreign technology. Mr. Armstrong argued against this notion, saying that he could see no basic differences between the standards of patentability in the two countries. He acknowledged, however, that the two nations approach IPR legal issues differently, a by-product of the fact that the Japanese system is "adopted" from Germany, whereas the U.S. system is rooted in the long history of common law.

One observer from the Arab Society for the Protection of Industrial Property noted that harmonization among the United States, Japan, and the EC seemed the most feasible, not because there are fewer disagreements or differences in their IPR systems, but because they collectively possess nearly all of the technological capacity in the world and have a natural, common interest in establishing strong protection for their assets. By the same logic, he argued, it should not be surprising that there is little incentive for the rest of the world to embrace this level of protection, since the majority of the world operates under completely different circumstances.

Many participants at the conference thought that the different technological and economic circumstances among countries must at least be acknowledged by the industrialized world as it pushes for a uniform system. To the lesser developed countries (LDCs), granting individuals or companies strong property rights to new varieties of grain seems counterintuitive when hunger is a basic problem for most of them. On the other hand, as LDCs have watched foreign pharmaceutical companies turn native germ plasm into valuable new medications beyond the purchasing power of their citizens, they have sought to gain recognition for the concept of "property rights" in local gene pools by the country of origin. As one audience member asked, "Can we distinguish between intellectual property rights for a razor blade and for a rice strain simply keeping in mind that food crops are closely linked to basic human rights, which are perhaps not an intellectual property right but are still a right?" The LDCs have difficulty taking the leap of faith needed to accept the logic that by extending private rights, temporary monopolies, and monetary incentives to individuals, the public ultimately will benefit.

The challenge of addressing LDC concerns in an international IPR dialogue prompted a discussion on the suitability of the GATT negotiations on Trade Related Aspects of Intellectual Property Rights (TRIPS) to serve as the forum in which to move toward uniformity. Given the origins of GATT as a mechanism for trade enhancement among industrialized countries, and the fact that the industrialized countries have controlled the agenda and rulemaking process of GATT ever since its establishment, it may not be a suitable forum for a North-South discussion. On the other hand, it was contended that because these multilateral talks address trade issues in a variety of sectors, the GATT presents an opportunity to make cross-sectoral agreements to the economic benefit of all parties.

The success of GATT is dependent, however, on agreement across all sectors. Given the difficulty of justifying to the LDCs on an empirical basis the benefits of specifics of the Dunkel draft in the TRIPS negotiation, the industrialized countries may have to be willing to make trade-offs in terms of market access or agricultural concessions to gain the acceptance of stronger IPRs by LDCs. The improbability of such concessions leaves open the prospect of losing all of the gains made toward agreement during the IPR negotiations.

A member of the academic community suggested that because the TRIPS proposal is based on the straightforward imposition of uniform standards, rather than on principles or stage-of-development considerations, its rigidity may be its weakest aspect. During the conference discussion, Paul David added that although there are theoretical arguments in favor of standardization, the creation of standards based on consensus alone may result in a regime that is inflexible and unadaptive to changes in technology and in international business relations. Moreover, he added, "the same laws in different cultural and economic environments do not imply the economic effects will be the same in all of those countries." He suggested that James Armstrong's assessment of the situation between the United States and Japan was a case in point, that is, similar laws have a different effect. As a possible alternative solution, he proposed a more flexible, constitution-like framework, which could be interpreted by the courts. Another option suggested was the adoption of a convention of "adequate and effective protection" for those issues not agreed to thus far in the TRIPS negotiations. The boundaries of what constitutes adequate and effective protection would be decided for each country by a special GATT committee, which would take into account stages of economic development.

The ambiguity that these various alternatives imply, however, may not be acceptable to the industrialized countries, whose principal goal is the assurance of stronger protection for IPRs internationally. To craft an agreement without the provisions of strong standards is, as Jacques Gorlin commented, "to put structure above substance."

The "fast-track" authority for U.S. congressional ratification of a GATT agreement will expire in 1993. If the GATT negotiations fail to produce a satisfactory TRIPS agreement, the United States will return to its plan of bilateral pressure, a method it has found successful but slow, and one that raises international resentments that may fuel further resistance to broad, multilateral solutions.

IV

Scientific and Technological Advance and Its Impact on the Role of Intellectual Property Rights

Introduction

As Paul David states in Chapter 2, intellectual property rights have traditionally been considered a stimulus to useful innovation and economic growth. The process of scientific and technological advance has, however, changed in ways that raise questions as to whether IPRs remain effective instruments for achieving these objectives. The papers in this section address the changing nature of scientific and technological advance and the implications of those changes for the role of IPRs in organizations in different industries, economic sectors, and countries.

With the global diffusion of technological capability, competition in technology-intensive goods and piracy of intellectual property have increased. The rising cost of scientific and technological development has increased the importance of IPR protection, as well as the incentive to engage in cooperative R&D. These and a variety of other factors associated with scientific and technological change that affect the need for and the effectiveness of IPR protection are discussed in this section.

Changes in the nature of scientific and technological advance and their effects on the role of IPRs have important implications for organizations involved with technology. These implications may vary depending on the industry, sector of the economy, or country. For example, now that universities are much more heavily involved in the commercialization of research, they are dealing with IPRs and are affected by international IPR issues. What are the implications for the role of IPRs in the corporate strategy of companies in different industries and in different countries? How do changing science and technology, and changing international IPR regimes, relate to the missions of the federal government?

The chapters in this section address these issues. In Chapter 8, John Armstrong discusses in detail the trends in global science and technology and what they mean for intellectual property systems. He identifies and describes three trends: (1) science will continue to provide an increasing number of discoveries; (2) research topics will continue to proliferate and the conduct of research will become globalized; and (3) short, quick steps to product application will be at a premium. From these trends he discerns two general principles: (1) in the face of continuing scientific and technological change, the best course of action will be to rely on flexible adaptation of existing, basic IPR concepts; and (2) since research, development, and invention are increasingly global, IPR systems should also be globalized.

In Chapter 9, three discussants describe the implications of the trends in IPRs for organizations in different economic sectors. John Preston describes the nature of his university's interest in intellectual property and how that drives the use of IPRs in licensing.

Bruce Merrifield addresses implications for the government sector. He discusses the opportunities and challenges that exist for using advanced technology to expand economies and improve the quality of life around the world. He argues that the federal government has an important role to play in providing incentives and that intellectual property protection is required for the entrepreneurial function to thrive.

George McKinney III discusses the gap between historical/theoretical models of IPRs and the real world in which U.S. corporations must operate. He raises several issues that he believes illustrate improvements that are needed in the IPR system.

In Chapter 10, representatives from companies in different industries and different countries describe the implications of recent changes in science, technology, and IPRs for corporate strategy. Otto Stamm attests to the importance of strong patent protection for continued development of new drugs. In the face of increasing development costs, government cost controls, and imitative products, he predicts that the pharmaceutical industry will be increasingly unable to afford a globally oriented marketing strategy for new products unless worldwide protection of pharmaceuticals is achieved.

Michiyuki Uenohara addresses the issue of the higher license fees being charged by some U.S. companies in those industries. He criticizes the higher fees on the basis that they obstruct advancement of the public welfare by contributing to higher product costs. He argues that it is time to restore the original purpose of intellectual property law, which he views as "to legally protect the inventor's right in order to promote the application of such valuable intellectual creations for the benefit of the public welfare."

Antonio Medina Mora Icaza discusses the implications of recent changes in the Mexican copyright law for the Mexican software industry. The Mexican software industry association sought modifications of Mexican copyright law to improve the protection of software, and in July 1991, those modifications were approved. In Mr. Medina Mora's view, strengthening intellectual property protection is a strategic move for developing countries such as Mexico, a move that can build the competitiveness of a nation's industries and gain the trust of foreign investors.

W.L. Keefauver describes how AT&T's use of intellectual property evolved over the life of the company. In the early years, AT&T used patents as entrepreneurial assets to establish markets. Later, as a heavily regulated enterprise, it used IPRs to obtain "freedom of design" through cross-licensing with other firms. With divestiture and the globalization of markets, AT&T has again entered international competition, and its IPR and licensing practices are being tailored to specific competitive strategies. This trend seems likely to continue, and IPRs in software have the potential of becoming the most important of all.

The following chapters illustrate that one of the principal effects of recent changes in the nature of scientific and technological advance has been to increase the importance of IPRs to organizations involved with technology, whether they be in industry, academia, or government. They also show that there are differences of opinion on the appropriate and adequate protection of intellectual property and on whether the present IPR system is functioning adequately, and that those differences do not correspond simply to sectoral, industrial, or national lines.

8

Trends in Global Science and Technology and What They Mean for Intellectual Property Systems

JOHN A. ARMSTRONG

INTRODUCTION

Powerful forces are altering existing patterns of global activity in science and technology. My task here is to look at some of those forces and characterize how they will influence the relationship between research and development activities and the world's intellectual property systems over the next few years.

I undertake this task as a scientist (who holds two patents), not as an expert on intellectual property. Yet from my experience at IBM it is clear that our level of R&D could not and would not be sustained without protection of the intellectual property that results from the $6 billion we spend annually on R&D around the world. We are interested in protecting our intellectual property rights to obtain freedom of action for future manufacturing and marketing, and to provide a level, competitive playing field between companies that perform R&D and those that do not. These dual interests are characteristic of the computer and electronics industries.

Without an intellectual property regime that provides an opportunity for us and for others to gain a return on our various investments, our R&D spending would be both less efficient and lower in absolute terms. Clearly the global and social business impact would be undesirable; less innovation would create less new wealth.

Because the evolution of technology is necessarily a global endeavor, worldwide consistency and predictability in protection of intellectual property are essential. It is hard to run a railroad efficiently if the track width differs from place to place. However, that is the situation we have now as we try to conduct research and move technology around the world. But I foresee an expanding realization that innovation and fair trade will move us toward the global approach to effective intellectual property protection that is required.

Since World War II, the world has seen two kinds of experimentation with intellectual property systems. In developing countries, we have seen attempts to diminish levels of protection radically. In the developed countries, we have seen attempts to create new sui generis forms of protection. In my view, both experiments have achieved far less than they promised. I see a growing awareness around the world that the basic concepts of intellectual property are sound and that flexible application of those concepts to new developments in science and technology makes sense. I return to these themes repeatedly in this chapter.

Why is the importance of intellectual property systems growing? Clearly it is because the role of knowledge, particularly technical knowledge, is becoming much more prominent in modern economic life. I hardly need to assert that to you or illustrate it for you. However, what this means for business and for nations is changing, which is really the theme of this chapter.

Twenty years ago, business seminars and academic research paid a lot of attention to investment. In the international setting, the focus was on foreign direct investment. That has changed. In recent years, we have all given a lot of attention to technology and, increasingly, to innovation and its role in the creation of wealth and the increased well-being of nations.

Let me give you some images to help make my point. In the 1970s it was as though we had a stage. At stage center sat an imposing figure called investment. The plot was all about attracting investment and calculating investment risk, country by country. Technology played a supporting role and intellectual property was in the program notes, but certainly not on stage.

Today, we have a video screen, not a stage. The actor *technology* is constantly the center of attention on this screen, and now we see investment pursuing technology, trying to keep up. As we look closely at the screen, we see the intellectual property system helping to guide investment in R&D and conditioning the origin, direction, and velocity of innovation.

Stated differently, classical and neoclassical economic theories have sought to depict productive human activity in terms of capital, labor, and resources (meaning things in the ground, such as oil and gold). This trio of factors has overlooked innovation as a primary factor. Only in the last few decades have economists begun to look hard at what happens when the results of scientific and technical research are injected into economic activity. This is preparing the way, let me suggest, for a serious scholarly examination of the role of intellectual property in conditioning the creation and diffusion of technology internationally. One of the exciting frontiers for this examination is in the developing countries, and I want to commend the organizers of this conference for pointing in that direction.[1]

Conventional wisdom in the business world says intellectual property is vital because it is a stimulus to innovation, a vehicle for technology transfers, and a magnet for financing. I would push that further. An intellectual property system is a crucial part of a country's economic infrastructure. It enhances the ability of any country to strengthen and advance its technological base in a sustained way. It helps to build human resources. It conditions priorities in allocating financial resources. It fosters the movement of technical knowledge across borders.

Again, although I am not expert on the fine points of intellectual property, it is clear to me from my experience and position that intellectual property is not simply a set of legal provisions. It is a system with integral parts. If parts are missing, less happens and the system is apt to fail. A system that works well will have laws that protect the full spectrum of technology, everything from patent and trade secret protection, to copyright and trademark protection, from semiconductor "chip" topographic design protection, to protection for living matter, computer programs, and more.

Just as important, a system that functions well also has reasonably predictable mechanisms for public administration of the various forms of protection and for judicially enforcing individual rights created by the system.

A system that functions well creates, within the country, a general public confidence that innovation and creative expression will, in fact, be protected.

Countries that have such systems will be able to do more with technology than countries that do not. From my experience around the world I know that there is a high correlation between private firm spending on research and the strength of intellectual property protection in country after country. It is private firm research, more than other activity, that converts scientific advances into useful products and services.

Today it is particularly important to think in terms of a comprehensive system because more and more technical activity requires several forms of intellectual property protection. In my industry, we rely on copyright, patent, trade secret and chip topography protection in a major way. People working in the biological sciences, particularly in biotechnology, increasingly

[1]For an elaboration of why economic theory has been slow to take up this subject, see Sherwood (1990a).

turn to both patents and trade secrets. Together, information technology and biotechnology underlie a large part of many countries' industrial base. So attention to the complete spectrum of intellectual property protection is very important.

Now that I have given you a context for my comments, let me tell you how this chapter is organized. First, I describe three trends that characterize global developments in science and technology. They each carry implications for intellectual property. Then, from these trends, two general principles are indicated. Finally, I return to the original context to reaffirm the importance of intellectual property for all countries.

Before describing the three trends in detail, I would like to preview the two principles I just mentioned: The first is that because activity in scientific and technological R&D will be increasingly global, it makes sense to globalize intellectual property systems at high levels of protection. The second is that because scientific and technological research will provide the world with constant surprises, it makes sense to keep intellectual property systems flexible.

THREE SELECTED TRENDS

As mentioned, I have selected three trends that characterize global developments in science and technology. Each carries implications for the ways we shape intellectual property systems in the international context.

Trend One—More Surprises from Science

The first trend I observe is that science continues to roll out surprises. It was never more true that we should expect the unexpected. Recent discoveries in high-temperature superconductivity, in nanotechnology, and in molecular and genetic engineering are clear evidence of this. I believe this means at least two things.

First, for companies and countries, the ability to support high-level scientific inquiry will be central to economic growth and success. The ability to convert scientific advances quickly into products and services for human use will be even more crucial to company and country success.

Second, market leadership positions will be eclipsed more quickly and widely than ever before. Indeed, entire industries can disappear when overtaken by new technology. The telegraph industry has almost vanished, the "public business" of mail delivery is rapidly changing character, and the list can easily be extended.

For intellectual property, this trend suggests that our legal systems will have more new fields of innovation to absorb and will need to do so quickly. This, in turn, suggests the avoidance of sui generis approaches, which are based on the assumption that new technological directions demand unique forms of protection. Sui generis laws—each unique by definition—fail to rely on or contribute to a body of consistent, developed principles of legal protection for intellectual property. Uncertainty is the result, and business risks are increased. Innovation is stifled because return on investment is jeopardized by unpredictable application of law worldwide. If we were to follow such a policy direction, we might soon have a bewildering array of legal tools that retard advances in applied science and technology, and delay the delivery of new products to market. The public good will be better served if we stick to basics.

By basics I mean that we will be well served by using the existing frameworks for patents, copyrights, trade secrets, and so forth, and by fostering their evolution to accommodate new forms of technology. The basic principles that underlie these forms of protection have proved flexible and served us well since at least the last century. Our reflex should be to expect that the existing, well-tested, basic principles will accommodate new forms of technology in the future, as they have in the past.

Let me illustrate my suggestion by referring to computer software. When software emerged as a separate area of innovation several decades ago, it seemed new and different. We were all uncertain about what form of intellectual property protection should be applied. After reflection, however, the fundamental principles of intellectual property were considered, and it could be seen that from an intellectual property point of view, software was not one thing but several.

It was seen that from a patent perspective, inventive activity could be involved. Thus, if the standard criteria of inventiveness were met, a patent could be granted for certain aspects of software. Similarly, it was seen that from a copyright perspective, creative expression was involved. Thus the standard criteria for copyright were present, and this form of protection could be applied. Indeed, this protection could be uniform from country to country through application of the Berne Convention. It was also seen that for certain kinds of software, particularly customized software, standard criteria for trade secret protection could be found. Broken down into its several component parts, we now see that software can be served by three forms of intellectual property protection, depending on various factors.

I want to conclude my reflection on this trend by stressing simply that when we are faced with new forms of technology there should be a strong preference for adaptation of existing intellectual property systems rather than a flight to exotic new mechanisms. I urge this primarily to achieve global congruence as rapidly as possible when new technology comes into play, chiefly through utilization of the international conventions of Berne and Paris, which have served well in this regard over the last hundred years.

Trend Two—Proliferation and Globalization of Research

The second trend I observe is a continuing expansion in the topics of research and an increasing globalization of the way research is conducted by private firms. I want to spend more time on this trend because it is less well known than the other two trends.

So much is happening at the frontiers of science that research can be pursued in an exponentially increasing number of directions. There is more than enough scientific advance to go around. This is true throughout the full range of research targets, from pure science, to applied technology, to process improvements.

At the same time, the ease with which firms can pursue research globally is increasing. I know this from travel to many parts of the world and from my association with many research managers here and abroad.

In the midst of this, I see the role of government changing in significant ways, both in conducting and in paying for research.

What are the implications of this second trend? The explosion of scientific advances will mean many things, but let me pick just three. First, with more to do, that which is most worth doing from the perspective of payoff is less obvious. Second, even the largest companies cannot alone keep up with everything that is coming into play in their fields. Third, the methods by which research is being pursued are changing.

About the first point, risk is increasing as those of us who manage research programs decide what is most worth doing in an expanding galaxy of options. Not every scientific advance rolls out startling new products. Not every new finding stimulates breakthrough technology. A small adjustment in scientific knowledge or in known technology may lead to major commercial consequences.

All of this implies risk. Risk is at the heart of business in open economies, of course, but as risk increases because of this explosion of research directions, we become even more sensitive to intellectual property protection. The research results we do achieve become particularly important to us. There is also a strong tendency today for us to want to share this risk.

As to the second point, I note that even the largest companies cannot keep up with all the new science that might bear on their field. This is creating opportunities for small, new entrants. Many large companies, including my own, allocate a significant portion of their R&D budgets to building relations with smaller firms. Although information technology industry has done this for some time, the emphasis now is increasing. There is an opportunity in this for smaller firms in developing countries. Given adequate intellectual property protection, there is no reason why they cannot participate in relationships with large companies. Although industry is accustomed to negotiating the treatment of intellectual property coming into and emerging from cooperative ventures, new complexities are introduced as government, industry, and academia explore joining forces as a catalyst to innovation. The varied cultures of each of these sectors will require flexibility to accommodate the interests of the others. Perhaps most important, a stable body of consistent law is required on which all parties can rely, thereby avoiding yet one additional complicating factor that would be introduced if sui generis laws were prevalent.

As to the third point, the catalogue of approaches to research is expanding: it includes large corporate programs, national laboratories, research done under contract, state-funded programs, university research, and individual research done "at the kitchen sink." We also see technology alliances and networks that do not necessarily depend on equity ownership or even contractual relationships. We see on-line computer-connected research done at two or more locations, even on opposite sides of the world. We see platform building through knowledge sharing that includes "casual" research relationships, in which one company or laboratory will volunteer information to another (on a confidential basis) with an expectation of some feedback. Also, there is more transient employment in laboratories as researchers move from place to place. More attention is being paid to how to conduct research to maximize creativity, and the computer is playing an ever greater role.

I have noted that one of the major ways R&D has changed is by being globalized. This is true of research being conducted by single companies as well as by research consortia. Computer linkages through satellites, data bases, and networks mean that research of all kinds can be conducted in different locations as a single effort. This is already far advanced. It is not unusual for teams operating in different time zones to relay information through electronic mail and other techniques so that research continues around the clock. These linkages are not confined to business. Some of the biggest networks are shared by private industry, universities, and government agencies. Some are public or quasi-public networks, whereas others are private or classified government communication channels.

There are several factors that are prompting research through global networks, now that the capability has been demonstrated. A desire for greater proximity to customers, suppliers, and university talent accounts for some global activity. In other areas, relatively lower personnel cost is a notable reason for going overseas. Another reason is the desire to tap talent found in other countries. A leading example is access to programmers in India through satellite links.

There are elements of risk sharing, cost sharing, and economies of scale in these arrangements, but the desire to pool knowledge is increasingly evident as a motivating factor. IBM's new arrangements with companies such as Siemens and Apple illustrate this. Whatever the mixture of motivation, shared research activity that crosses national boundaries is a growing phenomenon.

In developing countries, the concept of the state as mother of us all is in decline, with state enterprises being privatized or streamlined. Yet the importance of using technology as a tool for economic development is seen by all, and governments are spending accordingly. So there are some cross-currents at work in terms of the role of the state in research.

In the United States, we have seen the Carnegie Commission (1991) report suggesting that complex defense procurement procedures have built a wall between government-supported military and civilian research programs to the increasing detriment of the Pentagon. This report recommends that these two arenas be merged institutionally. In effect, the report signals a shift in defense procurement policy and practice, and this will carry intellectual property system implications as the private and public sector cultures attempt to mesh traditionally disparate views on the role and operation of intellectual property regimes (the former relying on laws to ensure return on investment, the latter relying on laws for national security and protection of the taxpayer's investment in public programs).

In nearly all countries, budgetary constraints are limiting government research expenditures. As a result, the private sector will be asked to pick up more of the national research bill, both for internal research and for public research in universities and public research institutes. This will mean that intellectual property systems will become a crucial supporting factor helping to induce private investment in research. In those cases where government pays more or does more, private industry will often be a companion.

I want to add that explicit government-to-government scientific cooperation is playing a role in research at the global level. For some time, the United States has entered into bilateral science and technology agreements with developing country governments, which are meant to foster good relations and encourage university professors to exchange information. The research funds offered by the United States are relatively modest, particularly when compared to those from Japan. Still, such cooperation has boosted university programs in some developing countries.

The 1984 reform of the U.S. trade laws added a requirement that bilateral science and technology agreements must have an ancillary intellectual property agreement. The requirement, which does not define the content of the agreement, was meant to leverage partner countries into adopting stronger intellectual property systems.

I propose an additional catalyst. Experts from both sides should be asked to devote structured time to a discussion of intellectual property. In most developing country universities today, there is intense curiosity about the growing emphasis on intellectual property in American universities.

Quite often the scientific leadership of developing countries is closer to government policymakers than is the local business community. Ultimately, the scientific leadership of a developing country could emerge as strong advocates of effective intellectual property protection and have greater political impact for reform than the current approach alone.

What will all of this mean for intellectual property systems? The explosion of research directions and the globalization of research mean that the contribution of intellectual property is being more widely felt. While some companies have long recognized its importance, others—including many smaller companies—are just beginning to realize that they can live or die by their ability to protect their intellectual property.

An offshoot of the globalization of R&D is the expanding opportunities this gives to researchers in many of the developing countries. In important areas of research, it does not require giant laboratory facilities to make significant advances. There are many excellent minds working in Third World universities who are capable of making contributions. For them, the issue is often how to move their work from the laboratory to the marketplace in the absence of strong intellectual property protection. Without protection, they typically have difficulty safeguarding their results and attracting needed start-up funds. Countries lacking adequate and effective systems will want to install strong legal protection so that their best minds will not be left out. It will be seen increasingly that more happens technologically in countries with effective protective systems.

As another aspect of the globalization of R&D, I have described the increase in research by alliances. When firms decide to jointly build knowledge platforms by sharing information and sharing risk, the ability to identify and protect both the information going in and the results emerging from that collaboration is crucial. Intellectual property plays this role. The various forms of intellectual property serve to define and "package" those results, to enhance their negotiability, and to defend them from loss to others who do not participate in the risk sharing.

To the extent that such research is conducted across borders—and this is happening more and more—the intellectual property systems of the participating countries need to be highly congruent. This means that protection increasingly needs to be achieved through similar and effective mechanisms. This is true, whether the project is small in terms of expenditure, or gigantic. For the participants, the ability to safeguard results is a necessary precondition to undertaking research in this way. Countries that want to give their scientific and technical people opportunities to be involved in such research alliances will want to be sure their intellectual property system meets the expectations of potential partners from other countries.

Let me emphasize that such research alliances are not restricted to large companies. The opportunity this presents for researchers in developing countries is significant, I believe, and there are already examples of research partnerships between large companies from developed countries and small companies from developing countries. This trend is, however, severely constrained by the typical weakness of intellectual property systems in many developing countries.

There is another relevant perspective based on the increasing tendency for researchers to change jobs. As they move from one company or institution to another, questions arise about what knowledge they are permitted to take with them to use in their new position. The tools of intellectual property, particularly the trade secret, are very helpful in defining and selecting the knowledge that is proprietary and therefore nontransferable. Countries without these tools will be at a disadvantage in building a research culture that can participate in research alliances.

I have noted some of the pressures that are building for greater congruence among the intellectual property systems of the world. Each country with a weak system will certainly feel these pressures as it contemplates the future of its own technology base and the various roles its own researchers are to play.

Let me clarify a point. I am not arguing that other countries adopt the particular intellectual property system of the United States. Within certain parameters, there are various ways that adequate and effective protection for inventions, technical knowledge, and creative expression can be provided. The point is that national systems for intellectual property protection will need to be sufficiently similar to the world norm if those nations are to participate in the globalization of research and in the wide range of shared research options that science is constantly opening up.

Much can be said for the U.S. system when you consider that the booming growth in the creation of computer software correlates with the early and strong protection available to software creators in this country. By the same token, I am told that, of the funds devoted to research on biotechnology on both sides of the Atlantic, 90 percent is spent in the United States because of the far stronger protection available here. It is no accident that an increasing number of European firms conduct a good portion of their research in the United States. All of this seems to indicate that the greater the degree of intellectual property protection, the higher is the level of research stimulation.

Trend Three—Short, Quick Steps at a Premium

The third trend I observe is that many industries place great emphasis on incremental refinements in technology and that in most cases there is great pressure to collapse the elapsed time from a discovery in science to resulting product application.

The reason is clear—competition. The constant pursuit of competitive advantage in our industry is driving down product lives, and shorter product lives demand—and are driven by—technological advance and the incremental improvement of products. That is the basis for competitive advantage.

Consumers are, of course, well served by such an emphasis. In market economies, consumers decide whether an improvement is worth their attention or not, and companies are accordingly kept on their toes. What does this mean for intellectual property systems?

In awarding the privilege of exclusivity, intellectual property systems must distinguish between that which is new and different and that which is simply an imitation. It is not always an easy task. Viewed from a distance, things can appear to be the same or closely similar. Viewed from closer range, things can be readily distinguished. The work of an intellectual property system is to make these fine distinctions. It is work that the well-established systems have performed reasonably well in the past.

This issue manifests itself primarily in the patent area where a basic concept of the law calls for inventions to be novel and nonobvious. When a patent application is filed it must describe what it "claims" as novel and nonobvious. Drafting claims can be crucial. If written too broadly, they can be attacked later as overreaching. If written too narrowly, the patent can be avoided by simply making superficial changes (Chisum, 1991).

The concepts of patent law must also address issues such as basic versus improvement inventions. Consider this example. If Jones invents the bicycle and obtains a patent, and later Smith invents the 10-speed bicycle gear shift, what can Smith do with his invention before Jones's patent expires? Will the basic patent preclude the improvement patent? In an age of incremental improvements and refinements, how can such inventiveness best be protected from imitation and how does it relate to more basic inventions?

Although the greater emphasis on incremental advances is new, the issue of how to deal with increments themselves is far from new for patent and copyright systems. Generally speaking, the improvement patent owner is asked to honor the basic patent. This forces the two parties to negotiate suitable arrangements so both inventions reach the market.

Trade secret protection also plays an important role in the context of incremental advances. Let me take a few moments to discuss trade secrets. They are quite important,[2] yet little known. This is because they are created by private action, not by a government office. There is neither a bureaucracy nor a cadre of specialist lawyers to attend to the trade secret. Those

[2]Perhaps two-thirds of the technology that moves from place to place does so under trade secret protection, see Sherwood (1990b).

who generate technical information that has valuable commercial or industrial application simply take reasonable precautions to keep competitors from learning such information. If the information is obtained by a competitor by unfair means, the courts will intervene to stop the competitor from using it. However, if the competitor develops the same information independently, he is free to use it. Trade secret protection is particularly appropriate in process technology. In my view, any country without strong trade secret protection today is severely handicapping itself in the global competition to improve commercial technology.

TWO GENERAL PRINCIPLES

I discern two general principles in these reflections on developments in science and technology around the world. The design of intellectual property systems will benefit from attention to both of them.

Principle One: Change Will Be Constant, So Keep Intellectual Property Flexible

Things we cannot foresee will emerge from science. This indicates reliance on the evolution of existing basic intellectual property concepts rather than resorting to novel new legal schemes. As a lesson, I point to the sui generis chip protection law of the United States. Rather than expanding patent and copyright concepts already functioning around the world, the United States saw fit in 1984 to create a new form of protection for the topographic layout of semiconductor chips. Enough time has passed since the enactment of this law to see at least two problems arising from this sui generis approach.

First, this experiment relied for its global reach on a unique reciprocity provision that says, in essence, that chips created in another country will receive protection in the United States only if that other country provides equivalent protection under its law. Japan reciprocated. However, many other countries did not. Instead, in 1989 they created a treaty[3] that has internationalized a lower order of protection for chips. The result is that two conflicting systems for chip protection are operating in the world. If

---

[3]The Treaty on Intellectual Property in Respect of Integrated Circuits (commonly referred to as the Washington Treaty) was adopted at an international diplomatic conference sponsored by the World Intellectual Property Organization (WIPO), held in Washington, D.C., in May 1989. The United States and Japan opposed the treaty based on an inadequate term of protection, unacceptable compulsory licensing provisions, failure to provide for remuneration to the right holder from innocent infringers, and failure to adequately protect rights holders from infringement through higher-level products incorporating pirated chips.

the current intellectual property negotiations in the General Agreement on Tariffs and Trade (GATT) Uruguay Round are successful, greater congruence would be achieved.

Second, a sui generis approach is by definition self-limiting. It is designed to apply only to a defined (new) technology. There has been interest recently in bioelectronic devices, parts of which might be produced by "natural" processes. It is by no means clear that U.S. or Japanese laws, or potentially the international treaty, will cover the new "biochip."[4] Likewise, these chip laws and treaty do not protect the masks used in producing micromotors and thin-film heads. Here we see how an attempt to create a new form of intellectual property protection has failed to predict the future. It will not be the last failure, given the surprises that science will surely produce.

A third objection to the sui generis approach, aside from the opportunity it provides for unintended mischief, is the delay in protection it causes. Exotic new technology is most vulnerable in its early stages for lack of protection. Imagine what will happen if research and development must be put on hold while our legal systems take several years to create a novel form of protection. Will this become the common reflex? Will we be taught to assume that new types of technology are not covered by existing forms of protection?

I urge, instead, that the common reflex be to assume that existing forms of protection can be adjusted and adapted to accommodate new technologies. Once this reflex is clearly in place, we will create more globally uniform protection more quickly than we would through sui generis approaches. In urging this, I am encouraged by the experience of the last two centuries which shows that traditional forms of protection have exhibited great flexibility in adjusting to new technology.[5]

Principle Two: Activity Will Be Global, So Make Intellectual Property Global

I suggest as the second principle that since research, development, and invention are all increasingly done globally, intellectual property systems

---

[4]The U.S. and Japanese mask work laws are restricted in application to semiconductor material. The WIPO Treaty on Intellectual Property in Respect of Integrated Circuits is broader in that it is limited to "elements, at least one of which is an active element . . . formed on a piece of material which is intended to perform an electronic function."

[5]The patents system has evolved to cover new technologies, from steam engines to electrical applications to radio devices, from computers to nuclear energy to biotechnology. Copyrights have expanded from novels and speeches to maps and charts, from sound recordings to computer programs, and there will certainly be new technologies in the future.

should be globalized. The generation of new scientific and technological insights and their diffusion and application are not confined to one or a few countries. If we wish to include all willing countries in this wealth-producing activity, and to catalyze global research and development activities, a greater degree of congruence among national intellectual property systems is obviously desirable.[6]

The boundaries of nations do not themselves impede the flow of technical knowledge. However, technical knowledge does not now flow equally into all countries (Sherwood, 1989). Just as electricity flows best through certain media and not through others, so too, technology flows best under certain conditions. Some countries, particularly those in the Third World, have built intellectual property system strategies on the supposition that technology flow is spontaneous and that weak protection increases the flow. As these strategies fail, I think it is getting clearer that this supposition is wrong.

Countries with weak intellectual property systems receive less technical knowledge. There are at least three reasons. First, even if proprietary knowledge can be "stolen" or "pirated," those who obtain it are denied associated knowledge from a willing source. So less is learned, and it is learned late. Those who pirate technology condemn themselves to perpetually catching up. Moreover, the skills learned from pirating are not the skills needed to conduct research and development.

Second, because such countries are hindered in building a knowledge infrastructure, they cannot even make good use of technology that is freely available. They do not develop people who can appreciate and work with such technology. This is true whether the freely available technical knowledge comes in the form of knowledge embodied in capital goods or through journals, magazines, conferences, and even newspapers. Moreover, without the means to protect innovation, there is little incentive to advance this technology even when it is freely available.

Furthermore, it will be increasingly difficult for countries with weak intellectual property systems to export products with "pirated" technology into markets that have strong protective systems. It will also be difficult for such countries to attract private foreign investment to supplement local support for research and development relevant to industrial development there.

The principle I distill from these observations is that if countries lack-

---

[6]Harmonization would be an additional degree of congruence. Congruence means that those active in science and technology would not have to give much thought to system differences, although specialist lawyers would be needed. Harmonization would mean that even the lawyers would not have to give much thought to system differences.

ing adequate and effective intellectual property systems wish to participate in global advances in science and technology, they will be well served by making their systems congruent with the many existing national systems that are adequate and effective. The more globally consistent the treatment of intellectual property is, the greater will be both the stimulation of research and the conductivity of technology across borders.

At the moment it is still not possible to know the outcome, but I hope the attention being given to intellectual property in the current GATT negotiations will confirm my observations. Although the December 20, 1991, Trade Related Aspects of Intellectual Property (TRIPS) text has its critics, it is a strong agreement in some important areas, including the protection of computer programs. It is clear that with additional improvements, particularly in the area of patent protection covering pharmaceuticals and chemicals, a TRIPS agreement could have a significant impact on improving the protection of intellectual property worldwide.

OBSERVATIONS

I have the following closing observations.

As Ideology Fades, Technology Will Drive Development

For much of this century, ideology has informed many aspects of public policy, particularly in developing countries. Today, however, pragmatism signals new approaches to many things as ideology fades rapidly (provided resurgent nationalism does not erect new barriers). Within one lifetime, ordinary people now see profound technical revolutions that change entire industries and countries. The deliberate quest to be part of such revolutions will drive government policy in many countries to an increasing degree, and this will in turn, I believe, encourage strong intellectual property systems for all countries. The presence or lack of strong intellectual property protection in developing countries will be a critical factor in their participation in the world's economic progress.

As Economies Open, Invention Will Flourish

The era of the closed economy and the import substitution model is rapidly closing. As economies open to join the global marketplace, exciting things are happening. Competition intensifies, putting a premium on innovation.

In this setting, and particularly as more of the global economy is directed by private rather than state decisions, the balancing of interests achieved by well-considered intellectual property systems, and the globalization of those balances, will serve both the generation and the diffusion of technology as it has in the past. Nations that take part in this globalization will participate in the resulting wealth.

As Research Grows, Everyone Will Benefit

Throughout the research chain, from basic science to incremental product improvements, the intellectual property system strongly conditions decision making. When those who make research decisions look across the globe, it is important for everyone that they see a landscape in which research is uniformly encouraged. This does not imply that intellectual property systems need to be uniform, only that the encouragement they offer needs to be uniformly adequate.

I firmly believe that given greater uniformity among intellectual property systems around the world, much more will happen at the international level. Large companies like mine operate widely already, but I foresee that smaller companies will link with counterparts in other countries to accelerate the advance of knowledge and technology in a great variety of special fields.

The trends in global science and technology indicate to me that the basic concepts of intellectual property, applied globally and flexibly, will be increasingly called on to serve research and development activity around the world.

ACKNOWLEDGMENT

I would like to acknowledge discussions with—and the assistance of—D.P. McCurdy, W.T. Ellis, and V. Siber, all of IBM.

REFERENCES

Carnegie Commission on Science, Technology, and Government. 1991. Technology and Economic Performance: Organizing the Executive Branch for a Stronger National Technology Base. New York.

Chisum, Donald S. 1991. Patents: A Treatise on the Law of Patentability, Validity, and Infringement. Matthew Bender, Chapters 8 and 18.

Sherwood, Robert M. 1989. New theory of conductivity in licensing. les Nouvelles: Journal of the Licensing Executives Society 24(4):186-189.

Sherwood, Robert M. 1990a. Intellectual property and economics. Chapter 4 in Intellectual Property and Economic Development. Boulder, Co: Westview Press.

Sherwood, Robert M. 1990b. The importance of trade secrets. Pp. 57-59 in Intellectual Property and Economic Development. Boulder, Colo.: Westview Press.

9

Sectoral Views

The Research University

JOHN T. PRESTON

I am going to discuss the Massachusetts Institute of Technology (MIT), which unlike many universities is interested in intellectual property for a slightly different reason than you might imagine. We are driven by one goal, and that goal is to use intellectual property to see our technology embedded in products, thereby serving the public good. A secondary goal is to make money from patents. Serving the public good is important because universities such as MIT are funded to a first approximation by the U.S. government. Therefore, MIT's image is closely tied to the public benefit it can demonstrate from its inventions and ideas. Let me give you a couple of examples. When we look at technology transfer there are two possible strategies. We can say, "Let us heavily protect the technology with patents and then license the patents exclusively," or "Let us ignore intellectual property or diminish the value of intellectual property and transfer it for the public good at a relatively low cost." I am of the opinion that the cheaper it is to take a technology and get it to the market, the better it is to lower intellectual property rights and grant low-cost nonexclusive licenses. The more expensive it is to get the technology to the market, the better it is to patent and grant an exclusive license.

For example, if I invent aspirin today, and I publish the formula for aspirin in *Science* magazine next month, without patenting, no company will spend $150 million going through Food and Drug Administration approval to commercialize aspirin knowing that the second company will not have to amortize the large investment. In the case of a novel pharmaceutical, one should generally patent inventions to commercialize them. However, if I develop a computer program such as X-Windows, where there is a relatively low cost to take that program from the point of development and get it embedded into products and out into industry, or when there is a desire to make it a standard, maybe the best thing to do is to license it for free. In fact, our license agreement for X-Windows has only one constraint. It costs nothing, by the way, but the one constraint is that you preserve MIT's name on the copyright notice. You can do anything else you want with it.

So I am coming from a slightly different point of view than most of the authors here, and I want to start by saying that I agree with about 90 percent of what John Armstrong says in Chapter 8, but I am going to discuss the 10 percent that I question, or disagree with.

Before I get into that, however, it is to look at some figures—important because they explain the economic incentives for commercializing technology. In other words, if we do not understand the driving motives for commercializing technology, we cannot create policy for intellectual property.

Figure 9-1 gives three scenarios for developing technology. In terms of the investment of money, the first scenario is curve A, which shows a very small investment of money over a long period of time to develop the technology. I call this the minimalist curve, and it is typical, quite frankly, of large U.S. companies. I am going to return to that point, but it is also typical of behavior when intellectual property is not a valuable asset to developing the technology.

The problem with curve A is (1) it gives a long window of opportunity for competition, and (2) the people running these businesses are spending too much time looking for money.

Curve B is the optimum curve in which significant resources are infused early on. If you have heavy protection of intellectual property, it drives your behavior more toward curve B. By the way, one thing this graph does not show is that if the total amount of investment to commercialize a technology is very small, it drives curves A and B closer together. This graph assumes that the total investment to get into the market is very large, which drives curves A and B apart. You can see that over a long period of time, curve B will make a lot more money than curve A. Japanese businesses are curve B companies because they sacrifice short-term profit for long-term gains and market share. Curve C is what happens when you throw too much money at a technology. You can actually corrupt the

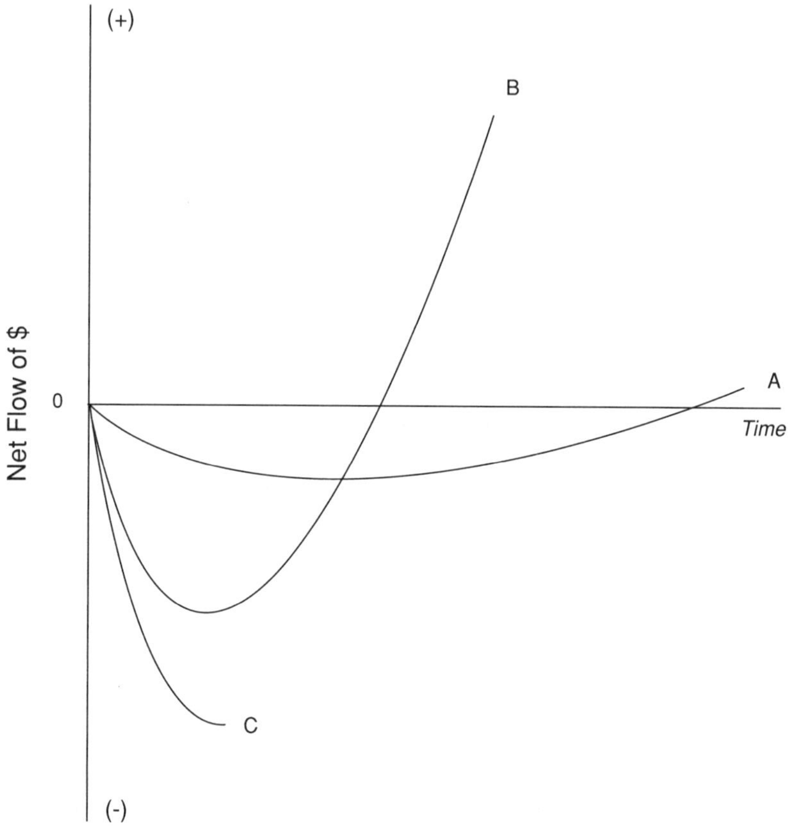

FIGURE 9-1 Three Scenarios for Developing Technology

management of a company by giving them too much money. Often this leads to spending money on the wrong things. I would sell stock in any company that buys a corporate jet, for example. I call this the Taj Mahal syndrome.

The point I am making is that what we want to do is use intellectual property to drive people to invest more rapidly in developing the technology in the short term, like curve B, to capture the market and to get products on the market more quickly.

If you think about the behavior mode of U.S. companies, one of the problems with curve B is that in the short term you are losing money relative to the companies on curve A.

In other words, if I am a manager of a company that is investing in capital equipment, technology, patents, and intellectual property along curve B, how can I look like a hero? I can look like a hero by cutting all curve B investments down to the least I can get away with and still look credible (curve A). I make more profit in the short term from a smaller investment, and I get promoted to destroy a bigger chunk of the company. I call this behavior the MBA syndrome. I confess that I am an MBA.

Figure 9-2 shows the stock performance of three groups of U.S. companies and demonstrates the importance of the curve B behavior mode. The lowest line is the stock performance of the Standard and Poor's 500 during the five years from 1984 through 1988. It went up 66 percent during this period. The group of companies right above it, called Universe, is a large group of companies chosen by a single criterion. Does one family own 10 percent or more of the shares of that company (e.g., Motorola, Corning, Ethyl, and DuPont)? IBM would have made it back in the 1970s when the Watson family owned more than 10 percent. The theory is that family companies are more willing to sacrifice short-term profitability for long-term gains, and in fact, the tax system in the United States drives them to sacrifice short-term profitability for long-term gains. Note that these companies outperformed the Standard and Poor 500 twofold during this period. I find that my license agreements tend to go to the Motorolas, the DuPonts, and the Ethyls, and not to the General Motors, and the reason is that they are willing to invest in intellectual property and commercializing technology more than General Motors and the broadly traded public companies.

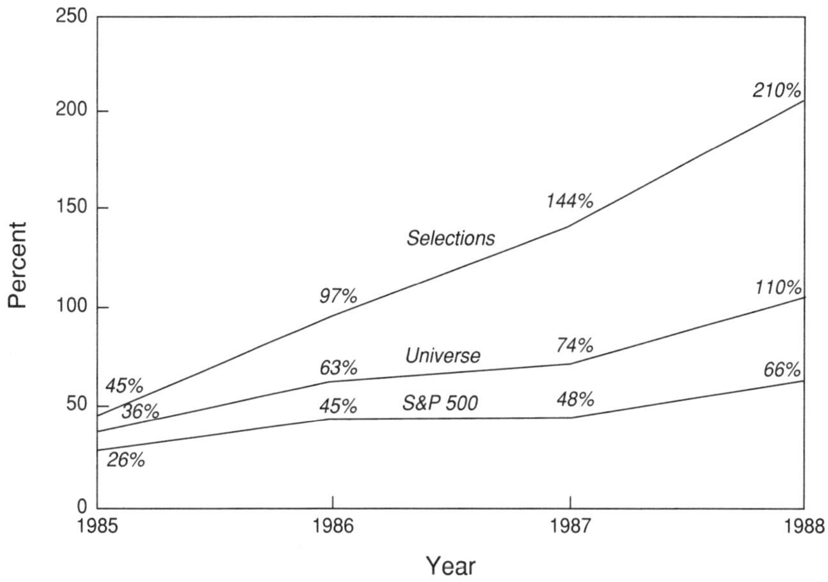

FIGURE 9-2 Stock Performance of Three Groups of U.S. Companies

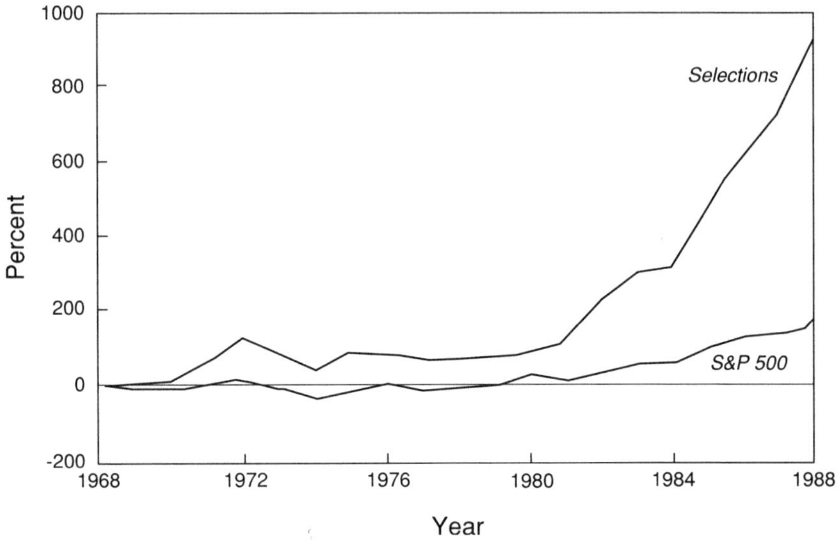

FIGURE 9-3 Cumulative Returns to Standard and Poor's 500 and Selected Family Managed Companies

Astonishingly, these data indicate stock performance twice as good as the Standard and Poor 500 in four years by that one single criterion. The "Selections" are publicly traded companies that (1) are 10 percent or more family owned like the Universe companies and (2) have significant family involvement in managing the company. There were actually 12 criteria for the Selections. I cannot give them to you because this work was done by an MIT alumnus who trades money for a living. He is Mark Cunningham, Vice President of Alliance Capital, and this is his competitive advantage. He has outperformed the stock market very regularly by a wide margin.

Figure 9-3 shows that the Selections outperformed the Standard and Poor's 500 by an order of magnitude (tenfold) over 20 years. When the market is doing well, they grow rapidly relative to the broadly publicly traded companies.

Now that we understand what is driving the development of intellectual property, let me discuss two of the things that John Armstrong mentions in Chapter 8. One is avoiding sui generis laws for intellectual property protection.

I disagree with this position because the crisper you make the laws, the easier it is for me to know whether I am infringing on an IBM product or IBM is infringing on one of mine.

The easier that is known or the crisper the laws, the better it is for the world economy because we spend less money litigating and hiring lawyers to work in a gray area. Anything that we can do to change this gray area into something that is black and white would be, in my opinion, really, really good.

The other point I want to make is that in considering sui generis, and our entire intellectual property system, one of the things that has not been discussed is the method used to enforce intellectual property right (IPR) laws. When you go into court you can have one of two scenarios: You are going to win or you are going to lose. Yet you may be going in with a very gray issue—for instance, look and feel in the software area is a very gray issue in my opinion. Yet one company is going to come out as winner and one as loser. What should at least be a topic of discussion is whether or not mediation or some other form of resolving disputes should be attempted before going to court. I have seen mediation work in billion-dollar settlements, and I think it is a much cheaper approach.

The second point I wanted to make about Chapter 8 actually dovetails with this. Armstrong favors using the existing framework. The existing framework in which you go to court to enforce your patents and copyrights again creates a problem from my point of view in that it favors large entities over small ones. In other words, as a university or a small start-up company going to court against IBM, you can almost predict who is going to win based on who has the deepest pockets. Who is going to be able to enforce its position better? Who is going to be able to hire the best lawyers? Mediation would perhaps help improve the quality of settlements.

Let me just make one final point. Rather than discussing only matters on which I disagree with John Armstrong, let me mention some about which I agree.

The general conference does not present a balanced view on intellectual property in the following sense. When the originators of intellectual property such as IBM, MIT, and AT&T enforce strong intellectual property laws, developing countries can make an argument that this is unfair. However, if you look at it on balance, we are transferring much more for free than we are protecting. In other words, if I purchase a new computer from IBM and design a competitive computer without infringing any of IBM's patents, the cost for me to design that competitive computer is cheaper than the cost IBM paid in the first place to design its computer. We are not looking at enforcement of intellectual property in a balanced way. When IBM enforces a patent or when anyone enforces a patent, the balanced way to look at it is that the company is transferring much more for free in that product than it is enforcing in intellectual property.

The Federal Government

BRUCE MERRIFIELD

Some of you have heard of Hernando DeSoto, a Peruvian who has written a book based on a seven-year study of Latin American economies. The bottom line of the book is that the definition of an underdeveloped country is one in which the entrepreneurial function has been made illegal. You can immediately translate that to the former Eastern Bloc countries and to a lot of other centrally controlled countries around the world.

Unfortunately, you can also translate it to some of our big Neanderthal companies in this country in which the entrepreneurial function virtually has been made illegal, but the important understanding there is that before entrepreneurial activity can thrive, intellectual property protection is required. Moreover, incentives that will allow this to happen also are needed.

Basically the world has changed. Ninety percent of everything we know in the sciences has been generated in the last 30 years. It will double again in the next 15. Product and process life cycles in electronics have collapsed to a few years, 2 to 3, rarely more than 5 to 10 years in most other industries, making existing facilities and equipment obsolete often long before their useful lives can be realized. This process really accelerated after World War II, in 1945, when Vannevar Bush wrote a report to the president identifying research as the endless frontier. The National Science Foundation was funded, and since then, we have pumped about a trillion dollars into our academic infrastructure in this country, more than any other nation can even begin to match. As a result of that, last year we spent about $25 billion on basic research, 10 times more than any other nation, but it is the source, the wellspring, for all the critical technologies—the 20 or so critical technologies that will dominate the twenty-first century. Moreover, any company that is not investing, not only in incremental improvements in its existing operations (just to maintain current cash flow) but also, simultaneously, in next-generation technology, may not survive the next decade.

There is a basic paradigm shift here in management strategy which says that wealth can no longer be measured in terms of ownership of fixed physical assets that can be obsolete in a few years. Wealth now has to be measured in terms of ownership or at least time critical access to proprietary, knowledge-intensive, high-value-added, technology-driven systems.

It is important, however, that we understand the significance of this paradigm shift for the developing countries. We now have a historically unprecedented opportunity to raise the quality of life of every nation in the world through expansion of their economies. Moreover, this is in the enlightened self-interest of the United States and all developed countries. For example, in the Marshall Plan after World War II, we pumped billions of dollars into Europe and Japan. Do you know who was the primary beneficiary of that investment? It was the United States because as those economies expanded, they offered tremendous opportunities for further investment and export.

As we help expand the developing country economies, the benefit will come back to us many times over. The problem is how to do that effectively. As we know, the communications revolution has created a situation in which capital can flow with the speed of light anywhere in the world. Technology also can follow rapidly wherever it is well treated, which means wherever intellectual property rights are respected.

Any rice paddy in the world can be transformed in 6 to 12 months by the big international construction people, to a state-of-the-art automated facility operated by $2 an hour labor. Nothing like this has ever happened before in history on this scale. Moreover, intellectual property protection can enhance foreign investment in these developing countries.

A model for doing this was developed at the U.S. Department of Commerce. We call it the Modern Marshall Plan. It is basically a joint venture arrangement in which a professionally trained function in each country first searches out emerging opportunities in the developing country. It is important to start with technology that interests the developing country. The second step is to match the foreign nation's company with a U.S. company to provide the missing skill, resources, and capabilities for jointly developing new technology. It is a win-win situation that multiplies the market potential, shares the risk in development, and accelerates development times.

The first pilot model with Israel has been remarkably successful. I think now there are more than 225 joint venture companies that have commercialized technology, with greater than $1 billion in sales, and something like a 90 percent success ratio. We have translated that model to a number of other countries. In India, for example, there have been remarkable results. Thirty-five out of thirty-six initiatives that were funded are now starting commercial operations. That is a pretty high success ratio. Basically every nation has latent entrepreneurial potential. India has 80 million educated people, and when that operation was established in Bombay (with some trepidation, admittedly), within about 6 weeks 300 proposals had swamped that office from all over India. Tremendously exciting things emerged, some of them low tech but many of them extraordinarily high tech in concept, and many of those are now going commercial.

This model, moreover, begins to create an incentive in the developing country for intellectual property rights because as indigenous companies develop their own technology, they understand the need for intellectual property protection, and put pressure on their own governments to provide that protection. This is important, because democracy cannot exist in a developing country without a small-business middle class that has a stake in political economic stability. Therefore, the first step in developing democracy is to help develop the middle class. That is really what we have here in the United States. Few of us probably realize what a remarkable small-business entrepreneurial culture we really do have. We have 15 million companies in this country. Of them 98 percent have less than 500 people; 90 percent fewer than 100 people. We have been, for the last decade, creating 650,000 to 700,000 new small businesses every year in the United States. Between 1980 and 1990, 20 million (net) new jobs were generated, 70 to 80 percent of them in these small businesses.

It is this bottom-up entrepreneurial revolution that creates much of the innovation and also the jobs. It is the small-business middle class that is needed to sustain democracy, and that is the model that can help us expand the economies of developing countries. It involves enlightened self-interest, and it is important that we provide incentives for this process.

The U.S. Agency for International Development (AID) provided the initial seed funding in India, and I think AID now sees this as a model for a much more extensive effort. The World Bank has yet to understand this model, but perhaps one day it will.

The important thing to understand, though, is that we now have historically unprecedented opportunities to raise the quality of life of all nations. For the first time in history we have point-to-point contact with any point on the surface of the earth. We can bring education through interactive video educational systems to any person on the earth. There are 4 billion people around the world just as smart as you and I who have never had access to education. We can change the global village in the next 25 years more than anyone might imagine, if we provide effective incentives and begin to develop the procedures that we well understand. We do not have to reinvent the wheel that has already been demonstrated. The Israel and India models can work with countries in the former Eastern Bloc, South Africa, and the Ivory Coast. It is currently operating in Chile and Finland. Even France has adopted the model, with rather remarkable success. It is called the FACET program over there.

To summarize, the opportunities we now have are historically unprecedented. We have the advanced technology that we need to share in international collaborative efforts, and as we do so, we will be the primary beneficiaries. I see a federal role here, which provides incentives, and a catalytic function that can help create such collaborative efforts on an international scale. I hope we can all work together to make that happen.

U.S. Industry

GEORGE W. McKINNEY III

I may be a relatively rare bird here in the sense that I have run three companies. I worked for a large company, Corning Glass Works, where I was in charge of business planning and corporate development; I worked for a major venture capital firm; and I was the first president of American Superconductor, a company that has been deeply involved in intellectual property with the high-temperature superconducting field. I currently spend 80 percent of my time doing venture activities and 80 percent running a company called Environmental Quality Corporation, which is involved in novel developments in the area of source reduction and pollution prevention.

In intellectual property I think there are significant differences between the mental images mentioned earlier that form the historical background, and the realities of the world we are in. I would like to raise five issues. I do not think anybody here is going to run out and cause change to occur, but these are areas in which we have to think about why change is not occurring. In that sense, I disagree with John Armstrong's position (Chapter 8) that we have a good system, it is working well, and we should not mess with it too much.

Let me raise the first issue. We talk about invention and intellectual property, and we think of the inventions of the world, which can be counted in the hundreds, in contrast to the vast majority of developments in intellectual property that are evolutionary.

Let us stop and consider the issue of true inventions. The reason I am concerned about this is that I deal with small companies, universities, national laboratories, and government laboratories, and I believe the majority of what is happening in this country is happening outside the large corporate structure. When we get big inventions, they do not necessarily happen in large companies. As somebody who lived through the early phases of high-temperature superconductivity in 1987, I was surprised to learn that there would not be a *basic* patent awarded to the Nobel Prize laureates from IBM who, in fact, invented oxide superconductors. No one, to my knowledge, questions that they invented a fundamental concept, but their application fails to meet somebody's definition.

I automatically say that this is a failure of the patent system because whether they would be willing, à la John Preston at MIT, to write an X-Window style (everybody is free to use) license is up to them. The idea that we have a system in which they could not achieve a fundamental patent is appalling.

I am concerned about this as we move to a first-to-file system. For a large company, the idea of $10,000 or $20,000 to file a patent is disruptive. I know at Corning when we had to file 50 to 60 patents to get into the optical waveguide business, that was a problem. For a small entrepreneur, an inventor, the idea of needing $15,000 or $20,000 up front, and taking all of this time (which should be going into research) to write an initial patent, is frightening.

I believe in first to file. I believe that we have to go to a global system that requires it, but I believe this country should lead in the development of—this is my first suggestion—a very simple, probably two-page description of an invention that will hold your place in first to file. This would cost nothing more than time, a notebook, and an application fee, and you would then have six months to finish up the due diligence. This application would be aimed primarily at true invention. One variant I would support is granting a window for filing to those who are "first to publish."

The second area I would like to discuss is our emphasis on invention rather than application. The original purpose of patents, as discussed earlier is in fact to encourage application, encourage commercialization, and the economic benefits.

Anybody who is familiar with the Small Business Innovation Research program knows that we have spent millions of dollars encouraging people in this country to do research that they have no motivation to commercialize. We know that one of the reasons large companies are moving away from internal research is precisely because a great deal of that research has not been commercialized. As a country, our concerns focus on commercialization.

I come out of a manufacturing background during my first few years of working. We do not emphasize cost-effectiveness enough. We emphasize in our intellectual research the idea of the invention that increases the performance by an ever so small amount. That idea has been driven in this country by the fact that so much of our research has come out of military and defense-supported areas in which it was critical to have the ability to get that last decimal point of performance accuracy.

I worry about cost-effectiveness. At Corning I was involved in the development of the emission control substrate that is in almost all cars. It is coated with a catalyst that is used to remove noxious fumes.

The fact of the matter is if you imagine a complex structure 2 feet long with 500 holes that extend the entire length, made out of ceramic, you would not expect it to cost about $1.00 to make. That cost is possible if it is made in quantities of 10 million to 20 million; you could never do it by manufacturing quantities of 100 to 1,000.

So many of our technologies in this country are oriented the other way.

I believe we have to redevelop in this country the passion to get things applied.

For that reason I would like to see a patent system that extends the time period of a patent for applications and, if the originator of a patent is able to *apply* it, grants those extensions. I think this is key in concept to problems in the drug industry where, in fact, it takes so many years to get the drug to commercialization and the period has been entirely used up by then.

The third idea I would like to discuss is the issue of globalization in fact, not just in word. At Corning there was no question from anybody as far as I can tell, that Corning Glass Works scientists, led by Bob Maurer, developed the glass that makes optical waveguides as we know them and optical communications possible. If someone wants to challenge me on that scientific fact, they are welcome to do so.

Those patents were issued in the United States seven years before they were issued in Japan. During those seven years they went unissued in Japan, Sumitomo was able to (1) go into manufacturing using an illegal process and (2) develop a competing process that allowed it to stay in business.

I consider that inexcusable—a failure of the patent system. I would recommend strongly, if anybody has the ability to make it happen, that there be a system in which, within two years of issuance by the first country, there is a presumption of issuance in all of the treaty countries.

The fourth concept I would like to discuss is the real confusion between evolutionary minor developments, which are so important to businesses, and inventions. The vast majority of these evolutionary developments are not inventions. They are things anybody who is an expert in the field could have done if he put his mind to it, and in some cases, they are being done simultaneously.

This is also an area in which small companies have problems in competing with big companies. When I was a managing partner of American Research and Development, one of our portfolio investments was a company called Fusion Systems. Those of you who know intellectual property will recognize Fusion as the company that makes ultraviolet (UV) lamps and has had a long-standing fight with Mitsubishi. There is *no* question again that Fusion invented the concept of the modern high-intensity UV lamp. Mitsubishi got an early example of it, proceeded to reverse engineer it, and then patented every variant imaginable in Japan. It went to Fusion and said, "If you want to sell in Japan you need to cross-license this—by the way, that includes rights to sell in the United States."

I consider that to be an inexcusable failure of the world patent system. I think the only benefit that should come from these minor variations is *permission* to compete, if you own the basic technology. I would like to see patents for minor developments done away with. It would certainly simplify the system.

The last issue is a deep concern as we have faster and faster paces of technology with flurries of inventions. I was in the superconducting field at its beginning. I can point to one intellectual property case in which four patents were filed within 30 days by four competing companies for the same precise invention, all of which were done independently. What benefit is served to anyone by awarding that patent to the person who happened to file it on March 28, versus the person who filed it on March 29, or the person who filed it on April 10 or April 15? I would strongly advocate that as long as independent invention is involved and there is no publication cycle in between, that there be a window in which simultaneous invention is presumed and awarded—that the inventors be forced to share the rights. During periods of rapid invention this will allow Digital to compete with IBM to compete with NEC, knowing full well that the cycle of technology invention may be days, not years.

10

Intellectual Property Rights and Competitive Strategy

A Multinational Pharmaceutical Firm

OTTO A. STAMM

A patent is something like an insurance policy against theft. One does not absolutely need it, for perhaps nothing will be stolen, but if one does need it and has not got it, then the experience is usually a painful one.

If the probability is great that something may be stolen from us, then it is best to insure oneself against the risk of loss. A good artist will therefore be well advised to insure his paintings against theft, for he has invested time, labor and hence money in producing a work that, after all, he wishes to sell.

To do so the artist must first make his work known, thereby risking that someone will fake it or copy his style. This means that it is not the paintings themselves (i.e., the artist's tangible assets) but the creative work the artist has put into them that will be stolen. Theft of this kind is worse than that of a tangible asset, because the theft of an idea that can be reduced to practice becomes common knowledge to which all can help themselves.

From this example of the artist, it can be said that a first object of patent protection is that it makes possible publication, avoids secrecy, and thus permits the spread of information without the creator being robbed of his or her intellectual property.

If, however, this theft is not prevented by legal protection of intellectual property, such as copyright or patent protection, then the painter or the inventor will lose the bulk of his expected return on investment, which he needs in order to survive. It is therefore a second object of patent protection to ensure that inventors receive an adequate remuneration for their creative achievement by preventing the theft of their intellectual property.

If the painter or inventor is no longer able to make a profit from his paintings or inventions, then he will stop painting or carrying out research and seek another means of earning an income. He has no further incentive to paint pictures or make inventions. Hence, it is finally a further object of patent protection to motivate people to engage in innovation.

If the manufacturer of a product that she has developed herself must reckon with the theft of that product, she will no longer make it because the money invested in development would be lost. The situation in the research-based pharmaceutical industry is by analogy the same as that just briefly outlined.

Can the products manufactured by the innovative pharmaceutical industry easily be stolen? The theft in question is not, of course, of the actual pills themselves (i.e., the tangible asset) but of the creative idea that produced them, in other words the invention, which is something intangible. It is precisely this intangible asset that a patent protects. The nub of the matter therefore is whether people have the incentive to copy an invention that they know, for example, from a publication.

It is a regrettable fact that of all branches of the chemical industry, the research-based pharmaceutical industry is the one most liable to theft of intellectual property. The reason is that the development of a new pharmaceutical product is extremely expensive and that the product has a number of typical properties: it is a specialty for which there is a need and which has a relatively long therapeutic life cycle. The product is one of high added value; it has consumer product characteristics (i.e., it is produced in a large number of units, each of which is relatively inexpensive and easily transportable). Last but not least, anyone with an elementary knowledge of chemistry and pharmacology can imitate most of these products at a fraction of the enormous costs incurred by the creator of the original product. For this reason, the pharmaceutical industry is a sector of industry that must have a key interest in securing protection for its innovations. A lead over the competition must be safeguarded by law; otherwise it is impossible to achieve an appropriate return on investment. Profits are necessary for reinvestment in future research. Without such a safeguard against imitation (i.e., without patent protection) there would be no research-based pharmaceutical industry.

The first basic point to be made is therefore that without patent protection, the pharmaceutical industry would make no investments in research.

The legitimacy of this is supported by the situation in countries that grant or granted no patent protection for pharmaceuticals. The fact is that no significant new drug has ever yet been developed in such countries as the former Soviet Union, Brazil, Argentina, and India, which have the necessary level of expertise to do so, but whose patent protection for pharmaceuticals is nonexistent or weak.

It may fairly be said that, without patent protection, there would be virtually no further progress in pharmaceutical research. Such a situation would be deplorable because only about 10,000 of the approximately 30,000 diagnosable diseases can actually be treated with drugs. Moreover, causal therapy is not available for most diseases; only the symptoms are treated. Nevertheless, it is noteworthy that those circles that condemn patents for pharmaceuticals or living organisms as being an unsocial monopoly are the very ones to insist categorically that the pharmaceutical industry should provide a drug to combat AIDS, cancer, or Alzheimer's disease. In a competitive environment, innovation and supportive patent protection are key both to the research-oriented firm and to society.

Drugs that were known not to be covered by patent protection would not be developed. That research is carried out *at all* is thus linked directly to patent protection. Patent protection must therefore be available in large markets, so that research-based private industry can exploit the possibility thereby provided of keeping ahead of the competition to make the profits it needs to finance research projects.

In the pharmaceutical sector there are few inventions that can be kept secret, perhaps only manufacturing processes, especially in the genetic engineering field. Yet protection for secrecy, even if it were legally ensured nationally, is of little use. Once the secret is out of the bag it cannot be put back in again, and it is known *worldwide*. Also, we have no warranty against this leakage of knowledge into the public domain, especially not in areas where government registration requires that all documents be handed over and where there is no absolute reliance on their actually being kept secret. In the case of self-reproducible matter, the loss of a single bacterium is theoretically enough to hand the factory over to a third party. For this reason, the deposit of cultures of microorganisms, if required for obtaining patent protection, should be better protected against usurpation by third parties.

What effect does it have that the level of patent protection varies from country to country? The kind of research being conducted depends far less on the local intellectual property situation than it does on the strategies for research planning. However, the pharmaceutical industry would not develop any product from which only a country such as Brazil, Argentina, India, Turkey, Thailand, or others that do not grant patent protection for pharmaceuticals, would profit.

To be competitive, it is important not only to develop new products but also to sell them (i.e., a strategy for marketing is also needed). How does patent protection influence the marketing strategy of an innovative pharmaceutical company?

When an artist has painted a picture, his primary aim is to sell it. What happens to the painting later probably interests him less, although it certainly will not leave him unmoved if an art dealer makes a much higher profit from his painting (i.e., his creative effort) at an auction. The same consideration applies to a product that a manufacturer has developed. She wants to sell it to make a profit. The sale is the primary object, but the later fate of the product on the market may also be important. Where there are substantial, usually government-dictated, price differences in different markets, as is the case with pharmaceuticals, it must be possible to prevent parallel imports (the catchword here is "international exhaustion"), otherwise competition will be distorted.

The second point to be made here is this: Patent protection tends to be rather of secondary interest for marketing strategy. The aim of a good marketing strategy is to bring a new product as rapidly as possible and as widely as possible onto the market, regardless of the quality of the patent protection in that market. If patents help to support an acceptable price level and to provide a lead over the competition, then they are naturally a welcome and important element in helping to achieve the desired profit optimization. Good patent protection also motivates marketing to speed up the introduction of a product in those countries so as to exploit the advantages conferred by patent protection for as long as possible. The efforts being made to extend the actual utilizable patent life by means of patent term restoration clearly demonstrate that these advantages are not to be disregarded. All available means for additionally safeguarding intellectual property are used to back up patent protection, but they do not have the absolute character of patent protection. These means include, in particular, trademark protection, which can very strongly influence the goodwill of the buyer. The trademark, once firmly anchored in the customer's mind, is a guarantee of quality and hence a sign of loyalty to the customer or patient. The trademark is also an aid to image cultivation. The advantage of the trademark is its unlimited life and the wide geographical spread of the protection accorded it. The drawback of the trademark is that it can be circumvented easily, and that the government regulations prevailing in many countries restrict its value. Registration protection (i.e., the ban on using registration data for a second registration before the expiration of a specific time limit) can also supplement patent protection. A certain marketing exclusivity is thereby also ensured, but only for a specific product. Finally, secrecy—especially for production processes—is another supplementary measure in certain cases for gaining the edge over one's competitors and thus for supporting a comprehensive marketing strategy.

Especially important, above all in countries without effective patent protection, are marketing steps that—varying from place to place—are independent of property rights. Such steps include customer orientation (e.g., the confidential relationship with doctors); satisfying real needs, which requires a true understanding of the customer's interests; promotion by means of medical-scientific and other explanatory literature; a reliable distribution organization; readiness to take up suggestions from the realm of practice; good contacts with government authorities; commercial probity; market-oriented price strategy; the guarantee of ongoing and consistent drug monitoring on an international basis; keeping ahead of competitors by rapid product launching; and a healthy capital structure of any company-owned production plants, which makes it possible to survive unfavorable economic conditions such as high inflation.

The unavailability of efficient patent protection in certain countries therefore does not keep a firm from marketing its products in these countries. Hence there are doubtless countries in which there is no patent protection—where it is not possible to take out risk insurance against theft, but where nonetheless original pharmaceutical products are sold.

Also, just *where* a product is first introduced does not depend primarily on whether efficient patent protection is available in a country. At present the wish and the need are to introduce new pharmaceuticals onto the market as quickly as possible and in as many countries as possible. The location of clinical testing and first market launching will be chosen in accordance with this principle, as will also the later sequence of local market launchings. The period of time until the first introduction onto the market will be influenced materially by government regulations and by the availability of doctors who carry out clinical tests. The desire for global market presence cannot therefore take into account the quality of local patent protection. This desire dictates solely the measures to be taken in addition to patent protection for successful marketing of the product. The sales in different countries can be effected by one's own sales organizations or, where this is economically or politically infeasible, through agents and outside distributors.

The best marketing strategy, however, cannot in the long run lead to the desired profits without new and improved products. Without new products a pharmaceutical company will ultimately lapse into trading in generics; that is, it will become a company that simply cannot afford to engage in such expensive research. In this case there would, of course, be no more progress in the pharmaceutical industry. Yet, as already mentioned, this is just what most people do not want. To be able to afford research, to be able to develop products at all, requires patent protection, especially in those countries that constitute large markets. Fortunately, in the United States, Western Europe, and Japan, patent protection for pharmaceuticals is good. Given a sufficiently long patent life and exclusivity, it is possible to achieve the kind of return on investment that will allow a pharmaceutical company to market new products in countries where the price level is inadequate. Naturally there are limits to this indirect subsidizing of market presence in countries with inadequate patent protection by countries with efficient patent protection. Since the prices of pharmaceuticals are usually government controlled even in countries with efficient patent protection, the research-based pharmaceutical industry can no longer afford the shortfall in sales resulting from the loss of markets caused by imitation products. The ever-spiraling costs of research into and development of a new drug also call for an ever-increasing return on investment. The newly industrialized countries must also make their contribution here if new drugs are to be made available to all of humankind in the future. Therefore, without worldwide patent protection, industry will be increasingly unable to afford a globally oriented marketing strategy for new products.

One concluding comment on the influence of efficient patent protection on the transfer of technology, and hence on corresponding new investments: It can quite confidently be ruled out that investments will be made in research into novel products in a country where patent protection does not exist. However, with regard to investments, the question of patent protection is only *one* factor among many for deciding on investment. Other factors, such as intellectual work potential, infrastructure, political stability, and economic structure must also be taken into account. The third important point is therefore that if a country is prepared to raise its patent system to a high level, the readiness to invest and transfer know-how will increase. For only through patent protection does a creative idea, an innovation, become a legally protected item of trade and thus of transnational transfer.

This point is best illustrated by citing the drift of investments away from Europe to the United States in the field of biotechnology. Table 10-1 shows the volume of investment made by European firms in biotechnology in Europe and in the United States for 1989 and the first half of 1990. Of the total investments made by European firms in this sector only three percent remained in Europe in 1989; in the first half of 1990, the figure was as low as 1.6 percent. Also, in this example, patent protection is only one part of the overall picture that influences the decision to make investments. Nevertheless, in view of the fact that in the United States, patent protection is fully available for the entire field of biotechnology (including plants and animals)—whereas in Europe effective protection is still lacking or explicitly not obtainable—it is very likely that the distribution of investments is not purely fortuitous and supports the argument that patent protection is a

TABLE 10-1 Investments in Biotechnology by European Firms in Million ECU

	1989	January-June 1990
In Europe	59.3	48.8
In USA	1925.2	2987.4
Percentage of total investment remaining in Europe	3%	1.6%

factor that *does* have an influence on the decision to make investments and therefore on technology transfer.

To sum up:

1. Without efficient patent protection in the most important trading areas, there can be no research and development of new pharmaceuticals.
2. For marketing strategy, rapid product launching and a worldwide presence (i.e., global sales with an edge over competitors) are more important than considerations of patent protection and its quality in different countries. These criteria influence the choice of countries in which the pharmaceutical industry is active first. Patent protection in big markets, however, is an important factor in securing a lead over competitors for a limited time and hence in optimizing profits. Industry therefore will not introduce important products onto the market if exclusivity, even if limited in time, is unavailable in key markets.
3. Without adequate patent protection, no investments will be made in pharmaceutical research, and consequently there will be no corresponding transfer of technology to countries that de facto grant no patent protection for pharmaceuticals.

One final comment, which has no bearing on the patent strategy of a multinational research-based pharmaceutical company but does concern the patent strategy of our politicians: The inventions of the pharmaceutical industry that will be on the market in 15 years' time are being made today. If awareness that no insurance is available for these innovations comes only in the next decade, then it will be too late. The entire investment of time, work, and money will benefit the imitator. For this reason, efficient patent protection without discrimination of individual technological fields, such as biotechnology in Europe and many other countries, is needed today and not 20 years from now. Otherwise the incentive will be lacking to pursue research whose results we need urgently. Therefore, patent protection or, more correctly, exclusion from patent protection does have an influence on the choice of research projects. This need not be if the politicians would finally realize that patent protection is there to protect intellectual property from theft and not to standardize ethical concepts; to act as an animal, plant or environmental protection law or as an instrument of price control; or finally to prevent any risks in industrial research, development, and production.

A Multinational Electronics Firm

MICHIYUKI UENOHARA

According to my historical observation, when manufacturing companies begin to need royalty income badly for the improvement of corporate financial performance, it is a strong sign of losing competitiveness. So I do not buy the stock of such companies, or I sell them if I own them.

Manufacturing companies should earn a profit as a result of good services, by providing a better product to the market. For this aim, intellectual property rights are a very important asset to pursue the corporate strategy of serving the market best without interference from other competitors.

My company, NEC, was established in 1899 jointly with Western Electric Company of Illinois and Japanese partners. We now manufacture computers, communications, semiconductors, and home electronics. These are the four core business areas. Hence, NEC strongly inherits Western Electric's corporate culture—better products and better services have been a solid corporate policy for more than 90 years. Executing corporate policy, we have secured patent rights as much as possible to maintain a healthy operation rather than to seek an extra royalty income. NEC's patent policy appears to be defensive, but I think it is the best offensive policy from a healthy business point of view.

We welcome cross-licensing agreements because they allow us to communicate freely and to concentrate fully on engineering efforts for better public services. When we transfer our technology to other companies, we select our partners and set the royalty in such a way that they are strongly motivated to apply the patent in marketable products and to secure reasonable profits. Technology and patents are mere knowledge. They cannot contribute to the advancement of public welfare unless they are applied, manufactured, and marketed as practical products.

A company should not be managed solely for the sake of profit; rather, profit is obtained as the result of better services for the customers. The marketable product cannot be produced by basic or central patents alone. Numerous peripheral technologies and inventions are necessary. Most customers appreciate peripheral inventions rather than basic inventions. This is market behavior. Only professionals appreciate basic inventions.

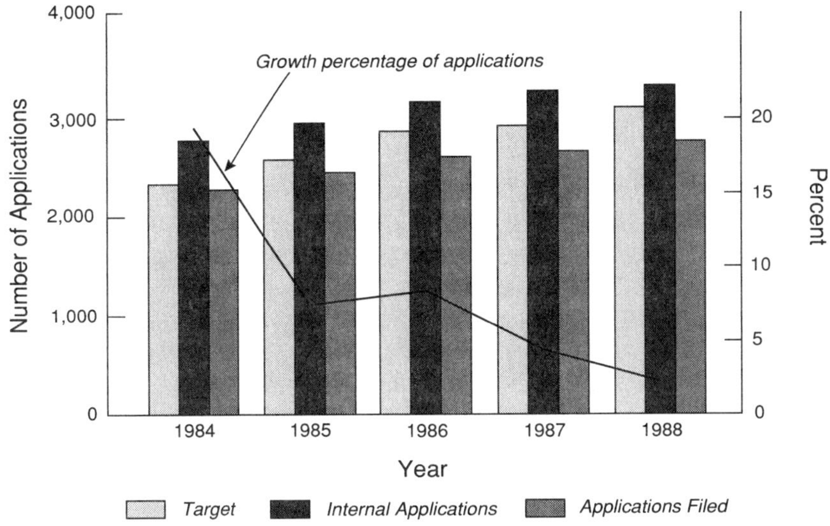

FIGURE 10-1 Patent Applications from NEC Central Research Laboratories

We have experienced severe business threats over peripheral patents of almost public knowledge quality. That is why we are encouraging our employees to reduce any idea to patents. Even researchers in central research laboratories are asked to file peripheral patents related to basic patents that they invented.

Figures 10-1 and 10-2 illustrate five years of statistics on the patent application process in our central research laboratories. In Figure 10-1, the left-most bar in each three-bar group shows the annual target that management suggests to our researchers. The central bar shows the total numbers of internal applications submitted and the right-most bar shows the total numbers submitted to the patent bureau after combining related claims.

As shown in Figure 10-2, the average number of patents filed by patenting researchers (some percentage of researchers are not asked to file any patents, just to do basic research) is about five.

Figure 10-3 compares the total number of patent applications disclosed for six competing companies over the same period from 1984 to 1988. Every year we issue extensive data showing the competitive situation to our employees. The average number of patents disclosed by patenting employees throughout the company is about 2.5, about half that of the researchers in the central research laboratory.

NEC markets products in more than 149 countries and manufactures in 15 countries. Many underdeveloped countries ask us to transfer our technology to develop their countries. When we negotiate a contract, we always base the terms not only on our interests but also on theirs. It is not easy to convince them that what they are asking is inappropriate for their development. We have recognized how important it is to educate their engineers and policymakers. Hence, we include education costs, actual costs without any profit, in the technology transfer fee. The patent royalty is very modest to motivate them to build up their businesses, since we can eventually get an appropriate profit by supplying them with the necessary components.

As a born engineer, I have devoted myself to advancing technology for the benefit of world welfare and am very much concerned about the current state of insufferable intellectual property right lawsuits. If this trend worsens, the cooperative spirit that is essential for the advancement of science and technology will be damaged, and the world welfare that has to be improved for reducing various conflicts will be degraded.

It is said that the basic idea concluding the Paris Treaty in 1883 on patent law was to legally protect the inventor's right in order to promote the application of such valuable intellectual creations for the benefit of the public welfare. Thomas Jefferson also stated a similar idea. Even though an exclusive right is granted to the inventor, he is also obliged to make its technical detail public so that it promotes research and development for public benefit.

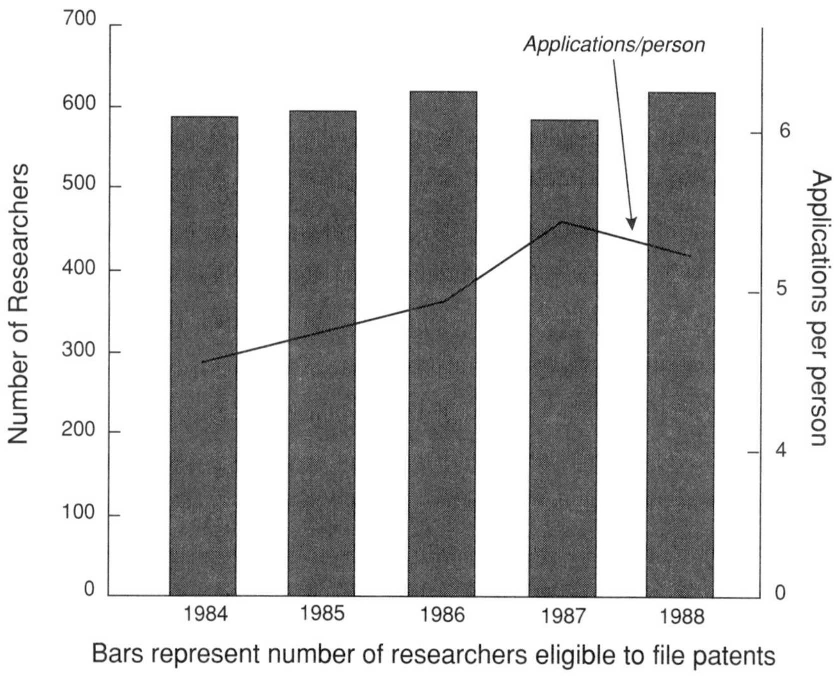

FIGURE 10-2 Patent Applications from NEC Central Research Laboratories

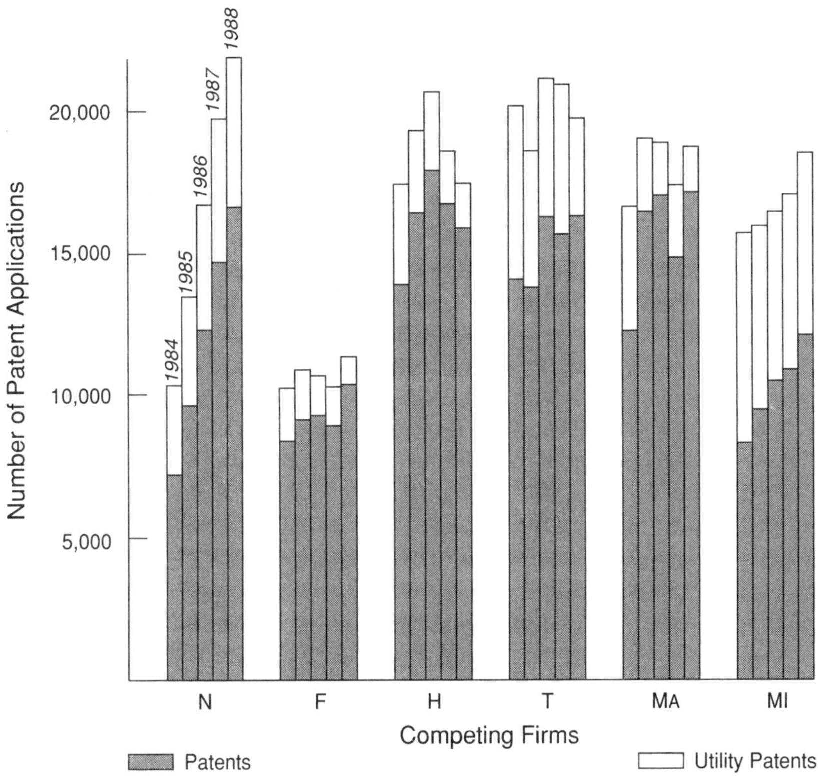

FIGURE 10-3 Annual Number of Patent Applications Disclosed by Six Competing Companies, 1984-1988

Much of the current litigation on intellectual property rights is based on selfish individual monetary interests and neglects the public interest. As the problems of global environmental energy and resources appear to be critical social issues, we are facing severe public criticism: "Is the advancement of science and technology really for the sake of human survival and happiness?" Modern industrial technology is more sophisticated and highly integrated. To solve future social problems such as the environmental problem, we not only must create new technologies but we must also utilize numerous existing generic technologies and inventions. We have to collaborate globally to solve many social problems and to help developing countries.

We have to respect intellectual property rights, and the beneficiary has to pay an appropriate license fee. However, except for very revolutionary new inventions, it is becoming difficult to survey all related patents, including those pending, before a new R&D program is initiated. Quite often, after a substantial investment has been made in R&D, manufacturing, and marketing, companies who are attempting to contribute to the public welfare are sued for patent infringement. Those patents are usually not basic, but rather are very minor and questionable from the viewpoint of Japanese patent law.

No matter how valid the claims are, the defendants have to expend substantial effort to respond to them. Many of the claimants have given up manufacturing and have no intention of applying their inventions. They are interested only in royalty or settlement payments, and their demand is unbearably high. Japanese like to settle outside court for cultural reasons, and most cases were settled out of court until the claims became unbearable.

However, many Japanese firms have had bitter experiences fighting in court, losing most cases at trial by jury. I understand American claimant tactics, but it is hard to understand why patent cases are allowed to be judged by juries who do not understand the technology or the value of the invention. If this trend gets worse, many manufacturing companies will lose interest in manufacturing. No matter how valuable inventions might be, if they are not applied and manufactured as marketable products, they do not contribute to the betterment of the general welfare. When I was still working at Bell Laboratories, I was told its basic patent policy. This agreed well with my belief that a patent right was secured to avoid paying royalties rather than for royalty income. To provide better products and better services had been AT&T's basic policy. Paying 3 percent royalty increases the cost of the product about 3 percent and degrades services.

Real technological progress is the popularization of frontier technology for providing appropriate hardware and software to help reduce handicaps resulting from mental, physical, financial, and social shortcomings, and environmental constraints. Without reducing inventions to a marketable product, the invention is mere knowledge and cannot contribute to the public. I respect the exclusive right of invention, but I question the principles of a claimant who demands a monopolistic royalty and obstructs the advancement of public welfare. This is the time to restore the original purpose of intellectual property law to avoid further confusing the current chaotic state.

The Mexican Software Industry

ANTONIO MEDINA MORA ICAZA

I was asked not only to speak from the point of view of my company, but also to discuss the Mexican software industry, the association of which I have the honor to be president, the status of software protection in Mexico, and finally, to say a little about the North American Free Trade Agreement (NAFTA) negotiations.

I will try to give you some data about the Mexican software industry, its size, and some other important aspects that are related to intellectual property rights protection. First of all, over the past 10 years, the Mexican software industry has grown at a rate of more than 20 percent. Those 10 years were considered tragic ones for the Mexican economy, so this was one of the most rapidly growing industries in Mexico.

In 1990-1991, according to studies done by the Asociacion Nacional de la Industria de Programas para Computadoras (ANIPCO), the size of the software industry was around $200 million. About 70 percent of wholesale items are imported software. Fifty percent of the market consists of payments to companies abroad, and the other 50 percent is in the form of services provided in Mexico. In 1980, the Mexican software industry represented about one-half of 1 percent of the global software industry and, by 1990, had fallen to one-third of 1 percent.

One of the reasons for this is clearly piracy. The year 1980 was the beginning of the personal computer (PC) era, so software was mainly for minicomputers or mainframes where piracy does not occur to the same extent as in PCs. By 1990, things had changed all over the world. Mexico was not an exception. Piracy is a very serious problem, even with the rate of growth we experienced over the past 10 years. It is a big problem and we consider piracy to be the cancer of the industry.

Some estimates based on studies we have made indicate that in Mexico we have five illegal copies of software per copy sold. That is across the board. However, calculations of the economic damage account for only those copies that would have been sold anyway (because there are people that collect software so they have hundreds of different software programs but never use them, just as a hobby). So economic damage is equated to be one illegal copy per copy sold, and that amounts to $200 million for 1991. You can assess the size of the problem.

There are basically two kinds of piracy that we account for and worry about. The first is industrial piracy, where somebody gets a copy, reproduces that copy, and then sells the product. The other is corporate piracy, which happens in the interior of organizations. Basically these are the same phenomena that we face all over the world.

For transnational companies it is clear that there has been severe damage to sales. If we recognize that for most of those companies, Mexico never accounts for more than 2 percent of the global sales, then the damage is not that important. However, if you look at the Mexican bloc of software industries, piracy is the difference between being solvent or being bankrupt. Why? Well, if you produce a program and put it on the market and someone copies it five times for each copy sold, you are soon out of the market.

This is illustrated by past experiences of my own company: We developed a word processor in 1983, and by 1985 we had to abandon that product and project because of piracy.

To complete the picture of the industry, we have around 400 to 500 software companies in Mexico. Most of them are small businesses with fewer than 10 employees. At least 10 percent of these companies are in good shape in terms of size and capacity for competing in a global environment. It is generally accepted in Mexico that there is a high capacity in the software industry basically because we have highly qualified individuals in terms of innovative, inventive, hard-working ability.

There are examples of companies that have successfully entered markets such as the American market. Maybe you have heard about our program, *DAC Easy*, which is an accounting system. The program was developed in Mexico and was very widely sold, accounting for company sales of more than $30 million a year.

Another company called Final Soft produces programs such as *Translate* and *Executive* that have also been sold in the United States. The group of companies with which I am involved created *The Coordinator*, software which is currently marketed in four different languages all over the world. Another example is a systems integration and software house called Soft Tec, which is involved in mainframe development and which has provided services in the United States.

Finally, there are several niches in which we see opportunity for the Mexican software industry. Basically these are not the spreadsheet, word processing, or rating systems niches, but rather those that are particular to our country—such as the administrative area, accounting, and inventory—because our fiscal laws differ from those of other countries, so products specifically adapted for our country are important. There are also some high-tech niches where we can take advantage of the capacity that we have.

Now, a few words about ANIPCO. ANIPCO is the national software association. It was founded in 1985. The people who got together to form it were concerned about the lack of representation of the software industry in Mexico. We spoke mainly of hardware and I see that this phenomenon is the same all over the world. We see the same story in Dr. Samuelson's discussion in Chapter 12, in terms of the history of software protection and the software industry in the United States. Software was bundled in with hardware at the beginning, and the same phenomenon occurred in Mexico.

The goals in creating the association were first to create the Mexican software industry, then to make it grow, and finally to take action over the main issues concerning the industry—which were piracy, financing, human resources, and information about the industry. Those key elements were in the minds of the people who created the association in 1985.

We have currently more than 200 members. We are an association not only of software developers, but also of hardware manufacturers, transnational companies associated with software in Mexico, hardware and software distributors, academic institutions, financial organizations, government agencies related to the industry, and users. One of our most important goals over the past years has been to seek legal protection for the industry by modification of the copyright law in Mexico.

What kind of protection do we have in Mexico? Basically, we are using the international consensus in terms of protection to software granted via the copyright law. However, I must note here that Mexican copyright law is not a copyright law—it is an author's rights law. It comes from the French tradition of regulating the rights of authors instead of the rights to copy. There is a basic difference. In the process of the NAFTA negotiation a lot of time has been consumed understanding the differences in the laws.

To give some history here, our copyright law—Ley Federal de Derechos de Autor—was created in 1963 when the software industry did not exist in Mexico. So there was no protection. In 1984, the government made a move similar to what occurred in the United States in allowing the registration of software programs. Yet that did not really motivate companies to seek protection. Why? Because this was not good in court. You could register your programs but you could not sue anyone.

In 1988 ANIPCO presented a proposal to the government for modification of the law. It was a very comprehensive idea for changing the law and looked to the law of Spain for the kind of modifications being suggested. It included a whole new chapter in the law to deal with computer software, in turn explaining what it is—all the rules, and so on.

Then in 1990, President Salinas sent a proposal for modification to the Congress and it was approved in July 1991. Now we are waiting for the regulation dealing with the registration of programs and that will complete the cycle. What is the current state of these modifications of the law? First of all, for the first time, protection is granted to software programs as one of the protected works of art under Article 7 of the law. Second, there is a clear and complete limitation in this law, regarding the right of a holder of a legal license of a program to make one copy for the purposes of backup and only one. Also we have clear sanctions for piracy in terms of jail and fines.

So we consider these to be a tremendous advance in protection for software because we can now go to court, have cases prosecuted, and so forth. At the same time, however, it is not enough; there are several areas in which the law is still lacking the strong protection we would like to see. These include enforcement, sanctions, and also, providing the equivalent protection given to literary works in terms of protecting the structure and organization of a work.

In terms of the industrial law in Mexico, no patents are granted to software. The industrial property law was also modified last year, and it is very clear that patents are not granted for software, but trade secret protection is granted under this new regulation that did not exist in the past. This industrial law, about which I am not going to be more specific, is considered a world-class law and has been adopted by other Latin American countries this past year.

In terms of how the protection mechanisms are used in Mexico, it is clear that when there was no protection, there was no incentive for companies to register programs and use this kind of protection. We expect that starting this year with the regulation for registry completed, we will have many companies registering software. At the same time, there is a committee that was formed within our association and is related to the Business Software Association of the United States, for developing cases and presenting them to the courts in Mexico. We expect to start testing the real benefit of the new law in the near future. We see that we have a huge educational process ahead of us and this mainly consists of changing the culture that says copying software is not bad. We need to teach people that it is a crime to copy software and to make them understand that.

I can make a few comments here regarding NAFTA. I am surprised that there was no reference to NAFTA during the conference, although some reference was made to bilateral agreements. I think this is a very good example of the kind of approach that can be used and what the future holds in store for us. In the case of Mexico, and also of Chile, the increase of intellectual property rights in these two countries is highly linked to their entering the global economy and to the open market.

Intellectual property rights protection in a country is a way to seek the trust of foreign investors in the country that will allow its economy to grow. I see that as a specific and very strategic move, one that explains why in countries such as Mexico and Chile, the intellectual property rights laws are changing at this particular time. Protection is a key to competitiveness, to building local industry, and to gaining the trust of foreign investors.

A Multinational Telecommunications Firm

WILLIAM L. KEEFAUVER

This section discusses how a large electronics-based firm such as AT&T has used intellectual property rights (IPRs) as part of its competitive strategy.

EVOLUTION

During the first several decades after the invention of the telephone, AT&T used its intellectual property rights—its patents in particular—to establish a market. The resulting litigation resulted in a complete volume of U.S. Supreme Court reports.

For the succeeding three-quarters of a century, its IPR practices became much more benign because of antitrust and regulatory considerations. As a regulated enterprise, AT&T had little reason to protect markets. Instead, it used technology, and the IPRs associated with it, to provide new services, reduce costs, and improve productivity. The operative phrase was "freedom of design," and the patent strategy was structured to acquire rights.

With Bell Labs as its principal technology engine, AT&T for the most part was able to acquire rights it needed from firms around the world on a royalty-free basis. Inventions were selected for patent protection based primarily on their potential interest to firms likely to generate technology of interest to AT&T. It was also recognized internally that patents were an extremely important factor in recognizing personnel and the professional contributions of individuals. The inventors themselves perceived patents to be very important.

This resulted in worldwide activity since many such firms were in Europe and, later, in Japan. For example, by acquiring patents in Germany, AT&T could exchange rights in Germany with Siemens for rights under the patents Siemens was acquiring in the United States.

Broad cross-licensing agreements did more than secure patent rights. They also provided an umbrella for informal technical discussions between Bell Labs researchers and those in other industrial firms. These firms, such as Siemens, Phillips, NEC, and IBM, were not then competitors of AT&T in any significant way. The patent umbrella made these discussions very easy to implement by reducing legal risk.

The invention of the transistor at Bell Labs led to a device whose ubiquity formed the base of an entire industry. This industry—semiconductors—developed its own brand of licensing. Because semiconductors were a critical technology for AT&T, semiconductor licensing became a major focus of AT&T licensing efforts.

The thing that was unique about the semiconductor industry is that its members did not want to worry about individual patents. They wanted to do what we called lump-sum licensing: I will give you $100,000 a year for the next five years (or $30,000 or whatever) and then you will not bother me with your patents, and you will give me $60,000 and I will not bother you. That is the way it worked for many companies in those peaceful years in the semiconductor industry.

Because of the complexity of interconnecting thousands of types of equipment in telecommunications networks, which are now international in scope, standards have always been of great importance. AT&T has long been an advocate of open networks with clearly defined and promulgated network interface standards.

It has also, however, been a staunch advocate of recognizing intellectual property rights arising from the extremely expensive R&D required to develop new standards, provided the owner of any patents or copyrights that apply to a proposed standard will agree to license anybody on reasonable terms. Of course, this was not difficult for AT&T but was its standard licensing policy.

Trade secret licensing similarly reflected the nature of the regulated U.S. business then engaged in by AT&T. For example, although major network elements used in the provision of telecommunications services were for the most part manufactured by Western Electric, its internal manufacturer, many materials and piece parts as well as systems were acquired from other manufacturers.

Many of these manufacturers used technology developed by AT&T but made available to them under "technology-information" agreements—the term used for trade secret licenses. The technology ranged from mainstream network equipment to highly specialized manufacturing and laboratory equipment.

Trademark strategy during this period was similarly more or less benign. Aside from its major corporate symbol—the Bell seal—and the product marks under which much of its hardware was sold (i.e., Western Electric and Teletype), AT&T made relatively little use of trademarks for a company of its size.

THE MODERN ERA

Two events caused major changes in the competitive strategy of AT&T: divestiture of the Bell operating companies and the globalization of markets. The first, in 1984, decreased the degree of AT&T regulation at both state and federal levels. It also terminated a consent decree that had codified AT&T's 1940s patent licensing practices. The second not only substantially increased the competition in telecommunications products in the United States but, together with the first, helped stimulate the transformation of AT&T once again into an international company.

As the business units of this new AT&T developed competitive strategies consistent with their new missions, IPR strategy was similarly revised. The result was that AT&T practices became more like those of other electronics-based firms.

The competitive characteristics of the various markets in which AT&T now operates differ, as do the specific practices of firms in these markets. Also, since AT&T is no longer subject to the restrictions of the now-terminated consent decree, the terms and conditions of licenses can be tailored to a greater degree to specific competitive strategies.

The new, more competitive, and more international nature of AT&T has also had a marked influence on its trademark strategy. This is reflected not only in the products and services for which marks are adopted, but also in how they are selected and how they are promoted. Trademarks obviously play a more important part in AT&T's competitive strategy than they did just a few decades ago.

Trade secret licensing has seen perhaps the most change, primarily due to the globalization of AT&T. The centerpiece of AT&T's offshore ventures in dozens of countries has often been a trade secret agreement, including a technology package. In fact, a technology package is a requisite part of the entry fee for doing business in many countries. These agreements usually end up being not just trade secret agreements. They are usually omnibus IPR agreements with both patents and trademark licenses involved.

Trade secret licensing has also been an important ingredient in the marketing of computer software. Initially, such licensing was primarily an adjunct to the sale and use of telecommunications network elements. In recent years, as AT&T's computer business expanded, it became important in that market sector as well.

The marketing of computer software has also increased the importance to AT&T of copyright protection. Historically, copyright was not of mainstream competitive importance to AT&T but was utilized to protect the many directories, books, documents, films, and other copyrightable works produced by the company. Now, however, with the increased recognition around the world that computer software is protected by copyright, it is an important element in the competitive strategy of AT&T.

In looking forward, it seems clear that computer software will continue to increase in competitive importance as hardware in both telecommunications and in computers becomes more and more a commodity. Thus, the intellectual property rights in software, whether protected by patents, trade secrets, or copyrights, have the potential of becoming the most important intellectual property elements for AT&T.

Because intellectual property has become such a significant part of the AT&T competitive strategy, negotiations such as patent harmonization and the General Agreement on Tariffs and Trade (GATT) Trade Related Aspects of Intellectual Property (TRIPS) agreements are of great interest to the company. AT&T itself does business in many countries around the world, and securing patent rights is very expensive even if you have a standardized system assumed. With nonstandardized systems, of course, the costs soar. So some level of patent harmonization would be highly desirable. The activities in GATT are also of great interest because a TRIPS agreement would both standardize and raise the level of protection in many countries around the world.

I really think that the Dunkel draft of the TRIPS agreement is much better than we had any reason to expect going in. It is certainly an excellent agreement for the final negotiations, so I hope that it will be given very serious consideration. Consider the alternatives. I, for one, do not consider many more bilaterals a very reliable alternative.

CONCLUSION

Intellectual property rights have been becoming an increasingly important part of AT&T's competitive strategy. With the continuing increase in worldwide competitiveness, this trend seems likely to continue.

I am going to take the liberty here of offering a few comments on some of the matters discussed earlier.

First let me make one other observation. AT&T does take into account the strength of intellectual property protection in making investment decisions in particular countries, and although the level of protection is not always a go-no-go decision maker in and of itself, it is an important element. In some instances it has been the sole determinant of an investment decision. When AT&T was asked to license UNIX in Brazil, we pointed out to the management of the company the level of software protection then available and the request was denied.

Of course, evaluation of the level of protection is heavily dependent on the age of the technology to be put into a particular country. That is alluded to by Mansfield in Chapter 5, who said you cannot simply look at gross investment numbers. You need to get underneath them and see what they represent. For example, in the semiconductor world, we measured the generation of technology by the width in microns of the lines etched on the chip. To go back a few years, if you were licensing 2.5-micron technology, you were going to have one degree of sensitivity to the level of IPR protection, but an entirely different sensitivity if you were licensing 2-micron technology.

I would like to clarify, if I may, several allusions that were made earlier in the conference to the European Community directive on computer software suggesting that it somehow gave recognition to or was a form of sui generis protection. It quite clearly is not. The EC directive, if you read it, says that it supports protection of computer software both by copyright under the Berne Convention and also under the patent laws. So instead of suggesting it provides support for a sui generis procedure, I think it is just the opposite.

Finally, one speaker suggested that since there have been some bad copyright decisions in the computer software area, and I fully agree, the only solution is sui generis protection, because these decisions show the copyright system does not work for software. To me, it only shows that some judges do not understand copyright law and I think we have a very effective appellate process for taking care of that. So, before we look to sui generis, let us first try the appellate process.

Discussion

Audience discussion of the papers and presentations in this part of the conference was divided into two sessions: one session dealt directly with Chapter 8 by John Armstrong and the remarks of the sectoral discussants; the other dealt with the other presentations on intellectual property rights (IPRs) and corporate strategy.

GLOBAL SCIENCE AND TECHNOLOGY TRENDS AND EFFECTIVENESS OF INTELLECTUAL PROPERTY RIGHTS

Much of the discussion focused on the effects of global scientific and technological (S&T) trends on the role of IPRs and, particularly, the effectiveness of current forms of IPRs. John Armstrong's first basic principle in Chapter 8—that the continuing evolution of science and technology will be best served by the flexible application of existing intellectual property concepts—stimulated considerable discussion. Some in the audience expressed the view that existing IPR concepts are incapable of providing the flexibility necessary for protecting new technologies.

It was suggested that the increasing prevalence of important technological breakthroughs that are technically obvious in the patent sense and very easy to reverse engineer is stimulating much of the debate over the effectiveness of current IPRs. This type of innovation, which has been termed *incremental innovation bearing know-how on its face* or *applied scientific know-how*, was said to constitute a gap between copyright and patents that cannot be filled by stretching these paradigms. Solving this problem does not necessarily lead to a plethora of sui generis laws; rather a single law could be developed to deal with this type of innovation.

Members of the audience clearly recognized that the sui generis issue is highly political. They also recognized that the passage of sui generis laws is likely to increase the level of IPR litigation for some time until the new legal issues and questions are settled. Nonetheless there seemed to be considerable opinion that new proposals for IPRs should be examined and that such examination should be conducted with scientific objectivity and impartiality.

A key question in the consideration of alternatives to current IPR forms is whether the current IPR system handles new technology properly. On the one hand, the current system was criticized on the grounds that it creates fuzziness in interpreting the laws. On the other hand, it was suggested that any attempt to define technologies crisply in IPR laws would fail because new technologies typically evolve more rapidly than the intellectual property system is capable of accommodating.

The optical fiber patent dispute between Corning Glass and Sumitomo (described in Chapter 9 by George McKinney) was discussed as an example of the debate over the adequacy of the current IPR system in dealing with new technology. In the opinion of one knowledgeable observer, Corning's optical fiber patent ran into trouble in Japan over a problem of language. This person thought that the invention was not described properly in the patent specification. Another view was also expressed, namely, that the Corning dispute is precisely the type of situation that exemplifies the failure of the system. In this view, if a team of qualified lawyers cannot describe a major invention in a manner that is satisfactory to patent authorities, it is clearly a failing of the patent system.

Another issue raised in the discussion was whether "use-it-or-lose-it" laws should apply more generally than they currently do. Unlike in some countries, U.S. patent law generally does not require the owner of an intellectual property right to exercise that right through manufacturing. However, U.S. law requires universities and not-for-profit research institutions that obtain IPRs stemming from government-funded R&D to exercise those rights within a certain period of time or risk the government taking them away. One discussant noted that the rationale for this provision is to promote the rapid diffusion of the benefits of government-sponsored R&D to the public at large. The point was also made, however, that manufacturing is not the only economically valuable use of patents. A corporation's use of patents for freedom of action and for access to other advances in technology—through cross-licensing, for example—is very important to that corporation. For this reason, it could be inappropriate and perhaps counterproductive to require firms to manufacture using their intellectual property.

The issue of software patents was also raised in discussion. Some aspects of software have been recognized as patentable. This is causing some confusion and difficulty, arising from the complexity of software and the lack of intellectual clarity in the industry as to its nature. One member of the audience commented that the recent increase in software patents has caused problems in managing small software operations because it seems to require a lot of defensive patenting. Another commented that several problems have arisen in the U.S. Patent and Trademark Office's examination of software patents; for example, there is neither standard terminology nor a standard of nonobviousness. It was suggested that this problem will probably be corrected as the software development community matures and as the Patent and Trademark Office gains experience in examining software patent applications.

Finally, a member of the audience made a plea for more focus on accelerating invention and innovation—that is, the discovery of new things and moving them from the mind to the marketplace. How can we reshape the U.S. system to do this? Speakers agreed that this is a major concern. It was pointed out that although improvements to speed up the patent process are necessary, they will not be sufficient to speed the innovation process. Accelerating the innovation process in the United States will also require companies to speed up their internal processes, as well as more collaborative efforts, low-cost capital, a stable monetary system, and an improved climate for investment, in addition to resolving IPR issues.

INTELLECTUAL PROPERTY RIGHTS AND COMPETITIVE STRATEGY

Discussion of the presentation on IPRs and competitive strategy focused on the issue of possible abuse of IPRs, among others. One member of the audience suggested that strong patent protection may be appropriate, as argued by Otto Stamm, but that it may lead to injustices, as pointed out by Michiyuki Uenohara, when, for example, firms misuse patent protection to avert competition. This person offered the view that it is entirely appropriate for firms that receive strong patent protection to be subject to consent decrees, such as that imposed on AT&T by the Department of Justice in 1956, which required AT&T to license anyone who wanted a license at a fair fee.

Others replied that the patent, like many other legal instruments, is subject to abuse and a regime is certainly needed to handle such abuse. There are in fact patent misuse doctrines and other laws to take care of the improper use of patents. Consent decrees may not be the most effective way of protecting against abuse because they are based on a particular set of technological conditions, which necessarily change over time, and it is difficult to get them modified to take into account changed conditions and circumstances.

Another person noted in reply that large companies, particularly in the pharmaceutical industry, are often criticized for using their patents to create monopolies and to harm small companies. This person argued that patents do not, in fact, create monopolies. Moreover, the same critics blame the pharmaceutical industry for not immediately creating drugs against cancer and other important diseases. It should also be pointed out, as George Rathmann does in Chapter 13, that the same patent system allows small biotechnology companies to protect their innovations and to license them.

As pointed out by a member of the audience, the legal framework of IPRs cannot be separated from issues of organizational and market structure when discussing competitive strategy, particularly in an industry such as telecommunications, which was heavily represented on the panel. The question was posed as to what market structure was favored by the panelists? Michiyuki Uenohara's reply indicated support for the current, very competitive market situation, which he noted is quite different from that previously enjoyed by AT&T. William Keefauver commented that in his opinion the U.S. structure for the industry is fine, Japan is moving in the right direction, but the European Community should push the members' Post, Telephone and Telegraph agencies to open up their cartels faster. Michiyuki Uenohara commented further that AT&T's near monopolistic position in the past helped considerably in achieving international communication interconnection standards. In his view, leadership is needed today, otherwise it will be very difficult to establish common standards for computer communication.

A member of the audience asked why software was excluded from patent protection under the recent changes in the Mexico's industrial property law. Antonio Medina Mora Icaza attributed it to the government's insistence that software is not patentable. International discussions on this issue have not yet been resolved. A further comment was made that the exclusion in Mexican law is the same as that in the laws of the European Community and the United States.

Another question from the audience concerned the practical effects on corporate competitive strategies of the differences in patent law and practice between Japan and the United States. Michiyuki Uenohara replied that until recently, Japan permitted only a single claim per patent application, but that has now been changed to multiple claims, as practiced by the United States. Nonetheless, Japanese companies believe it is important to apply for so-called peripheral patents, in addition to basic or central patents, to produce a marketable product. The huge amounts Japanese firms have paid in royalties to U.S. firms for basic patents have created real problems in serving the marketplace. He requested that U.S. firms also apply for peripheral patents, in addition to basic, central patents.

In a final comment, data were cited that appeared to dispute Otto Stamm's suggestion that countries without protection for pharmaceuticals have been unable to innovate in this area. It was also suggested that given the high concentration of pharmaceutical innovation among a small number of developed countries, patent protection for pharmaceuticals should be regarded as a necessary, but not a sufficient, condition for significant investments in R&D. Stamm replied that to evaluate the data it would be necessary to discuss further the nature of innovations made and whether they were of fundamental importance to the industry.

From the chapters and discussion in this section, it is clear that the process of scientific and technological advance is changing in ways that challenge the effectiveness of IPRs in stimulating economically valuable innovations. A number of central issues were raised in this discussion: Is the current IPR system capable of handling new technologies adequately? If not, is it preferable to modify existing IPR forms or to examine alternatives? What might be the nature of those alternatives? These issues were discussed from many perspectives, but not resolved. They are examined again in more depth in Section V.

V

Adapting Intellectual Property Rights to New Technologies

Introduction

FUNDAMENTAL ISSUES

New technologies are posing fundamentally new issues for intellectual property systems. The very nature of some new technologies requires a type and degree of protection to which the intellectual property system must adapt. This, in turn, may involve changes in legal doctrine, in rights granting, and in rights enforcement. In the United States, development of legal doctrine can take place directly, through legislation, or incrementally, through the courts; rights are granted through administrative organizations, such as the Patent and Trademark Office, and they are enforced through licensing agreements and, ultimately, through litigation. Because of the technical and legal subject matter involved and the far-reaching economic consequences, the process of adaptation is complex, involves many parties, and is inherently political. It is often a case of human institutions playing "catch-up" with advancing technology.

The fundamental question addressed in this section is how well the intellectual property system is adapting to the many new technologies that are emerging. Is it developing the appropriate new doctrines and mechanisms fast enough? Are they economically effective and efficient in encouraging innovation? Are some approaches to adaptation more successful than others?

EXISTING INTELLECTUAL PROPERTY PARADIGMS

One of the major issues is whether existing intellectual property right (IPR) paradigms, such as patent law and copyright law, can accommodate the new technologies. There are advantages to using existing legal frameworks to protect new technologies. The copyright and patent laws encompass an expansive and flexible subject matter and a historical body of principles and precedents. In the past, they have enabled the courts to meet the challenges of new technology by applying traditional principles without the need for repeated legislative action. Moreover, there are existing international treaties in both the patent and the copyright areas that, with modification, could be used to secure international protection for technologies.

The new technologies that pose the most troublesome issues for the intellectual property system are those that require protection but that do not fit easily within the existing intellectual property paradigms. For example, developers of computer programs and semiconductor chips, two of the most economically valuable modern technologies, have a great need for intellectual property protection because their efforts require considerable skill and creativity, but the results of their efforts can be appropriated easily once the product is marketed. It has been argued, however, that software and chips are generally too "functional" to fit in the copyright paradigm, as traditionally defined, and not "nonobvious" enough to be patentable. Similarly, incremental innovations in many technologies may be of critical economic importance, but they may not meet the requirements of intellectual property protection within existing paradigms.

Because of historical experience, many observers see it as natural for copyright law and patent law to expand beyond their traditional concepts to provide protection for the new technologies. Other observers, however, are concerned that there are limits beyond which the existing paradigms cannot be stretched without distorting the very purpose of the law. In this view, as pointed out by Pamela Samuelson in Chapter 12, the traditional purpose of copyright law, which is to promote dissemination of knowledge, is distorted when decompilation is an infringement of copyright. Similarly, Samuelson notes the concern that the overlap of patent and copyright protection in computer programs, especially as the scope of copyright is expanded, "would undermine important public policy goals of the patent system, which generally leaves in the public domain those innovations not novel or nonobvious enough to be patented."

In the United States, IPR law presents a problem for technologies such as computer programs and semiconductor chips because it basically assumes that something is either a writing (protectable by copyright) or a machine (protectable by patent), but cannot be both simultaneously. These new technologies challenge this fundamental assumption, because they have aspects of both—many see computer programs as involving both authorship and invention. This not only makes it difficult to draw meaningful boundaries between patents and copyright, it also raises the question of whether alternative intellectual property paradigms should be explored. In other countries, the dichotomy between copyright and patents is not so strong, and other intellectual property paradigms, such as utility models and industrial designs, exist and are widely used.

SUI GENERIS APPROACHES

Sui generis approaches are an alternative to existing intellectual property paradigms. Sui generis forms of IPR are unique property rights designed to protect a specific new technology that does not fit easily within existing IPR categories. They have the advantage of specifically recognizing the special issues posed by new technologies and tailoring protection to the specific problem. Moreover, developing a sui generis law allows the problem to be dealt with as a whole, rather than piecemeal in the courts. It can also avoid potential harm to the technology and to the law itself from applying an inappropriate legal framework.

Sui generis approaches have a number of disadvantages, however. They require legislative action, and, as pointed out by Morton David Goldberg in Chapter 14, it can be difficult to design a law that is flexible enough to maintain the desired balances and degree of protection in the face of continuing change in the technology and industrial structure. Moreover, both John Barton (see Chapter 11) and Goldberg note that it may be harder to undo errors in sui generis laws. Another factor weighing against a sui generis approach is the need to negotiate new treaties on emerging technologies for which there is little consensus as to the appropriate form or degree of protection. There is concern that sui generis laws make IPR law more complex, confusing, and uncertain.

INTERNATIONAL ADAPTATION

The international dimension adds another level of complexity to adapting intellectual property systems to new technology. A key difficulty is trying to achieve international consensus on an IPR approach to new technologies when there is often no consensus within countries on the appropriate approach and when the economic interests of different nations may dictate a different balancing of interests in the IPR law. These seem to be the issues that are holding up agreement on specific IPR rules in the Uruguay Round of the General Agreement on Tariffs and Trade (GATT) negotiations.

Computer programs may provide an example of the difficulties. The

United States has persuaded many countries to follow its lead in using copyright protection for computer programs. Yet according to Pamela Samuelson (see Chapter 12), the details of just what copyright protects beyond exact copying of program code in the various national laws remain unclear. Some countries are also following the United States on the patentability of software, but again standards of patentability are unclear.

Sui generis approaches pose their own issues when they move to the international arena. In Chapter 14, Goldberg notes that the sui generis treatment of semiconductor chips in the United states has been difficult to project into the international arena, and he points out that, if copyright had been used, existing treaties could have been employed to secure international protection. In the absence of a treaty, the United States has sought to internationalize the Semiconductor Chip Protection Act of 1984 through reciprocity with countries that protect U.S. mask works on the same basis. Goldberg indicates that this "arm-twisting" approach has brought some positive results, but it may have engendered resentment among other countries that now hinders cooperation in multilateral efforts to harmonize worldwide protection of chip topography.

SMALL BUSINESS

Small businesses pose special issues in IPR discussions because they are important sources of innovation in many rapidly moving fields and they have special needs with respect to IPRs. In biotechnology, for example, most of the major innovations have been made by small firms. The ability to get intellectual property protection for living organisms in the United States was a key factor in enabling new biotechnology firms to attract the investment necessary to grow and survive.

Some argue that the cost of IPR litigation often bars the use of the system by small companies; if they are able to—or must—use it, litigation diverts funds from research and innovation. Thus, it is argued that high litigation costs work against small businesses and hence new technologies. Although this may be more a result of the general character of U.S. litigation and not of IPRs in particular, the conditions surrounding the emergence of new technologies are probably conducive to high levels of IPR litigation. As Eugene Gordon shows in Chapter 15, when the economic stakes are high, there is more incentive for large firms to use IPRs to stifle small firms. Moreover, when a technology is new, the validity of IPRs may be uncertain and hence more likely to be litigated. Thus, it is important in designing IPRs for new technologies to consider their possible impact on small, innovative firms. It is also important to remember that solutions to IPR problems that require a long time may leave many failed start-up companies in their wake.

NEED FOR POLICY ANALYSIS

Thoughtful study and analysis can help in the process of adapting IPR systems to new technologies. Barton, in Chapter 11, suggests several instances in which analysis of the economic and innovation incentives of IPR laws and alternatives is needed—for example, when opposing doctrines are at a standoff, as in the controversy over whether a patent should extend to the progeny of patented life form. He also argues that identification of the point at which a gene becomes patentable should be subject to thoughtful analysis before decisions are made on the patenting of genome sequences. Similarly, in Chapter 12, Samuelson suggests that a better understanding of the nature of the innovation process in the software industry and the conditions that have promoted progress in that industry could provide a basis for policy decisions on software patenting.

Barton, in his examination in Chapter 11 of different approaches to adapting IPRs to new technologies, concludes that the European process for adapting IPRs to biotechnology, which involved standing study groups, was more effective at defining critical issues and subjecting them to debate than was the U.S. process, which involved primarily the courts and the Patent and Trademark Office. The keys to the European success were that repeated expert meetings produced substantive proposals that were then subjected to external criticism by all affected parties and the study groups were given enough time for staff work to help identify the new, "hard" issues.

ORGANIZATION OF THE SECTION

The chapters in this section discuss the special challenges to IPRs raised by new technologies and how the intellectual property system has adapted to those challenges. In Chapter 11, Barton evaluates the effectiveness of the IPR adaptation process in the case of three new technologies: biotechnology, computer programs, and integrated information networks. Barton suggests that routinized processes for the reform of IPR law are needed to deal with new technologies and that those processes should be international. He calls specifically for standing study and reform groups with an international composition to consider the issues and suggest appropriate responses on an ongoing basis.

In Chapter 12, Pamela Samuelson provides a detailed case study of the adaptation of intellectual property law to computer programs. After a historical review, she discusses the current legal approaches to protecting computer programs and the controversies surrounding the appropriate protection for the functional, as opposed to the written, aspects of computer programs. Samuelson argues that computer programs directly challenge existing IPR paradigms, and she expresses concern that attempts to protect computer programs under copyright and patent law may subvert the intent of the laws and harm the development of the technology.

In Chapter 13, George B. Rathmann presents a case study of the role of IPRs in the commercialization of biotechnology, which was accomplished in the United States through the formation and growth of many small, start-up companies. In this context of high technical risk and financial uncertainty, patents played an important role in attracting the large amounts of investment that were needed over many years. The *Chakrabarty* decision in the United States, which made living organisms patentable, was particularly important to launching the industry. Other patent issues that were more troublesome included the need for process protection and the slowness with which patents were issued. Rathmann warns that as rocky as the road was for biotechnology firms in the United States, what is coming up on the world scene may be much more difficult. He refers particularly to problems faced by U.S. biotechnology firms in protecting their technology in Japan.

In Chapter 14, Morton David Goldberg discusses the history of semiconductor chip technology and the sui generis intellectual property law Congress passed to protect the mask works used in making chips—the Semiconductor Chip Protection Act of 1984. He focuses on three serious shortcomings in the law: its narrow definition of the technology, its broad exception to proprietor's rights in the area of reverse engineering, and the difficulty of internationalizing protection. According to Goldberg, these shortcomings, together with the substantial change that has occurred in the technology and in the industry's structure, seriously weaken the protection given to innovators and potentially weakens the incentive for investment and innovation in the technology. Chapter 14 illustrates the concerns that many intellectual property practitioners have with sui generis approaches.

In Chapter 15, Eugene I. Gordon discusses IPR issues as they affect the field of optoelectronics. This field does not appear to pose fundamentally new issues for the intellectual property system in the way that biotechnology and computer programs do. Rather, it highlights IPR issues associated with fields that are rapidly advancing and exhibit high levels of entrepreneurship and new start-up companies. Gordon focuses on patents and proprietary information, an aspect of IPRs not covered elsewhere in this volume. By examining two case studies, Gordon illuminates the following problems: (1) the ability of large companies to use the threat of IPR litigation to stifle small companies and (2) the inefficiencies that result when the validity of patents is uncertain.

The chapters in this section describe the adaptation of IPRs to new technologies as a process fraught with complexities and challenges. The challenges posed by each new technology are different. Some technologies, such as optoelectronics, pose issues that apparently can be addressed and resolved within existing IPR regimes. Other technologies, including "hybrids" of writing and machines (such as computer programs and semiconductor chips), as well as economically valuable incremental innovations, pose a more fundamental challenge to the adequacy of the existing IPR paradigms. There are sharp differences on the advisability of modifying existing IPR laws versus creating sui generis IPR laws to accommodate new technologies.

As shown in this section, both approaches have their advantages and disadvantages. Unfortunately, there is a lack of policy analysis that would allow the effectiveness of the approaches developed so far to be evaluated more systematically. Will the intellectual property system adapt to new technology in the future? Can IPR reform processes be designed to achieve international consensus on IPRs and encourage the development and commercialization of new technologies? These and other fundamental questions remain in need of further thoughtful study.

11

Adapting the Intellectual Property System to New Technologies

JOHN H. BARTON

This chapter examines whether the intellectual property system is able to adapt to the current rate of change of technology.[1] As the intellectual property system attempts to adapt, it must respond to increased system load as, for example, the number of patents increases. Yet, far more important, it must deal with changes in the character of technology itself. New technologies may require fundamentally new encouragement mechanisms and pose fundamentally new issues for the intellectual property system. Thus the question here is whether the system develops the appropriate new doctrines and mechanisms at a rate adequate to maintain incentives to innovation.

This chapter uses as examples three of the fundamentally new technologies of our time: biotechnology, computer software, and computer-maintained and searched data bases. It describes the new issues posed by these technologies and reviews the approaches taken to adapt the intellectual property system in each case. It then evaluates the performance of this adaptation process, by looking at three levels: (1) the mechanisms for developing doctrine, (2) the systems (e.g., patent offices) that grant intellectual property rights, and (3) the formal systems (primarily courts) and informal systems (e.g., cross-licenses) that enforce intellectual property rights and shape

[1]This is, in a sense, the inverse of the usual question—whether the intellectual property system encourages technological innovation.

their practical economic implementation. In a number of the areas, the analysis is international and comparative; the conclusions are intended to be international as well.

THREE PARADIGM TECHNOLOGIES

The three paradigm technologies were chosen as those recent technologies that probably represent the most severe tests for the intellectual property system. Each requires new forms of expertise; for none can traditional legal principles be applied without substantial modification. Also, intellectual property protection is crucial for each of the three; for all, front-end costs are very large, copying of a marketed product is much easier than initial development, and product cycles are long enough that copying is a serious consideration.[2]

Biotechnology

Biotechnology is defined here as genetic engineering and particularly recombinant DNA manipulation. Although it is different from more traditional areas such as pharmaceutical technology in that living organisms are involved and can often reproduce themselves, what has most troubled the early evolution of the law is the fact that many of the products and processes being patented derive directly from natural products and processes.

Special Issues

The early generation of biotechnology products consists of proteins found in nature in very small quantities, but available in volume through cloning. Typically, the researcher started with the protein and then identified the sequences of its amino acids and of the corresponding nucleic acids in the gene that codes for the protein's production. The question, then, is whether such a protein or sequence can be patented; in common language, it has been discovered rather than invented. This question is usually answered in the United States on the basis that a purified natural product can be patented (because it is not found in purified form in nature),[3] but the

---

[2]Thus, this chapter does not face the question of the irrelevance of intellectual property to many industries, such as those in which the product cycle is much shorter than the time required to obtain a patent.

[3]In re *Bergy*, 563 F.2d 1031 (C.C.P.A. 1977), vacated sub. nom. *Parker* v. *Bergy*, 438 U.S. 902 (1978), on remand, In re *Bergy*, 596 F.2d 952 (C.C.P.A.), cert. granted sub nom. *Parker* v. *Bergy*, 444 U.S. 924 (1979), vacated and remanded with instructions to dismiss as moot sub nom. *Diamond* v. *Chakrabarty*, 444 U.S. 1028 (1980).

question is answered quite differently in the United Kingdom.[4] Moreover, the question raises important international policy issues—if natural genes are patentable, it is very hard to maintain that countries (or individuals) hosting organisms containing potentially important genes should make them freely available for scientific purposes.[5]

The issue is not just theoretical or political. A number of firms seeking therapeutic products have often pursued exactly the same proteins. In a context in which there is no single "Aha" of invention, but rather a process of identification and sequencing, the priority choice is extremely difficult to resolve. A court ends up having to decide, for example, between two different proposed inventors, one of which may have isolated the protein first, while another has sequenced it first.[6] The substantial delays that have marked the biotechnology patent area have intensified this problem, as competing firms have invested in research for a number of years without knowing which one will obtain the ultimate patent rights.[7]

This difficulty in identifying a specific point of invention is the fundamental problem underlying the current dispute over the National Institutes of Health patent application covering some 337 gene fragments sequenced as part of Craig Venter's cDNA approach to the human genome project.[8] Traditionally (i.e., for the last decade or so), identification of the therapeutic value of a protein arose before its sequence was known; identification of the sequence became a particularly important step in defining priority for gaining patent rights. Now with the human genome project, the process is reversed. This project is developing and publishing the nucleic acid se-

---

[4]Genentech Inc.'s Patent, [1989] R.P.C. 147 (Ct. of App. 1988). Note, however, that law in the United Kingdom will permit patenting of an organism containing an inserted gene (e.g., a yeast containing the sequence coding for a protein to be produced) on the grounds that the product is novel. Thus, there is still a form of protection available for the genetic engineer—genes in the abstract have little economic value.

[5]W. Tallent, Specific issues on proprietary rights, in Intellectual Property Rights Associated with Plants, Stuber et al., eds., Madison: Crop Science Society of America (1989), pp. 47-50 (discussion by U.S. Department of Agriculture official of impact of patenting on access to plant genetic resources); *Moore* v. *Regents of the University of California*, 51 Cal.3d 120 (1990).

[6]The example cited in the text is *Amgen, Inc.* v. *Chugai Pharmaceutical Co.*, 927 F.2d 1200 (CAFC 1991) (erythropoietin). For another example, involving a different factual situation, see Genentech Inc.'s Patent, supra (conflict with Wellcome), and *Scripps Clinic and Research Foundation* v. *Genentech*, 927 F.2d 1565 (CAFC 1991) (tissue plasminogen activase).

[7]For genetic engineering, these delays were as long as 47.4 months in 1989. See General Accounting Office, Processing Delays Continue for Growing Backlog of Patent Applications (1990), cited in Office of Technology Assessment, Biotechnology in a Global Economy 212 (October 1991).

[8]C. Anderson, U.S. patent application stirs up gene hunters, Nature 353:485-486 (October 10, 1991); L. Roberts, Genome patent fight erupts, Science 254:184-186 (October 11, 1991).

quences of a variety of genes; the proteins coded for and produced by these genes, however, have not yet been isolated and their function is unknown without further research. Admittedly, the utility requirement for granting a patent is not satisfied. Nevertheless, a cautious lawyer would wonder exactly what additional steps would be necessary (by the sequencer or by someone else) before patentability becomes possible and, in the face of this legal uncertainty, might wonder whether it would be malpractice not to file an immediate application.

The problems just identified are likely to change over the coming years. At some point, cloning and sequencing procedures will be regarded as obvious rather than novel, so that rights, if any, will be more likely to go to the discoverer of a pharmacologically active substance than to its cloner. The human genome project will move most human and some nonhuman research toward a pattern in which the researcher begins with a sequence and then looks for therapeutic activity (which may, in some cases, be suggested by the sequence's location on the gene). Because the sequence is known, it is obvious, and because the protein is directly coded, it may be obvious as well. Hence, the protein may be unpatentable (although its three-dimensional folding pattern may be difficult enough to reconstruct to provide a basis for nonobviousness). Sooner or later, the focus of biotechnology will move past natural proteins. Agricultural biotechnology has long been concentrating on transgenic plants and animals rather than therapeutic products; human biotechnology will probably look for new products not found in nature and for new ways to use them.

The fact that living organisms are involved in biotechnology (and especially in agricultural biotechnology) poses a different group of questions. Some are ethical. Some are technical—for example, defining the appropriate scope of intellectual property protection. The obvious example of such a technical question is whether a patent should be regarded as reaching the progeny of a patented life form. By traditional law, the seller of a patented item exhausts his or her rights in the item, so that the buyer is entitled to use it as he or she sees fit. By definition, however, the reproduction of a patented article is an infringement of the patent. Clearly, this is a logical standoff between two traditional doctrines that can only be resolved by a policy analysis exploring the economic and innovation incentives of the two alternatives. That policy analysis might reach one result for a yeast, which has to be multiplied for many generations as part of a single fermentation application, and a different result for a chick bred for sale for meat production.

Outline of Solutions, Thus Far

The intellectual property community has worked extensively—but almost entirely autonomously—in an effort to face these issues. It was a U.S.

Supreme Court decision, *Diamond* v. *Chakrabarty*, 477 U.S. 303 (1980), that launched the world's intellectual property system into the patenting of life forms, by authorizing the patenting of microorganisms. The U.S. Patent and Trademark Office (PTO) then, acting by internal decision, extended the Supreme Court's decision to plants and nonhuman animals.[9] With the assistance of the Industrial Biotechnology Association, the PTO worked hard to build up its capabilities to deal with the backlog that emerged in the area.[10]

Congress has played only a passive and responsive role in this sequence. The PTO's extension of patent rights to plants has raised some concern in the agricultural community; its extension to animals raised concerns in both the agricultural and the animal rights communities. The result has been hearings, proposals for moratoriums, and suggested legislation to restrict these patents—but no actual legislation.[11] Indeed the most relevant congressional action has been the passage of the 1983 Orphan Drug Act, which provides seven years of exclusivity, an equivalent of intellectual property protection, but enforced through the Food and Drug Administration's regulatory approval process.[12] As of 1991, 9 of the 15 biotechnology-derived drugs on the market had such status, as did some 19 under development.[13]

At the international level, the process of adapting intellectual property law has been somewhat more thought out. During a large portion of the decade of the 1980s, there was an expert study organized by the World Intellectual Property Organization (WIPO)[14] and, later on, a parallel study

---

[9]Ex parte *Hibberd*, 227 USPQ 443 (PTO Bd. App & Int. 1985) (plants); Ex parte *Allen*, 2 USPQ 2d 1425 (PTO Bd. App & Int. 1987) (animals). Note that there had also been special coverage procedures for many agricultural plants under the Plant Patent Act of 1930 (35 U.S.C. sec. 161-164) for most species of asexual plants and the Plant Variety Protection Act of 1970 (7 U.S.C. sec. 3231 et seq.) for most species of sexually propagated plants.

[10]Biotechnology in a Global Economy, supra at 210-214.

[11]Id. at 216.

[12]Pub. L. No. 97-414, 96 Stat. 2049, Jan. 4, 1983. There is also a Drug Price Competition and Patent Term Restoration Act of 1984, which authorized the extension of patent terms for pharmaceuticals to compensate partially for time lost during the regulatory process, as part of an elaborate rebalancing of the relationships of generic and pioneer pharmaceutical firms, Pub. L. No. 98-417, 98 Stat. 1585, Sept 24, 1984.

[13]Biotechnology in a Global Economy, supra at 92.

[14]This was the Committee of Experts on Biotechnological Inventions and Industrial Property, whose First Session was November 5-9, 1984; Second Session, February 3-7, 1986; Third Session, June 29-July 3, 1987; and Fourth Session, October 24-28, 1988; Committee of Experts on Biotechnological Inventions and Industrial Property, various documents in series BioT/CE. It is possible that there will be further meetings.

organized by the Union for the Protection of New Varieties (UPOV).[15] These are two international organizations that work with intellectual property; the former covers patents and a number of other forms of intellectual property, whereas the latter covers plant variety protection, a somewhat weakened form of intellectual property oriented toward the needs of the plant breeder.[16] These working groups were very successful in defining the various hard issues that needed to be faced in applying intellectual property to biotechnology (e.g., questions of the application of a patent to progeny, questions of the interpretation of the experimental use exemption in the biotechnological context, and questions of the relative rights of two parties who make different kinds of improvements on the same organism). By almost any measure, although one might disagree with the conclusions they reached, their analysis identified most of the difficult problems earlier than anything in the domestic process in the United States. Probably the only hard issue that they missed (at least of the issues that have already appeared) is that posed by the current proposal to patent raw sequences arising from the genome project.

The UPOV/WIPO work served as an intellectual (but not formal) basis for a European Community (EC) proposal for what amounts to sui generis protection for biotechnology.[17] (The proposal would deal formally with patent law, but would establish a number of very specific rules.) This proposal, issued in 1988, responds in large part to the fact that the European Patent Convention prohibits the patenting of "plant or animal varieties or essentially biological processes for the production of plants and animals"[18] and that there has been little uniformity in Europe in the national patent response to biotechnology.

The EC proposal has met severe resistance from those who oppose extension of the patent system to living organisms on ethical or environmental grounds; there is not yet a fixed date for it to enter into force. Some of the critics argue that life should not be made subject to a property system. Others fear that the extension of intellectual property to living organisms will harm the economic position of small farmers or encourage the development of environmentally dangerous products.

The European structure is not yet complete, and some of the technical issues that it faces will probably be solved in the United States through case

---

[15]Committee of Experts on the Interface Between Patent Protection and Plant Breeders' Rights, Geneva, January 29–February 2, 1990, Report, WIPO/UPOV/CE/I/4, February 2, 1990.

[16]This is the Plant Variety Protection Act of 1970, supra.

[17]Proposal for a Council Directive on the Legal Protection of Biotechnological Inventions, COM(88) 496 final – SYN 159, Brussels, October 17, 1988.

[18]Convention on the Grant of European Patents, Art. 53, signed October 5, 1973.

law. Nevertheless, it seems clear that the European process has moved more effectively than the U.S. process to define the critical issues and subject them to debate. The key is probably the WIPO/UPOV pattern of repeated expert meetings producing substantive proposals that are subjected to external criticism.

Computer Programs

Computer programs (software) pose an extremely different set of issues. In the first instance, they appear to be (and are) text, and one is concerned about direct copy of this text. Yet, as Randall Davis of the MIT Artificial Intelligence Laboratory stated, "Programs are not only text; . . . they also behave."[19] The type of protection to be provided that behavior is strongly debated, and given the extent to which computer and communications software (broadly conceived) is growing in market size and economic value compared to the corresponding hardware, the character of protection to be provided is extremely important economically.

Special Issues

The first special issue posed by software is its easy reproducibility. Discs can be copied cheaply and converted relatively easily from one computer language to another. This, of course, lends appeal to the copyright approach to software protection—and few would deny that copying or direct translation should in general be prohibited, because it fundamentally affects the incentives needed to develop software in the first place.

However, after this point, the rights that should be given to a software producer become unclear. Should a user be entitled to decompile the program as part of a reverse engineering process? Should software be given protection against other software that uses the same code (in places) or program outline, when one considers that the outline and parts of the code may be defined by the problem and that independent "clean-room" development may lead to the same outline and in some cases the same code? Should there be protection for the program's appearance to the user (the "look and feel")? What about protection for novel algorithms? Also, should standards and interfaces associated with programs be protectable?

These issues have arisen in an industrial structure that is, in general, much closer to that of engineering than to that of literature. Programs are

[19] R. Davis, Intellectual property and software: The assumptions are broken, in World Intellectual Property Organization, WIPO Worldwide Symposium on the Intellectual Property Aspects of Artificial Intelligence, Stanford University (March 25-27, 1991).

written by teams; they are constantly updated; consultants are used heavily; new programs build on previous programs. In a bizarre and wasteful response to the legal evolutions in the area, some are even written in clean rooms (i.e., written to specifications derived from an existing program but in a way designed to document that the writers of the new program had no knowledge of the detailed content of the first program). In an interesting new economic trend, programs are being written to common and evolving standards as part of "open systems."

The complications of today's industry are probably simple compared to those of tomorrow's. Programs for parallel processing and for artificial intelligence may prove to be logically reducible to the kinds of programs being published today, but there may be practical differences that have intellectual property implications. For example, what property rights should be assigned to the expert information that is inserted into an artificial intelligence shell through a series of interviews? Then too, the embedding of software in products may create even more difficulty—consider, for example, the proposals to couple computer chips and biological sensors in ways that might allow "intelligent" management of the construction of individual biological polymers.

Solutions, Thus Far

In contrast to biotechnology, this area started out with an early expert advisory study by the National Commission on New Technological Uses of Copyrighted Works (CONTU), which issued its final report in 1978.[20] This study explored the needs in the computer programming area and, without presenting any solid consideration of the implications, recommended use of the copyright pattern. Commissioner John Hersey's dissent warned of the point that programs act quite unlike traditional literary works and also foresaw the derivative works problem reflected in current look-and-feel debates.[21] Several factors, however, appeared particularly to move CONTU. The copyright pattern obviously fit the need to avoid direct copying. It provided coverage that is relatively inexpensive for the right holder (in contrast to patent coverage), and it appeared much more appropriate than the patent and trade secret alternatives that the study considered.[22] Although not explicit in the CONTU discussion, another important factor must

---

[20]National Commission on New Technological Uses of Copyrighted Works, Final Report, Library of Congress (July 31, 1978).

[21]Id. at 27-37.

[22]Id. at 14-38. The relation between copyright protection and trade secret protection for software is not yet resolved; it is hard to understand how one can combine the two protections in light of the tradition that copyright applies to material that is published.

have been that copyright permits use of the existing international copyright treaty network so that the right holder can obtain global coverage easily.[23]

CONTU went on to define the technical adaptations to copyright law needed to encompass computer programs, for example, to allow the "copying" of a program into computer memory as an essential part of the use of the program. Congress followed the CONTU recommendations in amending the Copyright Act in 1980 to define "computer program" and to authorize such copies.[24] Courts straightforwardly resolved a variety of similar, relatively technical issues (i.e., whether different principles should apply to software embedded in a microprocessor chip, on the grounds that such software is not published in the same sense as software sold on a disc.)[25]

However, the courts found it much more difficult to resolve the more fundamental questions noted above. In rather controversial decisions, the U.S. courts provided protection to structural features of a program[26] and to its look and feel,[27] but with at least one court taking a contrary opinion.[28] From the viewpoint of many critics, who include a large portion of the academic specialists in the area,[29] this extension of coverage provides the equivalent of patent protection (and for an irrelevantly long 75 years),[30] without requiring that a patent-quality innovation be achieved and disclosed. Moreover, the logic of the cases involves significant stretching of the distinction between idea and expression. It is understandable to seek to interpret the relevant law to provide some form of protection to the intellectual logic and structure of the program—a creation that may involve significant expense and creativity. Yet, as long as one is within the copyright tradition, it is hard to say that this logic and structure are not in fact ideas and therefore unprotectable. The courts' efforts to describe these features as

[23]Thus, the Berne Convention provides for automatic protection in all member countries, The Berne Convention for the Protection of Literary and Artistic Works, Art. 5 [Paris Act of July 24, 1971]. The Universal Copyright Convention, as revised at Paris, 1971, does have some formalities, but none requiring the types of filings that are typical of patent coverage.

[24]Pub. L. 96-517, 94 Stat. 3015.

[25]E.g., *Apple Computer, Inc.* v. *Franklin Computer Corp.*, 714 F.2d 1240 (3d Cir. 1983), cert. dismissed, 464 U.S. 1033 (1984) (operating system); *NEC Corp.* v. *Intel Corp.*, 10 USPQ 2d 1177 (N.D. Cal. 1989).

[26]*Whelan Associates, Inc.* v. *Jaslow Dental Laboratory, Inc.*, 797 F.2d 1222 (3d Cir. 1986), cert. denied, 479 U.S. 1031 (1987).

[27]*Lotus Development Corp.* v. *Paperback Software Int'l*, 740 F. Supp. 37 (D. Mass. 1990); *Broderbund Software, Inc.* v. *Unison World, Inc.*, 648 F. Supp. 1127 (N.D. Cal 1986).

[28]*Plains Cotton Cooperative Assn.* v. *Goodpasture Computer Service, Inc.*, 807 F.2d 1256 (5th Cir.), cert. denied, 484 U.S. 821 (1987).

[29]See, e.g., P. Menell, An analysis of the scope of copyright protection for application programs, 41 Stan. L. Rev. 1045 (1989); P. Samuelson, CONTU revisited: The case against copyright protection for computer programs in machine-readable form, 1984 Duke L.J. 663.

[30]17 U.S.C. sec. 302.

expression have significantly distorted copyright's distinction between idea and expression.

This difficulty is also shown by the question of decompilation of a program to understand its working. Were the program genuinely a literary work, such decompilation would be a normal form of study of the work. Were it genuinely a patented invention, such decompilation would be a normal step in reverse engineering or design improvement. Decompilation as a step toward design improvement is certainly a socially desirable activity (although there should be a reasonable way to sort out rights in the improved program). Yet the argument has been made (and accepted in Europe) that the reproduction of a program to study it is an infringement. Again, the problem is that the rights reasonably assigned to a software package do not match well with those defined by copyright law.

Perhaps in response to the difficulties posed by copyright law in the software area, a number of firms have been seeking software patents. The area is governed by one of the most opaque series of Supreme Court cases that can be found in any body of law: *Gottschalk* v. *Benson*, 409 U.S. 63 (1972); *Parker* v. *Flook*, 437 U.S. 584 (1978); and *Diamond* v. *Diehr*, 450 U.S. 175 (1981). As a measure of the confusion, in November 1989, two different panels of the Court of Appeals for the Federal Circuit (CAFC) issued diametrically opposed opinions on the issue.[31] Nevertheless, presumably in reflection of the fact that *Diehr* is relatively positive to patenting, the PTO has been granting a substantial number of software patents.

The problem lies in whether software amounts to patentable subject matter under the current statute, which provides for coverage of a process, a machine, a manufacture, a composition of matter, or an improvement of one of these.[32] The Court has long rejected the patenting of a "scientific truth or the mathematical expression of it."[33] Not only do such discoveries or inventions not fit the statute; they also pose questions as to the potential overbreadth of a monopoly right. Although the explicit tests are much more complex, the basic outcome of the case law is that a program or algorithm is patentable provided it is adequately embodied in a machine or adequately restricted to a particular range of applications. The difficulty, of course, is that the relevant innovation resides precisely in the program or algorithm, not the way it is embodied or restricted to a specified range of applications.[34]

---

[31] In re *Grams*, 12 USPQ 2d 1824 (CAFC 1989); In re *Iwahashi*, 12 USPQ 2d 1908 (CAFC 1989). For an effort to reconcile these opinions, see R. Laurie, The patentability of artificial intelligence under U.S. law, in WIPO Worldwide Symposium, supra at 121-150.

[32] 35 U.S.C. sec. 101.

[33] *Mackay Radio & Telegraph Co.* v. *Radio Corp.*, 306 U.S. 86 (1939).

[34] See P. Samuelson, *Benson* revisited: The case against patent protection for algorithms and other computer program-related inventions, 39 Emory L. J. 1025 (1990).

Other nations have gone through their own patterns, a number providing a U.S.-style mixture of patent and copyright law. A few have attempted sui generis proposals. Japan considered such an approach in the early 1980s and gave it up in the face of strong U.S. opposition.[35] The EC, after much U.S. lobbying and international discussion of the details of the proposal, adopted its own sui generis approach in a 1991 directive.[36] (As with the proposed EC biotechnology directive, this is technically an adaptation of existing law, copyright law in this case. The rules are so specific, however, that the directive can reasonably be called sui generis.) It is particularly interesting that each of these nation's laws have raised special issues that are both significant and different from those at issue in the United States. Thus, the Japanese law finally adopted as an amendment to the copyright law excludes "program languages, rules, or algorithms."[37] The EC directive debate extensively considered whether interfaces should be protectable. Protection of interfaces would economically strengthen those with a large installed software network by enabling them to restrict the ability of others to market interoperating equipment. The directive basically declined to create such protection, but it was difficult to devise an appropriate exception to the directive's general prohibition on decompilation of a program. The result is that some fear that they cannot understand a program's interface without inadvertently violating the directive's general prohibition against decompilation.

The legal difficulties in this area are an order of magnitude more severe than those in biotechnology. In biotechnology, one has a sense of new and difficult questions. In software, one has not only that sense, but also the more troublesome sense that the statutory models being used are fundamentally ill-adapted to the task and push the courts either to ignore important economic incentives or to twist the statutory language. The combination of troublesome questions and an ill-adapted statute suggests that CONTU was almost certainly wrong in its judgment that the copyright system should be used instead of a sui generis approach.[38]

---

[35]D. Karjala, Protection of computer programs under Japanese copyright law, [1986] 4 E.I.P.R. 105.

[36]Directive 91/250 on the legal protection of computer programs, May 14, 1991, OJ 1991 L122/42.

[37]Article 10(3) of Copyright Act, Law No. 48 (1970), as amended by Law No. 62, adopted June 7, 1985.

[38]For similar judgments, see Office of Technology Assessment, Intellectual Property Rights in an Age of Electronics and Information 88-94 (April 1986); and Prepared Statement of P. Samuelson before the House Subcommittee on Courts, Intellectual Property, and the Administration of Justice, November 8, 1989.

Note also that for each of the technical legal issues, there are relatively clear business winners and losers who have played an enormous role in the political debate. In general, up

Integrated Information Networks

The final working example is the integration of data networks. These networks, exemplified today by Prodigy, Compuserve, or Lexis, provide computer access to information from a variety of sources, ranging from published data to gene sequences and up-to-the-second financial information. They are only beginning to evolve—one can anticipate enormous expansion. The costs of reproducing, distributing, and searching material over a network are tending to zero, while the sophistication of the computer search systems grows rapidly. Precision in the ability to charge the information consumer for what is used is increasing rapidly. At the same time, the costs of producing information will remain as high as ever (and even increase with per capita income). Although it is not an intellectual property issue, the growth of these systems, and the likelihood that they will replace many current information distribution systems, may imply that important information will no longer be readily available to those who need it to be effective citizens, but who lack the ability to pay for it.[39]

Special Issues

These systems will pose at least two types of special intellectual property issues. One involves defining the rights to be held by the network; these rights will encourage development of the complex and sophisticated programs needed to assist in searching, linking, and translating individual data bases. Are there reasons—such as the possibility of monopoly—why such software should be treated differently from other software? How far should intellectual property law go to protect a network against use in ways not desired by its proprietor?

The other type of issue is that of protecting the information in data bases. It is the generation of this information that will be most expensive, and the network's computer capabilities will increasingly make these data bases more a source of information than a form of expression. Presumably, for example, any human language material in the data base will soon be automatically translated as necessary by the network. Likewise, statistics

until this point, these winners and losers have succeeded in maintaining the copyright approach with rather strong rights for the copyright holder; the difficulties of the approach may, however, be beginning to give rise to dissatisfaction in the business community as well as in the academic community.

[39]I have decided not to consider this type of problem in this chapter; nor do I consider the converse issue based on the fact that current technologies permit easy reproduction of printed or digital copyrighted materials. Nor are issues of collective authorship, which are posed by a different kind of computer network, considered here.

will probably be extractable from human language text for automatic assembly into a computer-generated total or time series.

Adaptations, Thus Far

These issues are just beginning to be faced, almost entirely by the courts alone, and it is impossible even to be confident that the hard problems will be those just enumerated. The first problem—that of adapting intellectual property protection, presumably copyright or a future sui generis protection, to the network software—may be only a special example of the admittedly difficult issue of software protection generally. Probably the critical issue here lies in defining the scope of the network's proprietary control over its materials. A leading case favoring the network is *Cable/Home Comm. Co.* v. *Network Prod. Inc.*, 15 USPQ 2d 1001 (11th Cir. 1990), which held that a computer chip designed to unscramble pay television programs transmitted via satellite was an infringement of the television system's copyrighted scrambling program. Although such unscrambling was also held to violate the communications laws,[40] this case rested in part on a judgment that the "pirate" chip infringed the copyright in the original scrambling chip. Sooner or later, someone will reverse engineer such a chip in a way that is legitimate under the copyright laws and perhaps in a circumstance in which the communications laws do not apply, and in at least one current reverse engineering case, infringement arguments have been rejected, *Vault Corp.* v. *Quaid Software Ltd.*, 847 F.2d 255 (5th Cir. 1988) (software designed to evade a software anticopying system).

However, the issue goes beyond "unauthorized access" type situations: Should a network, for example, have control over the ways in which its contents are further manipulated or combined? Suppose, for example, that the holder of a subscription to a data base wishes to resell access in a logically restructured form; for example, a subscriber to a national address and telephone number data base might combine that information with information available from census tapes and sell highly focused marketing information. If the copyright laws apply, should this be fair use? If some other form of law applies, should it be encouraged? The leading recent copyright precedent favors the user in the context of software designed to modify the characteristics of a Nintendo game, *Lewis Galoob Toys, Inc.* v. *Nintendo of America, Inc.*, 20 U.S.P.Q. 2d 1662 (N.D. Cal. 1991); the basic direction of European software protection law is similar. Nevertheless, there are strong pressures toward the opposite position on this issue of open versus closed

[40] 47 U.S.C. sec. 605.

systems, and the area is being shaped by patent and antitrust doctrines, as well as copyright doctrines.[41]

The question about protection of information in the data base may prove much more difficult, because the working line between information and expression is changing as new technologies permit more "intelligent" computer-based analysis of text. The most directly relevant case is *West Publishing Co.* v. *Mead Data Central, Inc.*, 799 F.2d 1219 (8th Cir. 1986), which found copyrightable expression in the page numbers and page breaks of West's reports. Presumably, then, one can use Lexis to find the citation but must still look at West's publication to obtain the citable page number for a quotation. This position appears to survive *Feist Publications, Inc.* v. *Rural Telephone Service Co., Inc.*, 111 S. Ct. 1282 (1991), in which the Supreme Court sharply upheld the proposition that the information in a data base (in this case a phone directory) was, because it was information, uncopyrightable.[42] The result, certainly correct from a copyright law and freedom of information perspective, but paradoxical for the future of the industry, is that the more expensive information input to integrated data bases is unprotectable, while the cheaper expression input is protectable.

Thus, although it may be a little early for making a judgment, the existing copyright protection system is likely to pose fundamental difficulties for data bases, just as it does for software. In this situation, however, there is no CONTU; U.S. courts are left on their own to struggle with the situation. The closest approach to official extrajudicial analysis has been the European discussions of the copyrightability of interfaces, an issue that is likely to have a substantial impact on the structure of the international information industry.

---

[41]For example, this line between open and closed systems was the underlying issue in *Digidyne* v. *Data General*, 734 F.2d 1336 (9th Cir. 1984), cert. denied, 473 U.S. 908 (1985), which dealt explicitly with whether antitrust principles prohibited sale of an operating system only with the basic computer. Congress has proposed greater freedom for such tying in the proposed Intellectual Property Antitrust Protection Act of 1989, S. 270, 101st Congress, 1st Sess. (1989); H.R. 469, 101st Congress, 1st Sess. See generally A. Silverman, Myth, empiricism, and America's competitive edge: The Intellectual Property Antitrust Protection Act, forthcoming Stan L. Rev. In *Allen-Myland Inc.* v. *IBM Corp.*, 746 F. Supp 520 (E.D. Pa. 1990), certain adaptation of copyrighted microcode was permitted on the basis of an interpretation of the Consent Decree in *United States* v. *IBM*, 1956 Trade Cases par. 68,245 (S.D.N.Y. 1956).

[42]Consider, to explore the scope of the issue, the Court's description of the source of originality in a data base: "The compilation author typically chooses which facts to include, in what order to place them, and how to arrange the collected data so that they may be used effectively by readers," 111 S. Ct. at 1289. All three of these factors change radically when access to the compilation is through a computer rather than a printed page.

EVALUATION: DOCTRINAL ISSUES

It is now possible to evaluate the performance of the system in developing the doctrines needed to resolve new intellectual property issues. Effective resolution requires that the new doctrines be economically effective in encouraging innovation. Legislators can take these economic aspects into account directly; courts must weave the economic factors into a preexisting network of doctrinal principles. In general, in the areas analyzed, doctrinal development has been left to the courts; there has been relatively little statutory activity.[43] Given this pattern (in the United States) one would anticipate a spurt of judicial activity as unresolved issues are explored on an incremental basis, followed by a settling down as stable doctrines are developed.

Case Law Processes

On the biotechnology side, however, there have been relatively few cases. A count of two volumes of the *United States Patents Quarterly Second* reveals 183 patent cases, of which only 4 involve biotechnology issues.[44] This probably suggests that the statutes need relatively little modification to respond to biotechnology. It should be noted, however, that certain of the court decisions, such as the new *Amgen* v. *Chugai* case, 927 F.2d 1200 (CAFC, 1991), have caused enormous confusion by leaving the principles unclear; the courts have almost certainly not done as well here as one would like.

The EC and WIPO activities in the biotechnology area have, however, been much more effective in locating the difficult issues early on; they suggest that a committee process can be more effective than case law in identifying new doctrinal issues. Although the EC proposals have not yet become law, the European discussions have produced quite thoughtful analyses of most of the hard issues and effectively illuminated the policy arguments on each side of these issues. Their result may be a sui generis EC law or modification of the existing patent law. Even in the absence of such a law, the analysis is one that is likely to be very helpful to courts (and it would be tragic if U.S. courts ignored these efforts).

On the software and network sides, case law efforts have been far less successful. Basically, it has proved impossible to modify the copyright concept to deal with software; the extensions needed to go beyond protection against literal copying have confused the doctrine and, in some cases,

---

[43]The arguable exceptions are the Orphan Drug Act and the copyright amendments of 1980.
[44]The count is based on volumes 15 and 16.

produced overprotection. The CONTU proposals failed to identify many of the hard issues and thus led the nation too quickly to a commitment to the copyright model. Moreover, in the case count (noted above) of two volumes of the *United States Patent Quarterly*, only 4 (and arguably fewer) of the 183 patent cases involved biotechnology. In contrast, of the 59 copyright cases, 7 involved software and 3 involved networks.[45] In short, the software area, in contrast to biotechnology, is in a state of substantive litigation chaos. Almost certainly, this reflects the doctrinal confusion in the area. Because of the close tie between software issues and network issues, one can expect the litigation chaos to extend to that area. Again, as for biotechnology, the EC processes appear to have been more effective in locating the difficult problems.

The Alternative: Sui Generis Approaches

Although the available examples skirt only the edges of the technologies examined here, sui generis approaches present a strikingly different picture. Two U.S. examples are available.

One is plant variety protection, the sui generis pattern for plant breeding.[46] This pattern allows for a certificate of protection on a new variety; the necessary showings are much simpler than those for a regular patent and the expense of obtaining protection is far smaller. The rights gained through protection are also smaller: farmers can reuse their harvested crop as seed, and breeders can use protected materials in competitive breeding programs without infringing the certificate holder's rights. Nevertheless, the system has worked; it is one of the few forms of intellectual property protection that has been shown to increase innovation,[47] and many firms in the traditional seed industry have urged use of this system for plants instead of the regular patent system. At the same time, there is forum shopping; some of the firms in the seed industry have attempted to obtain regular patents on materials that seem more appropriate to plant variety protection.[48]

---

[45]The difference in proportions of biotechnology and software cases (not including network cases) has a chi square of 7.53 and is significant at better than a 1 percent level.

[46]Plant Variety Protection Act of 1970 (7 U.S.C. sec. 3231 et seq.).

[47]See L. Butler and B. Marion, The Impacts of Patent Protection on the U.S. Seed Industry and Public Plant Breeding, North Central Regional Research Publication 304 (September 1985); R. Perrin, K. Kunnings, and L. Ihnen, Some effects of the U.S. Plant Variety Protection Act of 1970 (August 1983).

[48]See, e.g., Patent 4,812,600, issued March 14, 1989, on a Pioneer Hi-Bred International "inbred corn line PHK 29," and including a claim on use of the line as a hybrid parent. The plant variety system would appear more appropriate and provides comparable protection, 7 U.S.C. sec. 2541. Conceivably, the regular patent obtained in this case provides greater protection against reverse engineering or use of the material for breeding purposes—whether such greater protection should be available poses important policy issues.

The sui generis approach has apparently also been reasonably effective in the mask work context, with the Semiconductor Chip Protection Act of 1984.[49] This statute provides coverage for semiconductor mask works against reproduction by photographic means, but not against other forms of copying. Although the statute's allowance of reverse engineering can be criticized,[50] there are also arguments that such reverse engineering is economically desirable. Certainly there has been little litigation and substantial innovation in the area. The statute appears to have been successful enough in industry's eyes for the industry to urge continuation of its international provisions. Under these provisions, foreign firms can take advantage of the mask work act if their home nations offer comparable protection to U.S. firms. This reciprocity provision, which has played a role in enactment of comparable legislation in Japan and Europe, as well as other nations, provides a mechanism for obtaining international protection and thus serves as the equivalent of an international treaty.[51]

These approaches appear to have been quite successful in recognizing the special issues associated with a particular area of intellectual property law. In general, they allow tailoring of the protection to the specific problem. Nevertheless, as with case law, their doctrinal evolution may be unpredictable and slow, because they depend on legislative action or treaty negotiation (and Congress sometimes avoids the hardest recognized policy issues when it enacts such legislation). Moreover, they give the inventor an opportunity to engage in forum shopping and double coverage, choosing the form of protection that best satisfies the applicant's goals but may undercut an intended legislative balance. Also—although this point is far from clear—errors in creating sui generis forms of protection may be harder to undo than errors in adapting existing forms of protection.

Summary

In spite of these limitations, experience with sui generis approaches is relatively encouraging, compared to the ability of the courts or of short-term advisory panels to define ways to modify existing laws to meet new technologies. In addition, if an advisory or consultative process is used, it seems clear that the best approach is through a long-term continuing pro-

[49]Pub. L. No. 98-620, 17 U.S.C. sec. 901-914.

[50]M. Goldberg, Intellectual Property Rights and Technology—Semiconductor Chip Protection as a Case Study, Chapter 14.

[51]R. Stern, Chip topography protection in the USA, in The Law of Information Technology in Europe, A.P. Meijboom and C. Prins, eds., 153-168, Deventer (1991). For criticism of the statute, see T. Hoeren, Chip protection in Europe, Id. at 137-152.

cess open to outside criticism and lobbying (particularly by the firm or interest group that can point out an important missed issue).

These differences in timing and in opportunity for outside criticism are probably the important differences between the intellectually unsatisfying CONTU process and the more appealing WIPO expert pattern. CONTU was appointed in late July 1975 and first met in October of that year. Its first hearings were in May 1976, and its final report was filed in July 1978, three years after its creation.[52] Certain of the important issues were raised in Commissioner Nimmer's concurrence and Commissioners Hersey's and Karpatkin's dissents;[53] it is very plausible that the commission did not have adequate time to deal with the issues posed by these dissents. In contrast, the WIPO Committee of Experts, which published the results of its deliberations as it went along, met four times over the period November 1984 to October 1988 and may meet again.[54] The critical transition, at which consideration went from analysis based heavily on the previous state of the law to discussion of the new hard issues, appears to have been in the staff work between the second session in February 1986 and the third session in July 1987.[55] Thus, it took two years to identify these questions and to begin to deal with them; by that time, CONTU must have had to concentrate a significant portion of its attention on its final report.[56]

EVALUATION: THE RIGHTS-GRANTING PROCESS

Doctrinal changes are only part of the system's response; the operational performance of the system in terms of providing mechanisms for obtaining and enforcing intellectual property rights is just as important.

By way of introduction, one could hypothesize an intellectual property system's response to any technological perturbation. In the first instance, the system struggles with what rights to grant. If examination is needed as

---

[52]CONTU, supra at 3-8.

[53]Id. at 26-27, 27-37, and 37-38.

[54]Committee of Experts, supra.

[55]Compare Committee of Experts on Biotechnological Inventions and Industrial Property, Industrial Property Protection of Biotechnological Inventions; Report prepared by the International Bureau (prepared for the Second Session), WIPO BioT/CE/II/2 (November 5, 1985); with Committee of Experts on Biotechnological Inventions and Industrial Property, Industrial Property Protection of Biotechnological Inventions; Revised Report prepared by the International Bureau (prepared for the Third Session), WIPO BioT/CE/III/2 (April 8, 1987). The report of the Second Session reflects consideration of some of these issues. See Committee of Experts on Biotechnological Inventions and Industrial Property, Second Session, Report; Adopted by the Committee of Experts, WIPO BioT/CE/II/3 (February 7, 1986).

[56]The current Advisory Committee on Patent Law Reform has only two years from creation to final report.

it is for patents, the office hiring examiners will be in losing competition with industry for a small pool of capable specialists. Backlogs will build up and mistakes will be made. Ultimately, if the system can evolve, these problems will be solved. This process is unavoidable, no matter whether change occurs through statutory modification or through creation of a sui generis system; in either case there will be unanticipated issues to be faced.

The same kind of evolution will mark interpretation of the rights. At first, courts will have doubts about the scope of the rights—there will be lots of decisions, many in conflict, while these issues are worked out. Depending on the structure of the industry, new patterns of patent licensing may evolve. Again, however, if the system is working well, these disputes will settle down after a while.

As just suggested, the rights granting organization—e.g., the Patent and Trademark Office (PTO)—has to respond to new technologies by building new expertise. As an overall matter, the general increase in patent applications is taking place at a manageable rate. The number of patent applications filed increased from 106,295 in 1970 to 163,306 in 1989;[57] this corresponds to an increase of 2.3 percent per year, slightly less than the 3.1 percent annual increase of real R&D expenditures over the same period.[58] Also, if deviations in individual technology sectors are ignored, the PTO appears to have been quite successful in managing this case load. The backlog fell from 1.8 times the annual application rate in 1970 to 1.4 times that rate in 1989.[59]

There is, however, a difficulty in matching the personnel pattern to the needs of new areas of technology. As a new type of technology arises, the PTO must develop new expertise. As noted above, this requires education of existing employees or recruiting and holding of new employees in a market certain to be favorable to the employees. Data are available for the PTO backlog in the biotechnology sector; it is somewhat discouraging in that the average delay in the sector was three years in 1989—many years after the intellectual explosion in the area became obvious.[60] In some subsectors, the delay was nearly four years.[61] More recently, the biotechnology backlog has been brought down.

The process of adapting to new technologies leaves behind not only a

---

[57]Commissioner of Patents and Trademarks, Annual Report; Fiscal Year '89 (January 1990), Table 6.

[58]R&D estimates calculated from U.S. Department of Commerce, Statistical Abstract of the United States 1988 (December 1987), Table 950 (data for 1970-1987).

[59]Calculated from Commissioner of Patents and Trademarks, supra, Tables 6 and 8.

[60]Biotechnology in a Global Economy, supra at 212.

[61]Id.

delay (which is especially significant in biotechnology where a number of firms have been pursuing the same product)[62] but also errors. Almost every expert has a story of patents in new areas that were almost certainly granted only because the examiner was not really familiar with the state of the art in the technology or because there had not yet evolved a data base within which a search could be conducted.[63] These patents can severely complicate the economic development of the relevant field, although it is possible for a sensitive court to choose not to enforce them. Nevertheless, the overall evaluation of the rights-granting process has to be relatively favorable.

EVALUATION: THE RIGHTS-ENFORCING PROCESS

Intellectual property rights are meaningless unless enforced (and for software copyright, where the grant of rights is essentially automatic, enforcement is the only context in which litigation comes to the surface). Enforcement is, in the first instance, a litigation issue, but litigation is so expensive that its economics shapes the effective scope of intellectual property rights. A patent that its holder cannot afford to defend is worthless; likewise, a patent claim can be significantly stretched against a firm unable to afford defensive litigation. Equally important, intellectual property licenses—whose pattern differs radically from industry to industry—dramatically shape the real-world impact of these rights.

Judicial Enforcement

Patent litigation is extremely expensive; rumor suggests a cost of $0.5 million per claim litigated per side. There is little reason to anticipate a significantly different number for copyright, save perhaps for the possibility that some of the disputes are less fact intensive and more law intensive. (The costs of fact discovery and proof are especially high.) This is a nearly absolute bar to use of the system by small firms: where they must use the system, as in the biotechnology industry whose pharmaceutical products are

[62]There are sectors in which patentees would prefer delays in patent issuance because the market is growing faster than the interest rate so that a later monopoly is more valuable than an earlier one. This might even be true of biotechnology—except for the facts that development costs are so large, that development takes as long as the patent application processing time, and that these ongoing expenditures will be wasted if a competitor gains the patent rights.

[63]"Practically once a month, the nation's computer networks are abuzz with news of another patent issued on a fundamental concept that is widely used." S. Garfinkel, R. Stallman, and M. Kapor, Why patents are bad for software, 7 Issues in Science and Technology 50-55, 51 (1991).

easily imitated, the resulting costs are likely to drain away research funds. Some biotechnology firms are said to be spending more on litigation than on research.

The economics of litigation is thus likely to favor large firms at the expense of small ones. Large firms are more likely to be able to threaten litigation and to defend against litigation. There have been at least some cases of "strategic litigation" in which a large firm uses the threat of litigation costs to squash a start-up.[64]

The reality of such threats has evolved with recent strengthening of the intellectual property system. In the 1960s and 1970s, patents were generally considered nearly irrelevant, because they were so often found invalid.[65] Moreover, even if they were found valid, there were a number of patent misuse doctrines and antitrust doctrines that restricted the effective scope of patents. During the 1980s, nearly all these factors changed. The Court of Appeals for the Federal Circuit was created[66] which not only brought the body of patent law into uniformity but also substantively changed it. Congress, the courts, and the Reagan administration radically weakened the various defenses of an accused infringer and changed antitrust/patent perceptions.[67] There are signs of a change back, however, for example, in *Lasercomb America, Inc.* v. *Reynolds*, 911 F.2d 970 (4th Cir. 1990), in which the circuit court created a new copyright misuse concept analogous to the old patent misuse doctrines.

A separate group of enforcement issues arises at the international level and will probably be felt more strongly in the biotechnology sector than in other sectors discussed in this chapter. Many nations have hesitated to extend as wide a scope to intellectual property protection of biotechnology as does the United States, so that doctrines differ significantly from nation

---

[64]A. Silverman, Symposium report: Intellectual property law and the venture capital process, 5 High Technology L. Journal 157 (1990).

[65]Before the Court of Appeals for the Federal Circuit was created, appellate courts upheld only 30 to 40 percent of the patents found valid by trial courts, M. Adelman, The new world of patents created by the Court of appeals for the Federal Circuit, 20 U. Mich. J. L. Ref. 979 (1987). The new court upheld 89 percent of patents in such circumstances, D.Dunner, Special Comm'n. on CAFC, 1988 A.B.A. Sec. Pat., Trademark and Copyright L. 314, 325. Continuation of this trend would make it less likely that the courts would correct for patent office errors in new technological areas.

[66]Federal Courts Improvement Act of 1982, Pub. L. No. 97-164, sec. 127(a), 96 Stat. 37 (codified at 28 U.S.C. sec. 1295).

[67]For example, the CAFC rejected the idea that it was an antitrust violation for a patentee to claim infringement while knowing that there was not infringement. *Loctite Corp.* v. *Ultraseal, Ltd.*, 781 F.2d 861 (Fed. Cir. 1985). The Supreme Court and Congress strengthened the right of a patentee not to grant licenses, *Dawson Chemical* v. *Rohm and Haas Co.*, 448 U.S. 176 (1980); Pub. L. No. 100-703, sec. 201, 102 Stat. 4674, 4676 (1988) (codified at 35 U.S.C. sec. 271(d)).

to nation. Because of these differences and because the area offers the opportunity for substantial trade in intermediate and finished products, border restrictions are likely to be particularly important in this area.

The U.S. border restrictions are generally enforced through "section 337," which is administered by the International Trade Commission (ITC).[68] This procedure provides a U.S. intellectual property holder with extremely strong rights to bar the import not only of infringing products but also of the direct and indirect products of processes that would infringe if practiced within the United States. The biotechnology industry would like to extend these rights further through making essentially automatic the granting of process claims in a number of situations in which product claims are already granted.[69]

In a world that has not achieved patent law unification and an industry that is and must be internationally based, this kind of trade barrier will also be a barrier to progress by decreasing the extent to which product intermediates made legitimately in one nation can be exported to another. Significant extension of the section 337 process will lead to global suboptimization and enormous headaches for managers who are attempting to organize on an international level.[70] Nevertheless, current political trends in the U.S. Con-

---

[68]This is section 337 of the Tariff Act of 1930, ch. 479, 46 Stat. 590, 703 (codified as amended at 19 U.S.C. sec. 1337, amended by Omnibus Trade and Competitiveness Act of 1988, Pub. L. No. 100-418, sec. 1342, 102 Stat. 1107, 1212 (1988).

[69]E.g., S 654, the proposed Biotechnology Patent Protection Act of 1991. See Amended biotech process patent bill is cleared by Senate subcommittee, 42 BNA Patent, Trademark & Copyright J. 313 (August 1, 1991). The intended result would have been to change the outcome of the ITC phase of the Amgen-Chugai litigation, In re *Certain Recombinant Erythropoietin*, 10 USPQ 2d 1906 (ITC 1989). A Japanese firm held the product patent in the United States, while its U.S. competitor held a U.S. product patent on cells containing a gene inserted to manufacture the product. Because the foreign firm was able legally to use its cells with the inserted gene to produce the product abroad and sell it into the United States, the U.S. firm had little leverage to negotiate with its competitor. Had a process claim been available to the U.S. firm for its cells containing the gene, it would have been able to exclude the product under section 337 and would then have had a better bargaining position. The specific dispute was effectively resolved when, in parallel litigation, the Japanese firm's patent was unexpectedly found invalid, *Amgen, Inc.* v. *Chugai Pharmaceutical Co.*, 927 F.2d 1200 (CAFC 1991).

[70]See J. Barton, Testimony prepared for presentation before the Subcommittee on Intellectual Property and Judicial Administration of the House Judiciary Committee, May 16, 1991. It should also be noted that the existing section 337 provisions have been found to be in violation of the General Agreement on Tariffs and Trade (GATT), Re the DuPont de Nemours/AKZO Dispute: *European Economic Community* v. *United States of America*, reprinted at [1989] 1 C.M.L.R. 147, a problem that the administration suggested it would resolve as part of the implementing legislation for the Uruguay Round, [President Bush's] Memorandum for the United States Trade Representative; Subject Enforcement of Section 337 of the Tariff Act of 1930 (November 7, 1989). See generally, Barton, Section 337 and the international trading system, in Technology Trade and World Competition; Protecting Intellectual Property with

gress are probably toward such strengthening, presumably based on the judgment that it will increase the rents accruing to U.S. firms. Given that the U.S. share of all intellectual property is declining as non-U.S. research increases, this is probably a shortsighted position.

Thus, when one looks at the law as applied by courts, one is forced to retreat somewhat from the sense that the patent system is adapting relatively well in the biotechnology sector. The doctrine may be evolving rapidly enough to reduce the overall number of cases, but the expenses of these cases—and the still unsolved international issues—suggest a significant burden for the industry. It is at least possible that this burden falls disproportionately on the newer, smaller, more innovative firms.

Informal Approaches

Business has its ways of adapting to the difficulties of applying intellectual property systems. For example, the American Society for Composers, Authors, and Publishers is a privately created network that resolves the practical problems of collecting and distributing royalties in the musical area where there are so many individual performances of copyrighted works. Such systems are as much part of the effective body of law as are the formal statutes.[71]

On the basis of anecdotal evidence, the "normal" pattern of intellectual property enforcement is an industry-wide cross-license arrangement in which any payments by one firm to another are based on a very rough comparison of the relative value of the intellectual property contributed by each firm.[72] The firm with more patents collects from the firm with fewer patents, with little attention to the value of specific claims. The system thus rewards innovation while avoiding expensive litigation (which is saved for the cases in which a firm makes a serious challenge of the balance). This pattern was, for example, typical of the electronics industry prior to Texas Instrument's challenge of Japanese firms in 1986,[73] a challenge that triggered a round of litigation that may now be ending. The cross-license is a natural response

---

Trade Sanctions 1 (1990). The GATT panel's concerns go to certain procedural differences between ITC and District Court enforcement of patent rights against import; the problem discussed in the text is not yet on the reform agenda (and was not posed to the GATT panel), but is probably much more important than the issues identified by the panel.

[71]See Intellectual Property Rights in an Age of Electronics and Information, supra at 269-71.

[72]Consider, for example, the arrangements in the auto and the aircraft industries considered in *U.S.* v. *Automobile Manufacturers Assn., Inc.*, 1969 Trade Cases par. 72,907 (DC Cal. 1969), modified, 1982-83 Trade Cases par. 65,088; and *U.S.* v. *Manufacturers Aircraft Assn., Inc.*, 1976-1 Trade Cases par. 60,810 (DCNY 1975).

[73]See *Texas Instruments, Inc.* v. *United States International Trade Commission*, 871 F.2d 1054 (CAFC 1989).

to the limitations of litigation and a very effective response when each firm in the industry must use a variety of previously claimed inventions to produce a product.

There are two major exceptions to this "normal" pattern. One arises in those industries in which individual patents are the basis of specific products and each product has substantial market power. The obvious examples are chemistry and pharmaceuticals. Different chemicals and pharmaceuticals substitute only exceptionally for one another. Each patent, in effect, represents a market monopoly that can be practiced independently of competitors. The precise claims are taken quite seriously because the patent monopoly defines the returns from the product very precisely, serves as the basic competitive protection, and faces few countervailing claims that could be the basis for a cross-license. There are some rare parallel cases, such as fundamentally new technologies in which competitors can be excluded for a while (e.g., the early days of instant photography).

The medical biotechnology industry falls precisely within this exception.[74] It is an industry that anticipates producing a relatively small number of products, each of which requires enormous front-end research and regulatory investment, and each of which is likely to have a substantial product lifetime and to be readily imitated. In short, it is an industry exactly adapted to the patent system, because a period of monopoly is nearly essential as a mechanism of covering the front-end fixed costs. What is at stake in a patent suit is the possibility of access to this period of monopoly returns and the precise (and relatively easily evaluated) scope of these returns. There is no wonder that firms expend great efforts to protect their patent position. It is likely that the current litigation pattern will continue in this industry.

The other exception to the typical industry-wide cross-license is the occasional "flare-up" in an industry normally governed by a wide cross-license. This may typically be a response to an outsider who threatens to upset the competitive balance in an industry; this is certainly part of the explanation for Texas Instrument's use of patent law to protect its market against Japanese semiconductor competitors. (Texas Instruments has gone on, but without clearly improving its competitive position, to use patent royalties as a major source of income.) Given the international cross-flows of technology in the semiconductor sector, it is hard to imagine that the patent battle will not soon be settled.

---

[74]In contrast, agricultural biotechnology may well turn out to involve proprietary genes with a variety of applications marketed through licensees with expertise in particular seed markets; the licensing structure may not be a traditional cross-license pattern, but it is likely to be relatively stable.

One can anticipate, however, that the software and network areas will be within this exception for a long time. The economic structure of these areas is highly unsettled. Communications firms, computer firms, and software firms are all vying for additional influence and control, and are seeking to assert whatever intellectual property rights they can define or have defined as part of the jockeying. Along with the doctrinal confusion described above, this is undoubtedly a reason why litigation rates are so high in this industry. A cross-license may be the almost certain *ultimate* direction of software patents on algorithms and computational procedures; different claims are likely to be interlocked, and the task for a new competitor may prove to be to produce some useful algorithms of its own and then to enter the cross-license. On the broader software and interface issues, however, in the absence of a relatively stable industry structure and pecking order, continued intellectual property litigation is very likely.

In summary, in all three of the new areas, continued and expensive intellectual property litigation is likely. Given that any new technology upsets existing market structures, the pattern is likely to prevail for other technologies as well. This implies continued expenditures on litigation rather than on research. Perhaps more important, it will favor firms with strong economic positions at the expense of their challengers. If their most innovative challengers are, as may be suspected, small firms, the costs of litigation and the imperfections of the litigation system will cut against new technologies.[75]

A full response to intellectual property litigation costs would go far beyond the scope of this chapter, and the costs may well derive more from the general character of U.S. litigation than from the specific character of intellectual property. Nevertheless, three points can be suggested:

1. To the extent that legal uncertainty drives litigation, extensive use of expert studies and sui generis forms of coverage may help reduce costs.

[75]Economic evidence on whether small firms are more innovative than large firms is inconclusive. Gellman Research Associates (1975) and Acs and Audretch (1990) found that small firms produce more innovations per unit sales and per employee, respectively, than do large firms. Freeman (1982), however, found that the share of industrial innovations in Great Britain contributed by small firms was smaller than their share of production and employment. Despite inconclusive quantitative evidence, there are many historical and recent examples of small firms playing an important role in the establishment of new branches of industry and the rejuvenation of old ones. Gellman Research Associates, Indicators of International Trends in Technological Innovation, report prepared for the National Science Foundation (1975); Z. Acs and D. Audretch, Innovation and Small Firms, Cambridge, Mass.: MIT Press (1990); C. Freeman, The Economics of Industrial Innovation, 2nd ed., Cambridge, Mass.: MIT Press (1982).

2. Litigation can probably be discouraged somewhat if rights are defined less broadly.[76]

3. Again, if it is, as just suggested, desirable to favor smaller firms, the antitrust and intellectual property misuse doctrines may be helpful to assist them in their defense against litigation and thus to deter litigation against them.

OVERALL IMPLICATIONS

Ability of the System to Adapt to Increasing Innovation Rates

Nothing developed in this analysis provides any reason to believe that increased filing rates have posed fundamental problems for the system. To the contrary, the PTO has, overall, successfully kept up with the rate of innovation.

There are obvious qualifications. The PTO has not been able to hire new types of analytic capabilities rapidly enough; this has led to errors and delays in specific areas, and is likely to do so again and to cut against new technologies. In the biotechnology sector, in particular, the economic loss due to delay has been severe because firms are undertaking competing research projects and do not know which firm wins until a patent issues (and possibly until the patent is litigated). This is almost certainly a strong argument for switching to the global first-to-file system, with the applications made public a reasonable time after filing.[77]

Specific Doctrinal Implications of the Three Examples

The biotechnology case shows the system to be relatively successful in dealing with change. The critical pervasive problem is defining the point at which a gene becomes patentable; this is an issue that should ideally be

[76]Note that moderation is compelled by traditional economic analysis of intellectual property incentives. This analysis balances the incentive to innovation created by the monopoly right defined by intellectual property against the economic costs derived from the artificial prices created by the monopoly rent. An additional basis for moderation is suggested by the possibilities that smaller firms are better innovators and also victims of litigation costs. Put more broadly, under certain circumstances a first intellectual property monopoly right may in fact be exercised in a way that decreases incentives to subsequent innovation. For an argument in a similar direction, see Commissioner Hersey's dissent in CONTU, especially at pp. 35-36.

[77]With a first-to-file system, one can publish patent applications without fear of complicating priority disputes and thus speed the flow of scientific information. The typical European pattern, for example, is to require filing before publication, but then to publish the applications 18 months after filing (European Patent Convention, Articles 54 and 93). The publication conveys a form of interim protection (Article 67).

analyzed by a thoughtful study group before decisions are made on the patenting of genome sequences.

The software sequence, in contrast, shows up very badly. The basic CONTU decision to rely primarily on a modified copyright concept appears to have been wrong, and complementary efforts to apply patent law have been, at best, doctrinally confusing. This sector cries out for a new sui generis approach.

For the data bank context, it is too early to tell confidently, but if the Supreme Court's decision in *Feist* turns out to dominate the field, one can anticipate that there will be serious difficulties. This area also cries out for an early specific analysis.

Broader Implications

Technology is unpredictable; hence one must be careful about relying too heavily on the specific examples just discussed. Nevertheless, based on these examples (and on the other examples discussed in the analysis) one can suggest the following general conclusions:

1. Sui generis approaches are far more likely to be successful than one might have expected and should be utilized far more often. The European biotechnology and software directives and the plant variety protection and chip mask work statutes are all relatively encouraging.
2. It is essential to have routinized law reform processes to help deal with new technologies. The experience of the WIPO/UPOV panels is far more positive than that of CONTU. Although there may be a variety of reasons for this difference, one is that the WIPO/UPOV system was under much less time pressure, which enabled the expert panel and the staff to publish interim positions that could be broadly criticized and commented on. This is almost certainly better than the U.S. commission-style approach. It strongly suggests standing study groups, an approach that could be easily integrated into U.S. law. It might also suggest a pattern in which an expert panel is delegated the power actually to lay down and put in effect the rules for a sui generis system, leaving Congress the initiative to change the proposed rules if it wishes.[78]
3. The law reform process must be international. One of the reasons copyright has been pushed for software is that international coverage was relatively easy to achieve. In biotechnology, intellectual property protection is a possible trade barrier. The computer data bank issue is fundamen-

[78]See R. Stern, The bundle of rights suited to new technology, 47 U. of Pittsburgh L. Rev. 1229 (1986).

tally international. In short, future sui generis rights (and many proposals for interpretation of existing conventions) should be negotiated on an international basis,[79] and standing reform and study groups should be international.

4. It is crucial to take into account the role of small firms in innovation. Such firms may be prime sources of innovation in areas of new technology and there may be a serious risk that intellectual property rights can be used to stifle them. This means that such firms must be represented on study groups; it also means that intellectual property rights must be defined with consideration for their real-world impact on industrial structure.

---

[79]Note that in the absence of international agreement, one can use statutory reciprocity in the pattern of the chip mask work arrangement. Even so, it would be best to coordinate such efforts with as many other nations as possible.

Note also that the politicization of the international intellectual property system may make internationalization difficult. Yet, this politicization is primarily along a North-South axis, whereas the key negotiations in these new technology areas will generally be among the developed nations or with developing nations who have an interest in being included in a special regime.

12

A Case Study on Computer Programs

PAMELA SAMUELSON

HISTORICAL OVERVIEW

Phase 1: The 1950s and Early 1960s

When computer programs were first being developed, proprietary rights issues were not of much concern. Software was often developed in academic or other research settings. Much progress in the programming field occurred as a result of informal exchanges of software among academics and other researchers. In the course of such exchanges, a program developed by one person might be extended or improved by a number of colleagues who would send back (or on to others) their revised versions of the software. Computer manufacturers in this period often provided software to customers of their machines to make their major product (i.e., computers) more commercially attractive (which caused the software to be characterized as "bundled" with the hardware).

To the extent that computer programs were distributed in this period by firms for whom proprietary rights in software were important, programs tended to be developed and distributed through restrictive trade secret licensing agreements. In general, these were individually negotiated with customers. The licensing tradition of the early days of the software industry has framed some of the industry expectations about proprietary rights issues, with implications for issues still being litigated today.

In the mid-1960s, as programs began to become more diverse and complex, as more firms began to invest in the development of programs, and as some began to envision a wider market for software products, a public dialogue began to develop about what kinds of proprietary rights were or should be available for computer programs. The industry had trade secrecy and licensing protection, but some thought more legal protection might be needed.

Phase 2: Mid-1960s and 1970s

Copyright law was one existing intellectual property system into which some in the mid-1960s thought computer programs might potentially fit. Copyright had a number of potential advantages for software: it could provide a relatively long term of protection against unauthorized copying based on a minimal showing of creativity and a simple, inexpensive registration process.[1] Copyright would protect the work's "expression," but not the "ideas" it contained. Others would be free to use the same ideas in other software, or to develop independently the same or a similar work. All that would be forbidden was the copying of expression from the first author's work.

In 1964, the U.S. Copyright Office considered whether to begin accepting registration of computer programs as copyrightable writings. It decided to do so, but only under its "rule of doubt" and then only on condition that a full text of the program be deposited with the office, which would be available for public review.[2]

The Copyright Office's doubt about the copyrightability of programs

---

[1]Under the federal copyright statute in effect in 1964, the law was mainly used to protect published works from unauthorized copying and distribution. Protection lasted 28 years, but could be renewed for an additional 28 years. After that, the work could be freely copied. Upon publication, a copyright notice had to appear on each copy of the work (or else the work would be considered to have been dedicated to the public domain). The author would generally register the work with the Copyright Office upon publication. The Copyright Office would give the work a cursory examination to determine that it met copyright's modest substantive standards. After this examination, and upon payment of a modest fee, the office would issue a certificate of registration. Unpublished works were generally protected by state law.

Under the Copyright Act of 1976, the rights of copyright attach to original works of authorship from the moment of their first fixation in a tangible medium and last for the life of the author plus 50 years. Registration with the Copyright Office remains a simple inexpensive process; registration is necessary to bring an action for copyright infringement, but not for rights to attach. See 17 U.S.C. sec. 101 et seq. (1988), and 17 U.S.C. sec. 1 et seq. (superseded).

[2]See Samuelson, CONTU revisited: The case against copyright protection for computer programs in machine-readable form, 1984 Duke L.J. 663 (1984). The Copyright Office will deny registration to works that are clearly uncopyrightable but, on occasion, will register works about whose "copyrightability" the office has some doubt, leaving to the courts the ultimate question as to whether the work qualifies for copyright protection. The registration certificate issued for such a work will reflect that its issuance was under the rule of doubt.

arose from a 1908 Supreme Court decision that had held that a piano roll was not an infringing "copy" of copyrighted music, but rather part of a mechanical device.[3] Mechanical devices (and processes) have traditionally been excluded from the copyright domain.[4] Although the office was aware that in machine-readable form, computer programs had a mechanical character, they also had a textual character, which was why the Copyright Office decided to accept them for registration.

The requirement that the full text of the source code of a program be deposited in order for a copyright in the program to be registered was consistent with a long-standing practice of the Copyright Office,[5] as well as with what has long been perceived to be the constitutional purpose of copyright, namely, promoting the creation and dissemination of knowledge.[6]

Relatively few programs, however, were registered with the Copyright Office under this policy during the 1960s and 1970s.[7] Several factors may have contributed to this. Some firms may have been deterred by the requirement that the full text of the source code be deposited with the office and made available for public inspection, because this would have dispelled its trade secret status. Some may have thought a registration certificate issued under the rule of doubt might not be worth much. However, the main reason for the low number of copyright registrations was probably that a mass market in software still lay in the future. Copyright is useful mainly to protect mass-marketed products, and trade secrecy is quite adequate for programs with a small number of distributed copies.

Shortly after the Copyright Office issued its policy on the registrability of computer programs, the U.S. Patent Office issued a policy statement concerning its views on the patentability of computer programs. It rejected the idea that computer programs, or the intellectual processes that might be embodied in them, were patentable subject matter.[8] Only if a program was

---

[3]*White-Smith Music Co.* v. *Apollo*, 209 U.S. 1 (1908).

[4]See infra notes and accompanying text.

[5]The collection of the Library of Congress was built in substantial part with copies of copyrighted works deposited with the Copyright Office. The office remains a subunit of the Library of Congress.

[6]See, e.g., Office of Technology Assessment, Intellectual Property Rights in an Age of Electronics and Information (1986).

[7]National Commission on New Technological Uses of Copyrighted Works, Final Report (1979) (cited as "CONTU report") (reporting about 1,200 registrations of computer programs between 1964 and 1978).

[8]Only machines, manufactures, compositions of matter, and processes are patentable subject matter under 35 U.S.C. sec. 101 (1988). Processes that can be carried out in one's head or with aid of pen and paper, even when embodied in a writing, have been considered unpatentable as "mental processes." See Samuelson, Benson revisited: The case against patent protection for algorithms and other computer program-related inventions, 39 Emory L.J. 1025, 1033-1040 (1990).

claimed as part of a traditionally patentable industrial process (i.e., those involving the transformation of matter from one physical state to another) did the Patent Office intend to issue patents for program-related innovations.[9]

Patents are typically available for inventive advances in machine designs or other technological products or processes on completion of a rigorous examination procedure conducted by a government agency, based on a detailed specification of what the claimed invention is, how it differs from the prior art, and how the invention can be made. Although patent rights are considerably shorter in duration than copyrights, patent rights are considered stronger because no one may make, use, or sell the claimed invention without the patent owner's permission during the life of the patent. (Patents give rights not just against someone who copies the protected innovation, but even against those who develop it independently.) Also, much of what copyright law would consider to be unprotectable functional content ("ideas") if described in a book can be protected by patent law.

The Patent Office's policy denying the patentability of program innovations was consistent with the recommendations of a presidential commission convened to make suggestions about how the office could more effectively cope with an "age of exploding technology." The commission also recommended that patent protection not be available for computer program innovations.[10]

Although there were some appellate decisions in the late 1960s and

---

[9]Id. But see Chisum, The patentability of algorithms, 47 U. Pitt. L. Rev. 959 (1986).

[10]Report of the President's Commission on the Patent System, "To Promote the Progress . . . of the Useful Arts" in an Age of Exploding Technology 13 (1966), quoted in *Gottschalk* v. *Benson*, 409 U.S. 63, 72 (1972):

> Uncertainty now exists as to whether the statute permits a valid patent to be granted on programs. Direct attempts to patent programs have been rejected on ground of nonstatutory subject matter. Indirect attempts to obtain patents and avoid the rejection, by drafting claims as a process, or a machine, or components thereof, programmed in a given manner rather than as a program itself, have confused the issue further and should not be permitted.
>
> The Patent Office now cannot examine applications for programs because of a lack of classification technique and the requisite search files. Even if these were available, reliable searches would not be feasible or economic because of the tremendous volume of prior art being generated. Without this search, the patenting of programs would be tantamount to mere registration and the presumption of validity would be all but nonexistent.
>
> It is noted that the creation of programs has undergone substantial and satisfactory growth in the absence of patent protection and that copyright protection for programs is presently available.

Some would argue that the report's conclusions are as valid today as they were 25 years ago. See, e.g., Kahin, The software patent crisis, Technology Review (April 1990) at 53.

early 1970s overturning Patent Office rejections of computer program-related applications, few software developers looked to the patent system for protection after two U.S. Supreme Court decisions in the 1970s ruled that patent protection was not available for algorithms.[11] These decisions were generally regarded as calling into question the patentability of all software innovations, although some continued to pursue patents for their software innovations notwithstanding these decisions.[12]

As the 1970s drew to a close, despite the seeming availability of copyright protection for computer programs, the software industry was still relying principally on trade secrecy and licensing agreements. Patents seemed largely, if not totally, unavailable for program innovations. Occasional suggestions were made that a new form of legal protection for computer programs should be devised, but the practice of the day was trade secrecy and licensing, and the discourse about additional protection was focused overwhelmingly on copyright.

During the 1960s and 1970s the computer science research community grew substantially in size. Although more software was being distributed under restrictive licensing agreements, much software, as well as innovative ideas about how to develop software, continued to be exchanged among researchers in this field. The results of much of this research were published and discussed openly at research conferences. Toward the end of this period, a number of important research ideas began to make their way into commercial projects, but this was not seen as an impediment to research by computer scientists because the commercial ventures tended to arise after the research had been published. Researchers during this period did not, for the most part, seek proprietary rights in their software or software ideas, although other rewards (such as tenure or recognition in the field) were available to those whose innovative research was published.

Phase 3: The 1980s

Four significant developments in the 1980s changed the landscape of the software industry and the intellectual property rights concerns of those who developed software. Two were developments in the computing field; two were legal developments.

The first significant computing development was the introduction to the market of the personal computer (PC), a machine made possible by improvements in the design of semiconductor chips, both as memory storage

---

[11] *Gottschalk* v. *Benson*, 409 U.S. 63 (1972) and *Parker* v. *Flook*, 437 U.S. 584 (1978). These cases are discussed at length in Samuelson, supra note 8.

[12] Id.

devices and as processing units. A second was the visible commercial success of some early PC applications software—most notably, Visicalc, and then Lotus 1-2-3—which significantly contributed to the demand for PCs as well as making other software developers aware that fortunes could be made by selling software. With these developments, the base for a large mass market in software was finally in place.

During this period, computer manufacturers began to realize that it was to their advantage to encourage others to develop application programs that could be executed on their brand of computers. One form of encouragement involved making available to software developers whatever interface information would be necessary for development of application programs that could interact with the operating system software provided with the vendor's computers (information that might otherwise have been maintained as a trade secret). Another form of encouragement was pioneered by Apple Computer, which recognized the potential value to consumers (and ultimately to Apple) of having a relatively consistent "look and feel" to the applications programs developed to run on Apple computers. Apple developed detailed guidelines for applications developers to aid in the construction of this consistent look and feel.

The first important legal development—one which was in place when the first successful mass-marketed software applications were introduced into the market—was passage of amendments to the copyright statute in 1980 to resolve the lingering doubt about whether copyright protection was available for computer programs.[13] These amendments were adopted on the recommendation of the National Commission on New Technological Uses of Copyrighted Works (CONTU), which Congress had established to study a number of "new technology" issues affecting copyrighted works. The CONTU report emphasized the written nature of program texts, which made them seem so much like written texts that had long been protected by copyright law. The CONTU report noted the successful expansion of the boundaries of copyright over the years to take in other new technology products, such as photographs, motion pictures, and sound recordings. It predicted that computer programs could also be accommodated in the copyright regime.[14]

Copyright law was perceived by CONTU as the best alternative for protection of computer programs under existing intellectual property regimes. Trade secrecy, CONTU noted, was inherently unsuited for mass-marketed products because the first sale of the product on the open market would dispel the secret. CONTU observed that Supreme Court rulings had cast

---

[13]Whether computer programs were protectable under copyright law prior to the 1980 amendments has been the subject of some dispute. See Samuelson, supra note 2.

[14]See CONTU report, supra note 7.

doubts on the availability of patent protection for software. CONTU's confidence in copyright protection for computer programs was also partly based on an economic study it had commissioned. This economic study regarded copyright as suitable for protecting software against unauthorized copying after sale of the first copy of it in the marketplace, while fostering the development of independently created programs. The CONTU majority expressed confidence that judges would be able to draw lines between protected expression and unprotected ideas embodied in computer programs, just as they did routinely with other kinds of copyrighted works.

A strong dissenting view was expressed by the novelist John Hersey, one of the members of the CONTU commission, who regarded programs as too mechanical to be protected by copyright law. Hersey warned that the software industry had no intention to cease the use of trade secrecy for software. Dual assertion of trade secrecy and copyright seemed to him incompatible with copyright's historical function of promoting the dissemination of knowledge.

Another development during this period was that the Copyright Office dropped its earlier requirement that the full text of source code be deposited with it. Now only the first and last 25 pages of source code had to be deposited to register a program. The office also decided it had no objection if the copyright owner blacked out some portions of the deposited source code so as not to reveal trade secrets. This new policy was said to be consistent with the new copyright statute that protected both published and unpublished works alike, in contrast to the prior statutes that had protected mainly published works.[15]

With the enactment of the software copyright amendments, software developers had a legal remedy in the event that someone began to mass-market exact or near-exact copies of the developers' programs in competition with the owner of the copyright in the program. Unsurprisingly, the first software copyright cases involved exact copying of the whole or substantial portions of program code, and in them, the courts found copyright infringement. Copyright litigation in the mid- and late 1980s began to grapple with questions about what, besides program code, copyright protects about computer programs. Because the "second-generation" litigation affects the current legal framework for the protection of computer programs, the issues raised by these cases will be dealt with in the next section.

As CONTU Commissioner Hersey anticipated, software developers did not give up their claims to the valuable trade secrets embodied in their programs after enactment of the 1980 amendments to the copyright statute.

---

[15]See, e.g., Raskind, The uncertain case for special legislation protecting computer software, 47 U. Pitt. L. Rev. 1131 (1986).

To protect those secrets, developers began distributing their products in machine-readable form, often relying on "shrink-wrap" licensing agreements to limit consumer rights in the software.[16] Serious questions exist about the enforceability of shrink-wrap licenses, some because of their dubious contractual character[17] and some because of provisions that aim to deprive consumers of rights conferred by the copyright statute.[18] That has not led, however, to their disuse.

One common trade secret-related provision of shrink-wrap licenses, as well as of many negotiated licenses, is a prohibition against decompilation or disassembly of the program code. Such provisions are relied on as the basis of software developer assertions that notwithstanding the mass distribution of a program, the program should be treated as unpublished copyrighted works as to which virtually no fair use defenses can be raised.[19]

Those who seek to prevent decompilation of programs tend to assert that since decompilation involves making an unauthorized copy of the program, it constitutes an improper means of obtaining trade secrets in the program. Under this theory, decompilation of program code results in three unlawful acts: copyright infringement (because of the unauthorized copy made during the decompilation process), trade secret misappropriation (because the secret has been obtained by improper means, i.e., by copyright

---

[16]Shrink-wrap licensing agreements are printed forms inserted between the box containing the discs on which the software is loaded and the clear plastic covering the box. The forms typically have boldface instructions to read the form before opening the package, and inform the consumer that by opening the package, the consumer agrees to all of the terms contained in the form, some of which pertain to warranties, and others of which pertain to uses that can and cannot be made of the software. Shrink-wrap agreements typically inform the consumer that he or she is not the owner of a copy of the software, but a licensee of the software, and that breach of any term of the licensing agreement will terminate all of his or her rights to use of the program. Depriving the consumer of the status of an "owner of a copy" of the software is said to be the way to avoid the provisions of 17 U.S.C. sec. 117 that grant certain rights to modify and make backup copies of the software—rights that are granted only to "owners" of copies.

[17]See, e.g., Hazen, Contract principles as a guide for protecting intellectual property rights in computer software, 20 U.C. Davis L. Rev. 105 (1986).

[18]One appellate court decision has struck down portions of a state law purporting to validate shrink-wrap restrictions as a matter of contract law because the court thought the statute was in conflict with policies underlying the federal copyright law. See *Vault Corp.* v. *Quaid Software, Ltd.*, 847 F.2d 255 (5th Cir. 1988). See also Rice, Licensing the use of computer program copies and the Copyright Act's first sale doctrine, 30 Jurim. J. 157 (1990).

[19]*Harper and Row Publishers, Inc.* v. *Nation Enterprises*, 471 U.S. 539 (1984) (fair use rarely available as to unpublished works). But see LaST Frontier Conference Report on copyright protection for computer software, 30 Jurim. J. 15 (1989) (reporting a consensus of intellectual property scholars that unless there are valid contractual provisions to the contrary, decompiling a program to study its contents should be treated as a "fair use").

infringement), and a breach of the licensing agreement (which prohibits decompilation).

Under this theory, copyright law would become the legal instrument by which trade secrecy could be maintained in a mass-marketed product, rather than a law that promotes the dissemination of knowledge. Others regard decompilation as a fair use of a mass-marketed program and, shrink-wrap restrictions to the contrary, as unenforceable. This issue has been litigated in the United States, but has not yet been resolved definitively.[20] The issue remains controversial both within the United States and abroad.

A second important legal development in the early 1980s—although one that took some time to become apparent—was a substantial shift in the U.S. Patent and Trademark Office (PTO) policy concerning the patentability of computer program-related inventions. This change occurred after the 1981 decision by the U.S. Supreme Court in *Diamond* v. *Diehr,* which ruled that a rubber curing process, one element of which was a computer program, was a patentable process. On its face, the *Diehr* decision seemed consistent with the 1966 Patent Office policy and seemed, therefore, not likely to lead to a significant change in patent policy regarding software innovations.[21] By the mid-1980s, however, the PTO had come to construe the Court's ruling broadly and started issuing a wide variety of computer program-related patents. Only "mathematical algorithms in the abstract" were now thought unpatentable. Word of the PTO's new receptivity to software patent applications spread within the patent bar and gradually to software developers.

During the early and mid-1980s, both the computer science field and the software industry grew very significantly. Innovative ideas in computer science and related research fields were widely published and disseminated. Software was still exchanged by researchers, but a new sensitivity to intellectual property rights began to arise, with general recognition that unauthorized copying of software might infringe copyrights, especially if done with a commercial purpose. This was not perceived as presenting a serious obstacle to research, for it was generally understood that a reimplementation of the program (writing one's own code) would be

---

[20]*Compare Hubco Data Products Corp.* v. *Management Assistance Corp.*, 219 U.S.P.Q.(BNA) 450 (D. Idaho 1983) (decompilation was infringement) and *E.F. Johnson Co.* v. *Uniden Corp.*, 623 F. Supp. 1485 (D. Minn. 1985) (decompilation was not infringement). However, the Ninth Circuit Court of Appeals' decision to dissolve an injunction against Accolade Software in a suit brought by Sega Enterprises, which had alleged infringement on the sole ground that a copy had been made to make a compatible program, may be influential in persuading other courts to treat decompilation as fair use.

[21]*Diamond* v. *Diehr*, 450 U.S. 175 (1981). For a discussion of the PTO's change in interpretation of the patentability of program-related inventions after *Diehr*, see Samuelson, supra note 8.

noninfringing.[22] Also, much of the software (and ideas about software) exchanged by researchers during the early and mid-1980s occurred outside the commercial marketplace. Increasingly, the exchanges took place with the aid of government-subsidized networks of computers.

Software firms often benefited from the plentiful availability of research about software, as well as from the availability of highly trained researchers who could be recruited as employees. Software developers began investing more heavily in research and development work. Some of the results of this research was published and/or exchanged at technical conferences, but much was kept as a trade secret and incorporated in new products.

By the late 1980s, concerns began arising in the computer science and related fields, as well as in the software industry and the legal community, about the degree of intellectual property protection needed to promote a continuation of the high level of innovation in the software industry.[23] Although most software development firms, researchers, and manufacturers of computers designed to be compatible with the leading firms' machines seemed to think that copyright (complemented by trade secrecy) was adequate to their needs, the changing self-perception of several major computer manufacturers led them to push for more and "stronger" protection. (This concern has been shared by some successful software firms whose most popular programs were being "cloned" by competitors.) Having come to realize that software was where the principal money of the future would be made, these computer firms began reconceiving themselves as software developers. As they did so, their perspective on software protection issues changed as well. If they were going to invest in software development, they wanted "strong" protection for it. They have, as a consequence, become among the most vocal advocates of strong copyright, as well as of patent protection for computer programs.[24]

---

[22]Samuelson and Glushko, Comparing the views of lawyers and user interface designers on the software copyright "look and feel" lawsuits, 30 Jurim. J. 121 (1989) (reporting the results of a survey reflecting this view).

[23]Id.

[24]IBM Corp., Apple Computer Corp., and Digital Equipment Corp. have been especially prominent advocates on these intellectual property issues. IBM, several other computer manufacturers, and a computer manufacturers association argued to the U.S. Supreme Court during the *Benson* appeal in the early 1970s that patent protection should not be available for computer program innovations. This is not their position today.

CURRENT LEGAL APPROACHES IN THE UNITED STATES

Software developers in the United States are currently protecting software products through one or more of the following legal protection mechanisms: copyright, trade secret, and/or patent law. Licensing agreements often supplement these forms of protection. Some software licensing agreements are negotiated with individual customers; others are printed forms found under the plastic shrink-wrap of a mass-marketed package.[25] Few developers rely on only one form of legal protection. Developers seem to differ somewhat on the mix of legal protection mechanisms they employ as well as on the degree of protection they expect from each legal device.

Although the availability of intellectual property protection has unquestionably contributed to the growth and prosperity of the U.S. software industry, some in the industry and in the research community are concerned that innovation and competition in this industry will be impeded rather than enhanced if existing intellectual property rights are construed very broadly.[26] Others, however, worry that courts may not construe intellectual property rights broadly enough to protect what is most valuable about software, and if too little protection is available, there may be insufficient incentives to invest in software development; hence innovation and competition may be retarded through underprotection.[27] Still others (mainly lawyers) are confident that the software industry will continue to prosper and grow under the existing intellectual property regimes as the courts "fill out" the details of software protection on a case-by-case basis as they have been doing for the past several years.[28]

---

[25]One thing that distinguishes software from other commodities distributed in the general marketplace is the rarity of outright sales of software to customers. Most commodities in the marketplace are distributed on a sale basis. Publishers sell copies of copyrighted books; manufacturers sell machines embodying patented components; but software developers overwhelmingly distribute programs on what purports to be a licensing-only basis. Even shareware and freeware are distributed on a conditional basis.

Another thing that distinguishes software from other commercial products is that so many different legal mechanisms seem to be available to it. Even after enactment of the Copyright Act of 1976, which for the first time extended federal copyright protection to unpublished works, copyright is still—software excepted—largely utilized by those who commercially distribute their works in a manner that inevitably forecloses trade secret protection for the work (since publication discloses the contents of the work). Nor has patent protection previously been available to the intellectual processes embodied in a copyrighted writing.

[26]Samuelson and Glushko, supra note 22 (reporting the results of a survey of user interface designers).

[27]See, e.g., A. Clapes, Software, Copyright, and Competition (1989).

[28]Maier, Software protection—integrating patent, copyright, and trade secret law, 28 Idea 113 (1987).

What's Not Controversial

Although the main purpose of the discussion of current approaches is to give an overview of the principal intellectual property issues about which there is controversy in the technical and legal communities, it may be wise to begin with a recognition of a number of intellectual property issues as to which there is today no significant controversy. Describing only the aspects of the legal environment as to which controversies exist would risk creating a misimpression about the satisfaction many software developers and lawyers have with some aspects of intellectual property rights they now use to protect their and their clients' products.

One uncontroversial aspect of the current legal environment is the use of copyright to protect against exact or near-exact copying of program code. Another is the use of copyright to protect certain aspects of user interfaces, such as videogame graphics, that are easily identifiable as "expressive" in a traditional copyright sense. Also relatively uncontroversial is the use of copyright protection for low-level structural details of programs, such as the instruction-by-instruction sequence of the code.[29]

The use of trade secret protection for the source code of programs and other internally held documents concerning program design and the like is similarly uncontroversial. So too is the use of licensing agreements negotiated with individual customers under which trade secret software is made available to licensees when the number of licensees is relatively small and when there is a reasonable prospect of ensuring that licensees will take adequate measures to protect the secrecy of the software. Patent protection for industrial processes that have computer program elements, such as the rubber curing process in the *Diehr* case, is also uncontroversial.

Substantial controversies exist, however, about the application of copyright law to protect other aspects of software, about patent protection for other kinds of software innovations, about the enforceability of shrink-wrap licensing agreements, and about the manner in which the various forms of legal protection seemingly available to software developers interrelate in the protection of program elements (e.g., the extent to which copyright and trade secret protection can coexist in mass-marketed software).

---

[29]A "translation" of a program from one programming language to another in a related family of programming languages would likely be regarded as an infringement of the program copyright, even though there might be no literal similarities between the two programs. See, e.g., LaST Frontier Conference Report, supra note 19.

Controversies Arising From *Whelan* v. *Jaslow*

Because quite a number of the most contentious copyright issues arise from the *Whelan* v. *Jaslow* decision, this subsection focuses on that case. In the summer of 1986, the Third Circuit Court of Appeals affirmed a trial court decision in favor of Whelan Associates in its software copyright lawsuit against Jaslow Dental Laboratories.[30] Jaslow's program for managing dental lab business functions used some of the same data and file structures as Whelan's program (to which Jaslow had access), and five subroutines of Jaslow's program functioned very similarly to Whelan's. The trial court inferred that there were substantial similarities in the underlying structure of the two programs based largely on a comparison of similarities in the user interfaces of the two programs, even though user interface similarities were not the basis for the infringement claim. Jaslow's principal defense was that Whelan's copyright protected only against exact copying of program code, and since there were no literal similarities between the programs, no copyright infringement had occurred.

In its opinion on this appeal, the Third Circuit stated that copyright protection was available for the "structure, sequence, and organization" (SSO) of a program, not just the program code. (The court did not distinguish between high- and low-level structural features of a program.) The court analogized copyright protection for program SSO to the copyright protection available for such things as detailed plot sequences in novels. The court also emphasized that the coding of a program was a minor part of the cost of development of a program. The court expressed fear that if copyright protection was not accorded to SSO, there would be insufficient incentives to invest in the development of software.

The Third Circuit's *Whelan* decision also quoted with approval from that part of the trial court opinion stating that similarities in the manner in which programs functioned could serve as a basis for a finding of copyright infringement. Although recognizing that user interface similarities did not necessarily mean that two programs had similar underlying structures (thereby correcting an error the trial judge had made), the appellate court thought that user interface similarities might still be some evidence of underlying structural similarities. In conjunction with other evidence in the case, the Third Circuit decided that infringement had properly been found.

Although a number of controversies have arisen out of the *Whelan* opinion, the aspect of the opinion that has received the greatest attention is the test the court used for determining copyright infringement in computer

---

[30] *Whelan Associates, Inc.* v. *Jaslow Dental Laboratories, Inc.* 797 F.2d 1222 (3d Cir. 1986).

program cases. The "*Whelan* test" regards the general purpose or function of a program as its unprotectable "idea." All else about the program is, under the *Whelan* test, protectable "expression" unless there is only one or a very small number of ways to achieve the function (in which case idea and expression are said to be "merged," and what would otherwise be expression is treated as an idea). The sole defense this test contemplates for one who has copied anything more detailed than the general function of another program is that copying that detail was "necessary" to perform that program function. If there is in the marketplace another program that does the function differently, courts applying the *Whelan* test have generally been persuaded that the copying was unjustified and that what was taken must have been "expressive."

Although the *Whelan* test has been used in a number of subsequent cases, including the well-publicized *Lotus* v. *Paperback* case,[31] some judges have rejected it as inconsistent with copyright law and tradition, or have found ways to distinguish the *Whelan* case when employing its test would have resulted in a finding of infringement.[32]

Many commentators assert that the *Whelan* test interprets copyright protection too expansively.[33] Although the court in *Whelan* did not seem to realize it, the *Whelan* test would give much broader copyright protection to computer programs than has traditionally been given to novels and plays, which are among the artistic and fanciful works generally accorded a broader scope of protection than functional kinds of writings (of which programs would seem to be an example).[34] The *Whelan* test would forbid reuse of many things people in the field tend to regard as ideas.[35] Some commentators have suggested that because innovation in software tends to be of a more incremental character than in some other fields, and especially given the long duration of copyright protection, the *Whelan* interpretation of the scope of copyright is likely to substantially overprotect software.[36]

One lawyer-economist, Professor Peter Menell, has observed that the model of innovation used by the economists who did the study of software for CONTU is now considered to be an outmoded approach.[37] Those econo-

---

[31]*Lotus Development Corp.* v. *Paperback Software Int'l*, 740 F. Supp. 37 (D. Mass. 1990).

[32]See, e.g., *Plains Cotton Cooperative Assn.* v. *Goodpasture Service*, 807 F.2d 1256 (5th Cir. 1987) and *Computer Associates International, Inc.* v. *Altai, Inc.*, 23 U.S.P.Q. 2d 1241 (2d Cir. 1992).

[33]See, e.g., LaST Frontier Conference Report, supra note 19.

[34]Id.

[35]See, e.g., Samuelson and Glushko, supra note 22.

[36]Menell, An analysis of the scope of copyright protection for computer programs, 41 Stan. L. Rev. 1045 (1989).

[37]Remarks of Peter Menell, Annenberg Washington Program, Symposium on Intellectual Property Rights in Software, May 20-21, 1991.

mists focused on a model that considered what incentives would be needed for development of individual programs in isolation. Today, economists would consider what protection would be needed to foster innovation of a more cumulative and incremental kind, such as has largely typified the software field. In addition, the economists on whose work CONTU relied did not anticipate the networking potential of software and consequently did not study what provisions the law should make in response to this phenomenon. Menell has suggested that with the aid of their now more refined model of innovation, economists today might make somewhat different recommendations on software protection than they did in the late 1970s for CONTU.[38]

As a matter of copyright law, the principal problem with the *Whelan* test is its incompatibility with the copyright statute, the case law properly interpreting it, and traditional principles of copyright law. The copyright statute provides that not only ideas, but also processes, procedures, systems, and methods of operation, are unprotectable elements of copyrighted works.[39] This provision codifies some long-standing principles derived from U.S. copyright case law, such as the Supreme Court's century-old *Baker* v. *Selden* decision that ruled that a second author did not infringe a first author's copyright when he put into his own book substantially similar ledger sheets to those in the first author's book. The reason the Court gave for its ruling was that Selden's copyright did not give him exclusive rights to the bookkeeping system, but only to his explanation or description of it.[40] The ordering and arrangement of columns and headings on the ledger sheets were part of the system; to get exclusive rights in this, the Court said that Selden would have to get a patent.

The statutory exclusion from copyright protection for methods, processes, and the like was added to the copyright statute in part to ensure that the scope of copyright in computer programs would not be construed too broadly. Yet, in cases in which the *Whelan* test has been employed, the courts have tended to find the presence of protectable "expression" when they perceive there to be more than a couple of ways to perform some function, seeming not to realize that there may be more than one "method" or "system" or "process" for doing something, none of which is properly protected by copyright law. The *Whelan* test does not attempt to exclude

---

[38]Other speakers at the Annenberg Symposium referred to in note 37, including Lewis and Anne Wells Branscomb, expressed the view that the incremental nature of innovation in software was pertinent to the degree of intellectual property protection that should be available to software developers. See also Reichman, Computer programs as applied scientific know-how: Implications of copyright protection for commercialized university research, 42 Vand. L. Rev. 639 (1989).

[39]17 U.S.C. sec. 102(b) (1988).

[40]101 U.S. 99 (1879).

methods or processes from the scope of copyright protection, and its recognition of functionality as a limitation on the scope of copyright is triggered only when there are no alternative ways to perform program functions.

Whelan has been invoked by plaintiffs not only in cases involving similarities in the internal structural design features of programs, but also in many other kinds of cases. SSO can be construed to include internal interface specifications of a program, the layout of elements in a user interface, and the sequence of screen displays when program functions are executed, among other things. Even the manner in which a program functions can be said to be protectable by copyright law under *Whelan*. The case law on these issues and other software issues is in conflict, and resolution of these controversies cannot be expected very soon.

Traditionalist Versus Strong Protectionist View of What Copyright Law Does and Does Not Protect in Computer Programs

Traditional principles of copyright law, when applied to computer programs, would tend to yield only a "thin" scope of protection for them. Unquestionably, copyright protection would exist for the code of the program and the kinds of expressive displays generated when program instructions are executed, such as explanatory text and fanciful graphics, which are readily perceptible as traditional subject matters of copyright law. A traditionalist would regard copyright protection as not extending to functional elements of a program, whether at a high or low level of abstraction, or to the functional behavior that programs exhibit. Nor would copyright protection be available for the applied know-how embodied in programs, including program logic.[41] Copyright protection would also not be available for algorithms or other structural abstractions in software that are constituent elements of a process, method, or system embodied in a program.

Efficient ways of implementing a function would also not be protectable by copyright law under the traditionalist view, nor would aspects of software design that make the software easier to use (because this bears on program functionality). The traditionalist would also not regard making a limited number of copies of a program to study it and extract interface information or other ideas from the program as infringing conduct, because computer programs are a kind of work for which it is necessary to make a copy to "read" the text of the work.[42] Developing a program that incorporates interface information derived from decompilation would also, in the traditionalist view, be noninfringing conduct.

---

[41]Reichman, supra note 38.

[42]LaST Frontier Report, supra note 19.

If decompilation and the use of interface information derived from the study of decompiled code were to be infringing acts, the traditionalist would regard copyright as having been turned inside out, for instead of promoting the dissemination of knowledge as has been its traditional purpose, copyright law would become the principal means by which trade secrets would be maintained in widely distributed copyrighted works. Instead of protecting only expressive elements of programs, copyright would become like a patent: a means by which to get exclusive rights to the configuration of a machine—without meeting stringent patent standards or following the strict procedures required to obtain patent protection. This too would seem to turn copyright inside out.

Because interfaces, algorithms, logic, and functionalities of programs are aspects of programs that make them valuable, it is understandable that some of those who seek to maximize their financial returns on software investments have argued that "strong" copyright protection is or should be available for all valuable features of programs, either as part of program SSO or under the *Whelan* "there's-another-way-to-do-it" test.[43] Congress seems to have intended for copyright law to be interpreted as to programs on a case-by-case basis, and if courts determine that valuable features should be considered "expressive," the strong protectionists would applaud this common law evolution. If traditional concepts of copyright law and its purposes do not provide an adequate degree of protection for software innovation, they see it as natural that copyright should grow to provide it. Strong protectionists tend to regard traditionalists as sentimental Luddites who do not appreciate that what matters is for software to get the degree of protection it needs from the law so that the industry will thrive.

Although some cases, most notably the *Whelan* and *Lotus* decisions, have adopted the strong protectionist view, traditionalists will tend to regard these decisions as flawed and unlikely to be affirmed in the long run because they are inconsistent with the expressed legislative intent to have traditional principles of copyright law applied to software. Some copyright traditionalists favor patent protection for software innovations on the ground that the valuable functional elements of programs do need protection to create proper incentives for investing in software innovations, but that this protection should come from patent law, not from copyright law.

---

[43]See especially *Lotus Development Corp.* v. *Paperback Software Int'l*, 740 F. Supp. 37 (D. Mass. 1990); but see Brief Amicus Curiae of Copyright Law Professors, *Lotus Development Corp.* v. *Borland Int'l, Inc.*, Civ. No. 90-11662-K (filed Oct. 3, 1991).

Controversy Over "Software Patents"

Although some perceive patents as a way to protect valuable aspects of programs that cannot be protected by copyright law, those who argue for patents for software innovations do not rely on the "gap-filling" concern alone. As a legal matter, proponents of software patents point out that the patent statute makes new, nonobvious, and useful "processes" patentable. Programs themselves are processes; they also embody processes.[44] Computer hardware is clearly patentable, and it is a commonplace in the computing field that any tasks for which a program can be written can also be implemented in hardware. This too would seem to support the patentability of software.

Proponents also argue that protecting program innovations by patent law is consistent with the constitutional purpose of patent law, which is to promote progress in the "useful arts." Computer program innovations are technological in nature, which is said to make them part of the useful arts to which the Constitution refers. Proponents insist that patent law has the same potential for promoting progress in the software field as it has had for promoting progress in other technological fields. They regard attacks on patents for software innovations as reflective of the passing of the frontier in the software industry, a painful transition period for some, but one necessary if the industry is to have sufficient incentives to invest in software development.

Some within the software industry and the technical community, however, oppose patents for software innovations.[45] Opponents tend to make two kinds of arguments against software patents, often without distinguishing between them. One set of arguments questions the ability of the PTO to deal well with software patent applications. Another set raises more fundamental questions about software patents. Even assuming that the PTO could begin to do a good job at issuing software patents, some question whether

---

[44]See, e.g., Chisum, supra note 9. The PTO position is that it will not issue patents for computer programs as such. Notwithstanding that program instructions are just as much processes as the algorithms they implement, the PTO regards program instructions as running afoul of the "printed matter" rule. See Samuelson, supra note 8; but see also Chisum, supra note 9, for a critique of this aspect of the PTO's position. Because it will not issue patents for program instructions, the PTO insists there are no "software patents." The PTO says it issues patents only for computer program processes. See U.S. Patent and Trademark Office and U.S. Copyright Office, Patent-Copyright Overlap Study, prepared for the House Subcommittee on Intellectual Property and the Administration of Justice (May 1991) (referred to hereinafter as "overlap study").

[45]See, e.g., Garfinkel, Stallman, and Kapor, Why patents are bad for software, Issues in Science and Technology (Fall 1991).

innovation in the software field will be properly promoted if patents become widely available for software innovations. The main points of both sets of arguments are developed below.

Much of the discussion in the technical community has focused on "bad" software patents that have been issued by the PTO. Some patents are considered bad because the innovation was, unbeknownst to the PTO, already in the state of the art prior to the date of invention claimed in the patent. Others are considered bad because critics assert that the innovations they embody are too obvious to be deserving of patent protection. Still others are said to be bad because they are tantamount to a claim for performing a particular function by computer or to a claim for a law of nature, neither of which is regarded as patentable subject matter. Complaints abound that the PTO, after decades of not keeping up with developments in this field, is so far out of touch with what has been and is happening in the field as to be unable to make appropriate judgments on novelty and nonobviousness issues. Other complaints relate to the office's inadequate classification scheme for software and lack of examiners with suitable education and experience in computer science and related fields to make appropriate judgments on software patent issues.[46]

A somewhat different point is made by those who assert that the software industry has grown to its current size and prosperity without the aid of patents, which causes them to question the need for patents to promote innovation in this industry.[47] The highly exclusionary nature of patents (any use of the innovation without the patentee's permission is infringing) contrasts sharply with the tradition of independent reinvention in this field. The high expense associated with obtaining and enforcing patents raises concerns about the increased barriers to entry that may be created by the patenting of software innovations. Since much of the innovation in this industry has come from small firms, policies that inhibit entry by small firms may not promote innovation in this field in the long run. Similar questions arise as to whether patents will promote a proper degree of innovation in an incremental industry such as the software industry. It would be possible to undertake an economic study of conditions that have promoted and are promoting progress in the software industry to serve as a basis for a policy decision on software patents, but this has not been done to date.

Some computer scientists and mathematicians are also concerned about patents that have been issuing for algorithms,[48] which they regard as dis-

---

[46]See Office of Technology Assessment, Finding a Balance: Computer Software, Intellectual Property and the Challenge of Technological Change, 8-12 (May 1992).

[47]See, e.g., Kahin, supra note 10.

[48]Newell, The models are broken! The models are broken!, 47 U. Pitt. L. Rev. 1023 (1986).

coveries of fundamental truths that should not be owned by anyone. Because any use of a patented algorithm within the scope of the claims—whether by an academic or a commercial programmer, whether one knew of the patent or not—may be an infringement, some worry that research on algorithms will be slowed down by the issuance of algorithm patents. One mathematical society has recently issued a report opposing the patenting of algorithms.[49] Others, including Richard Stallman, have formed a League for Programming Freedom.

There is substantial case law to support the software patent opponent position, notwithstanding the PTO change in policy.[50] Three U.S. Supreme Court decisions have stated that computer program algorithms are unpatentable subject matter. Other case law affirms the unpatentability of processes that involve the manipulation of information rather than the transformation of matter from one physical state to another.

One other concern worth mentioning if both patents and copyrights are used to protect computer program innovations is whether a meaningful boundary line can be drawn between the patent and copyright domains as regards software.[51] A joint report of the U.S. PTO and the Copyright Office optimistically concludes that no significant problems will arise from the coexistence of these two forms of protection for software because copyright law will only protect program "expression" whereas patent law will only protect program "processes."[52]

Notwithstanding this report, I continue to be concerned with the patent/copyright interface because of the expansive interpretations some cases, particularly *Whelan*, have given to the scope of copyright protection for programs. This prefigures a significant overlap of copyright and patent law as to software innovations. This overlap would undermine important economic and public policy goals of the patent system, which generally leaves in the public domain those innovations not novel or nonobvious enough to be patented. Mere "originality" in a copyright sense is not enough to make an innovation in the useful arts protectable under U.S. law.[53]

A concrete example may help illustrate this concern. Some patent lawyers report getting patents on data structures for computer programs.

---

[49]Mathematical Programming Society, Report of the Committee on Algorithms and the Law, 33 Optima 2 (June 1991).

[50]See Samuelson, supra note 8.

[51]In no other domain than software can one infringe a patent by writing a copyrightable text. Algorithm patents, depending on how the claims are written, may even be infringed by reading or writing articles about them.

[52]Overlap study, supra note 44.

[53]Other nations have copyright-like industrial design protection laws. The United States has such a law only as to semiconductor chip designs.

The *Whelan* decision relied in part on similarities in data structures to prove copyright infringement. Are data structures "expressive" or "useful"? When one wants to protect a data structure of a program by copyright, does one merely call it part of the SSO of the program, whereas if one wants to patent it, one calls it a method (i.e., a process) of organizing data for accomplishing certain results? What if anything does copyright's exclusion from protection of processes embodied in copyrighted works mean as applied to data structures? No clear answer to these questions emerges from the case law.

Nature of Computer Programs and Exploration of a Modified Copyright Approach

It may be that the deeper problem is that computer programs, by their very nature, challenge or contradict some fundamental assumptions of the existing intellectual property regimes. Underlying the existing regimes of copyright and patent law are some deeply embedded assumptions about the very different nature of two kinds of innovations that are thought to need very different kinds of protection owing to some important differences in the economic consequences of their protection.[54]

In the United States, these assumptions derive largely from the U.S. Constitution, which specifically empowers Congress "to promote the progress of science [i.e., knowledge] and useful arts [i.e., technology], by securing for limited times to authors and inventors the exclusive right to their respective writings and discoveries."[55] This clause has historically been parsed as two separate clauses packaged together for convenience: one giving Congress power to enact laws aimed at promoting the progress of knowledge by giving authors exclusive rights in their writings, and the other giving Congress power to promote technological progress by giving inventors exclusive rights in their technological discoveries. Copyright law implements the first power, and patent law the second.

Owing partly to the distinctions between writings and machines, which the constitutional clause itself set up, copyright law has excluded machines

[54]The interchangeability of software and hardware is a commonplace notion in the computing field. There are often engineering and other reasons why one might prefer to implement certain functionalities in one form or another. From the standpoint of the technical community, it seems to make no sense to have intellectual property rules that are dramatically different depending on which choice is made for the form of implementation. If a hardware implementation is chosen, no copyright protection will be available and patent protection will be available only for inventive new components. If a software implementation is chosen, copyright protection arises automatically, and patents, at least under the existing PTO practice, would seem also available for inventive components.

[55]Article I, sec. 8, cl. 8.

and other technological subject matters from its domain.[56] Even when described in a copyrighted book, an innovation in the useful arts was considered beyond the scope of copyright protection. The Supreme Court's *Baker* v. *Selden* decision reflects this view of the constitutional allocation. Similarly, patent law has historically excluded printed matter (i.e., the contents of writings) from its domain, notwithstanding the fact that printed matter may be a product of a manufacturing process.[57] Also excluded from the patent domain have been methods of organizing, displaying, and manipulating information (i.e., processes that might be embodied in writings, for example mathematical formulas), notwithstanding the fact that "processes" are named in the statute as patentable subject matter. They were not, however, perceived to be "in the useful arts" within the meaning of the constitutional clause.

The constitutional clause has been understood as both a grant of power and a limitation on power. Congress cannot, for example, grant perpetual patent rights to inventors, for that would violate the "limited times" provision of the Constitution. Courts have also sometimes ruled that Congress cannot, under this clause, grant exclusive rights to anyone but authors and inventors. In the late nineteenth century, the Supreme Court struck down the first federal trademark statute on the ground that Congress did not have power to grant rights under this clause to owners of trademarks who were neither "authors" nor "inventors."[58] A similar view was expressed in last year's *Feist Publications* v. *Rural Telephone Services* decision by the Supreme Court, which repeatedly stated that Congress could not constitutionally protect the white pages of telephone books through copyright law because to be an "author" within the meaning of the Constitution required some creativity in expression that white pages lacked.[59]

Still other Supreme Court decisions have suggested that Congress could not constitutionally grant exclusive rights to innovators in the useful arts who were not true "inventors."[60] Certain economic assumptions are connected with this view, including the assumption that more modest innovations in the useful arts (the work of a mere mechanic) will be forthcoming without the grant of the exclusive rights of a patent, but that the incentives of patent rights are necessary to make people invest in making significant technological advances and share the results of their work with the public instead of keeping them secret.

---

[56] Samuelson, supra note 2.

[57] Samuelson, supra note 8.

[58] The Trademark Cases, 100 U.S. 82 (1879).

[59] *Feist Publications, Inc.* v. *Rural Telephone Service Co.*, 111 S. Ct. 1282 (1991).

[60] *Graham* v. *John Deere Co.*, 383 U.S. 1 (1965).

One reason the United States does not have a copyright-like form of protection for industrial designs, as do many other countries, is because of lingering questions about the constitutionality of such legislation. In addition, concerns exist that the economic consequences of protecting uninventive technological advances will be harmful. So powerful are the prevailing patent and copyright paradigms that when Congress was in the process of considering the adoption of a copyright-like form of intellectual property protection for semiconductor chip designs, there was considerable debate about whether Congress had constitutional power to enact such a law. It finally decided it did have such power under the commerce clause, but even then was not certain.

As this discussion reveals, the U.S. intellectual property law has long assumed that something is either a writing (in which case it is protectable, if at all, by copyright law) or a machine (in which case it is protectable, if at all, by patent law), but cannot be both at the same time. However, as Professor Randall Davis has so concisely said, software is "a machine whose medium of construction happens to be text."[61] Davis regards the act of creating computer programs as inevitably one of both authorship and invention. There may be little or nothing about a computer program that is not, at base, functional in nature, and nothing about it that does not have roots in the text. Because of this, it will inevitably be difficult to draw meaningful boundaries for patents and copyrights as applied to computer programs.

Another aspect of computer programs that challenges the assumptions of existing intellectual property systems is reflected in another of Professor Davis's observations, namely, that "programs are not only texts; they also behave."[62] Much of the dynamic behavior of computer programs is highly functional in nature. If one followed traditional copyright principles, this functional behavior—no matter how valuable it might be—would be considered outside the scope of copyright law.[63] Although the functionality of program behavior might seem at first glance to mean that patent protection would be the obvious form of legal protection for it, as a practical matter, drafting patent claims that would adequately capture program behavior as an invention is infeasible. There are at least two reasons for this: it is partly because programs are able to exhibit such a large number and variety of states that claims could not reasonably cover them, and partly because of

[61]See R. Davis, Intellectual property and software: The assumptions are broken, in Proceedings of WIPO's Worldwide Symposium on Legal Aspects of Artificial Intelligence, Stanford University (March 1991).

[62]Id.

[63]See *Computer Associates, Inc.* v. *Altai, Inc.*, 23 U.S. P.Q. 2d (BNA) 1241 (behavior of programs not protectable by copyright law).

the "gestalt"-like character of program behavior, something that makes a more copyright-like approach desirable.

Some legal scholars have argued that because of their hybrid character as both writings and machines, computer programs need a somewhat different legal treatment than either traditional patent or copyright law would provide.[64] They have warned of distortions in the existing legal systems likely to occur if one attempts to integrate such a hybrid into the traditional systems as if it were no different from the traditional subject matters of these systems.[65] Even if the copyright and patent laws could be made to perform their tasks with greater predictability than is currently the case, these authors warn that such regimes may not provide the kind of protection that software innovators really need, for most computer programs will be legally obvious for patent purposes, and programs are, over time, likely to be assimilated within copyright in a manner similar to that given to "factual" and "functional" literary works that have only "thin" protection against piracy.[66]

Professor Reichman has reported on the recurrent oscillations between states of under- and overprotection when legal systems have tried to cope with another kind of legal hybrid, namely, industrial designs (sometimes referred to as "industrial art"). Much the same pattern seems to be emerging in regard to computer programs, which are, in effect, "industrial literature."[67]

The larger problems these hybrids present is that of protecting valuable forms of applied know-how embodied in incremental innovation that cannot successfully be maintained as trade secrets:

> [M]uch of today's most advanced technology enjoys a less favorable competitive position than that of conventional machinery because the unpatentable, intangible know-how responsible for its commercial value becomes embodied in products that are distributed on the open market. A product of the new technologies, such as a computer program, an integrated circuit

[64]Those who regard software as a "literary work," such as the authors of the "Silicon Epics and Binary Bards" article published a couple of years ago, tend to be proponents of the business-oriented approach to interpreting copyright law to programs, rather than legal scholars. Compare LaST Frontier Report, supra note 19 (describing computer programs as functional works), and Clapes, Lynch, and Steinberg, Silicon epics and binary bards, 34 UCLA L. Rev. 1493 (1987). The latter three authors are attorneys who worked for IBM in some software copyright litigation.

[65]Reichman, supra note 38; Samuelson, supra notes 2 and 8.

[66]*Feist Publications, Inc.* v. *Rural Telephone Service Co.*, 111 S. Ct. 1282 (1991); Reichman, Goldstein on copyright law: A realist's approach to a technological age, 43 Stan. L. Rev. 943, 966-976 (1991).

[67]Reichman, Design protection and the new technologies: The United States experience in a transnational perspective, 1991 Industrial Property 220, 269, 271 (1991).

> design, or even a biogenetically altered organism may thus *bear its know-how on its face*, a condition that renders it as vulnerable to rapid appropriation by second-comers as any published literary or artistic work.
>
> From this perspective, a major problem with the kinds of innovative know-how underlying important new technologies is that they do not lend themselves to secrecy even when they represent the fruit of enormous investment in research and development. Because third parties can rapidly duplicate the embodied information and offer virtually the same products at lower prices than those of the originators, there is no secure interval of lead time in which to recuperate the originators' initial investment or their losses from unsuccessful essays, not to mention the goal of turning a profit.[68]
>
> From a behavioral standpoint, investors in applied scientific know-how find the copyright paradigm attractive because of its inherent disposition to supply artificial lead time to all comers without regard to innovative merit and without requiring originators to preselect the products that are most worthy of protection.[69]

Full copyright protection, however, with its broad notion of equivalents geared to derivative expressions of an author's personality is likely to disrupt the workings of the competitive market for industrial products. For this and other reasons, Professor Reichman argues that a modified copyright approach to the protection of computer programs (and other legal hybrids) would be a preferable framework for protecting the applied know-how they embody than either the patent or the copyright regime would presently provide. Similar arguments can be made for a modified form of copyright protection for the dynamic behavior of programs. A modified copyright approach might involve a short duration of protection for original valuable functional components of programs. It could be framed to supplement full copyright protection for program code and traditionally expressive elements of text and graphics displayed when programs execute, features of software that do not present the same dangers of competitive disruption from full copyright protection.

The United States is, in large measure, already undergoing the development of a sui generis law for protection of computer software through case-by-case decisions in copyright lawsuits. Devising a modified copyright approach to protecting certain valuable components that are not suitably protected under the current copyright regime would have the advantage of allowing a conception of the software protection problem as a whole, rather than on a piecemeal basis as occurs in case-by-case litigation in which the

---

[68]Id. at 269.
[69]Id. at 271.

skills of certain attorneys and certain facts may end up causing the law to develop in a skewed manner.[70]

There are, however, a number of reasons said to weigh against sui generis legislation for software, among them the international consensus that has developed on the use of copyright law to protect software and the trend toward broader use of patents for software innovations. Some also question whether Congress would be able to devise a more appropriate sui generis system for protecting software than that currently provided by copyright. Some are also opposed to sui generis legislation for new technology products such as semiconductor chips and software on the ground that new intellectual property regimes will make intellectual property law more complicated, confusing, and uncertain.

Although there are many today who ardently oppose sui generis legislation for computer programs, these same people may well become among the most ardent proponents of such legislation if the U.S. Supreme Court, for example, construes the scope of copyright protection for programs to be quite thin, and reiterates its rulings in *Benson*, *Flook*, and *Diehr* that patent protection is unavailable for algorithms and other information processes embodied in software.

INTERNATIONAL PERSPECTIVES

After adopting copyright as a form of legal protection for computer programs, the United States campaigned vigorously around the world to persuade other nations to protect computer programs by copyright law as well. These efforts have been largely successful. Although copyright is now an international norm for the protection of computer software, the fine details of what copyright protection for software means, apart from protection against exact copying of program code, remain somewhat unclear in other nations, just as in the United States.

Other industrialized nations have also tended to follow the U.S. lead concerning the protection of computer program-related inventions by patent

[70]For example, some questionable conduct by lawyers for Atari Games in obtaining a copy of the source code of a Nintendo program that was on deposit with the Copyright Office (which was used by Atari Games' engineers to figure out how to make a compatible program) strongly influenced the trial judge who ruled that Nintendo had made a strong enough showing of copyright infringement to get a preliminary injunction against Atari Games' distribution of Nintendo-compatible cartridges. See *Atari Games Corp.* v. *Nintendo of America, Inc.*, 18 U.S.P.Q.2d (BNA) 1935 (N.D. Cal. 1991). The future of interoperable software under U.S. copyright law may well hang in the balance of the resolution of this case.

law.[71] Some countries that in the early 1960s were receptive to the patenting of software innovations became less receptive after the *Gottschalk* v. *Benson* decision by the U.S. Supreme Court. Some even adopted legislation excluding computer programs from patent protection. More recently, these countries are beginning to issue more program-related patents, once again paralleling U.S. experience, although as in the United States, the standards for patentability of program-related inventions are somewhat unclear.[72] If the United States and Japan continue to issue a large number of computer program-related patents, it seems quite likely other nations will follow suit.

There has been strong pressure in recent years to include relatively specific provisions about intellectual property issues (including those affecting computer programs) as part of the international trade issues within the framework of the General Agreement on Tariffs and Trade (GATT).[73] For a time, the United States was a strong supporter of this approach to resolution of disharmonies among nations on intellectual property issues affecting software. The impetus for this seems to have slackened, however, after U.S. negotiators became aware of a lesser degree of consensus among U.S. software developers on certain key issues than they had thought was the case. Since the adoption of its directive on software copyright law, the European Community (EC) has begun pressing for international adoption of its position on a number of important software issues, including its copyright rule on decompilation of program code.

There is a clear need, given the international nature of the market for software, for a substantial international consensus on software protection issues. However, because there are so many hotly contested issues concerning the extent of copyright and the availability of patent protection for computer programs yet to be resolved, it may be premature to include very specific rules on these subjects in the GATT framework.

---

[71]See, e.g., Dreier, Patent protection for computer programs in Europe, and Ozawa, Patent protection for program-related inventions in Japan, Proceedings of the SOFTIC Third International Symposium on Legal Protection of Software (1991) (cited hereinafter as "SOFTIC proceedings").

[72]H. Hanneman, The Patentability of Computer Software (1985). Many European nations and the European Patent Convention have rules against the patenting of computer programs per se (as does the PTO) but will issue patents for processes that utilize computer programs as long as the processes achieve a "technical effect."

[73]See generally Trade-related aspects of intellectual property, 22 Vand. J. Transnat'l L. 223-384 (1989).

Europe

Prior to the adoption of the 1991 European Directive on the Protection of Computer Programs, there was general acceptance in Europe of copyright as a form of legal protection for computer programs. A number of nations had interpreted existing copyright statutes as covering programs. Others took legislative action to extend copyright protection to software. There was, however, some divergence in approach among the member nations of the EC in the interpretation of copyright law to computer software.[74]

France, for example, although protecting programs under its copyright law, put software in the same category as industrial art, a category of work that is generally protected in Europe for 25 years instead of the life plus 50-year term that is the norm for literary and other artistic works. German courts concluded that to satisfy the "originality" standard of its copyright law, the author of a program needed to demonstrate that the program was the result of more than an average programmer's skill, a seemingly patentlike standard. In addition, Switzerland (a non-EC member but European nonetheless) nearly adopted an approach that treated both semiconductor chip designs and computer programs under a new copyright-like law.

Because of these differences and because it was apparent that computer programs would become an increasingly important item of commerce in the European Community, the EC undertook in the late 1980s to develop a policy concerning intellectual property protection for computer programs to which member nations should harmonize their laws. There was some support within the EC for creating a new law for the protection of software, but the directorate favoring a copyright approach won this internal struggle over what form of protection was appropriate for software.

In December 1988 the EC issued a draft directive on copyright protection for computer programs. This directive was intended to spell out in considerable detail in what respects member states should have uniform rules on copyright protection for programs. (The European civil law tradition generally prefers specificity in statutory formulations, in contrast with the U.S. common law tradition, which often prefers case-by-case adjudication of disputes as a way to fill in the details of a legal protection scheme.)

The draft directive on computer programs was the subject of intense debate within the European Community, as well as the object of some intense lobbying by major U.S. firms who were concerned about a number of issues, but particularly about what rule would be adopted concerning decompilation of program code and protection of the internal interfaces of

---

[74] See, e.g., Verstrynge, Protecting intellectual property rights within the new Pan-European framework: Computer software, SOFTIC proceedings, supra note 69.

programs. Some U.S. firms, among them IBM Corp., strongly opposed any provision that would allow decompilation of program code and sought to have interfaces protected; other U.S. firms, such as Sun Microsystems, sought a rule that would permit decompilation and would deny protection to internal interfaces.[75]

The final EC directive published in 1991 endorses the view that computer programs should be protected under member states' copyright laws as literary works and given at least 50 years of protection against unauthorized copying.[76] It permits decompilation of program code only if and to the extent necessary to obtain information to create an interoperable program. The inclusion in another program of information necessary to achieve interoperability seems, under the final directive, to be lawful.

The final EC directive states that "ideas" and "principles" embodied in programs are not protectable by copyright, but does not provide examples of what these terms might mean. The directive contains no exclusion from protection of such things as processes, procedures, methods of operation, and systems, as the U.S. statute provides. Nor does it clearly exclude protection of algorithms, interfaces, and program logic, as an earlier draft would have done. Rather, the final directive indicates that to the extent algorithms, logic, and interfaces are ideas, they are unprotectable by copyright law. In this regard, the directive seems, quite uncharacteristically for its civil law tradition, to leave much detail about how copyright law will be applied to programs to be resolved by litigation.

Having just finished the process of debating the EC directive about copyright protection of computer programs, intellectual property specialists in the EC have no interest in debating the merits of any sui generis approach to software protection, even though the only issue the EC directive really resolved may have been that of interoperability. Member states will likely have to address another controversial issue—whether or to what extent user interests in standardization of user interfaces should limit the scope of copyright

[75]IBM was among the U.S. firms that joined the Software Action Group for Europe, which lobbied during the time the EC directive was pending. This group lobbied against decompilation and for the protection of interfaces. Sun Microsystems was among the American firms that were members of the European Committee for Interoperable Systems that lobbied in favor of a decompilation privilege and against protection of interfaces. Sun Microsystems is among the founding members of the American Committee for Interoperable Systems which recently filed an amicus brief in *Computer Associates, Inc.* v. *Altai* litigation in support of Altai's position on the copyright issues; the district court's decision in favor of Altai is on appeal to the Second Circuit Court of Appeals.

[76]For programs authored by individuals, the EC directive provides that protection should last for the life of the author plus 50 years. If member states adopt a "work for hire" provision giving employers ownership of software developed by employees, the firm is to be given 50 years of protection from the time the program is first made available to the public.

protection for programs—as they act on yet another EC directive, one that aims to standardize user interfaces of computer programs. Some U.S. firms may perceive this latter directive as an effort to appropriate valuable U.S. product features.

Japan

Japan was the first major industrialized nation to consider adoption of a sui generis approach to the protection of computer programs.[77] Its Ministry of International Trade and Industry (MITI) published a proposal that would have given 15 years of protection against unauthorized copying to computer programs that could meet a copyright-like originality standard under a copyright-like registration regime. MITI attempted to justify its proposed different treatment for computer programs as one appropriate to the different character of programs, compared with traditional copyrighted works.[78] The new legal framework was said to respond and be tailored to the special character of programs. American firms, however, viewed the MITI proposal, particularly its compulsory license provisions, as an effort by the Japanese to appropriate the valuable products of the U.S. software industry. Partly as a result of U.S. pressure, the MITI proposal was rejected by the Japanese government, and the alternative copyright proposal made by the ministry with jurisdiction over copyright law was adopted.

Notwithstanding their inclusion in copyright law, computer programs are a special category of protected work under Japanese law. Limiting the scope of copyright protection for programs is a provision indicating that program languages, rules, and algorithms are not protected by copyright law.[79] Japanese case law under this copyright statute has proceeded along lines similar to U.S. case law, with regard to exact and near-exact copying of program code and graphical aspects of videogame programs,[80] but there have been some Japanese court decisions interpreting the exclusion from protection provisions in a manner seemingly at odds with some U.S. decisions.

---

[77]See Karjala, Lessons from the computer software protection debate in Japan, 1984 Ariz. St. L. J. 53 (1984).

[78]MITI, for example, thought that "moral rights" were inappropriate for programs because of their technical nature; MITI was also concerned about the market power that might be conferred as a result of the long term of copyright for what was essentially a functional work. Id.

[79]Karjala, Copyright protection of computer software in the United States and Japan: Part II [1991] E.I.P.R. 231 (1991). Among the other special provisions applicable to programs are ones limiting "moral rights" protection to authors of copyrighted programs, making rules about programs written in an employment setting, and giving users certain modification rights in programs. Id.

[80]Id. at 232.

The Tokyo High Court, for example, has opined that the processing flow of a program (an aspect of a program said to be protectable by U.S. law in the *Whelan* case) is an algorithm within the meaning of the copyright limitation provision.[81] Another seems to bear out Professor Karjala's prediction that Japanese courts would interpret the programming language limitation to permit firms to make compatible software.[82] There is one Japanese decision that can be read to prohibit reverse engineering of program code, but because this case involved not only disassembly of program code but also distribution of a clearly infringing program, the legality of intermediate copying to discern such things as interface information is unclear in Japan.[83]

Other Nations

The United States has been pressing a number of nations to give "proper respect" to U.S. intellectual property products, including computer programs. In some cases, as in its dealings with the People's Republic of China, the United States has been pressing for new legislation to protect software under copyright law. In some cases, as in its dealings with Thailand, the United States has been pressing for more vigorous enforcement of intellectual property laws as they affect U.S. intellectual property products. In other cases, as in its dealings with Brazil, the United States pressed for repeal of sui generis legislation that disadvantaged U.S. software producers, compared with Brazilian developers. The United States has achieved some success in these efforts. Despite these successes, piracy of U.S.-produced software and other intellectual property products remains a substantial source of concern.

FUTURE CHALLENGES

Many of the challenges posed by use of existing intellectual property laws to protect computer programs have been discussed in previous sections. This may, however, only map the landscape of legal issues of widespread concern today. Below are some suggestions about issues as to which computer programs may present legal difficulties in the future.

---

[81]Id.

[82]Id. at 232-237.

[83]Id. at 235. It is worth noting, however, that Japanese copyright law does not have a fair use provision; but see Durney, Reverse engineering under Japanese law, IP ASIA, 2-6 (March 15, 1990).

Advanced Software Systems

It has thus far been exceedingly difficult for the legal system to resolve even relatively simple disputes about software intellectual property rights, such as those involved in the *Lotus* v. *Paperback Software* case. This does not bode well for how the courts are likely to deal with more complex problems presented by more complex software in future cases. The difficulties arise partly from the lack of familiarity of judges with the technical nature of computers and software, and partly from the lack of close analogies within the body of copyright precedents from which resolutions of software issues might be drawn. The more complex the software, the greater is the likelihood that specially trained judges will be needed to resolve intellectual property disputes about the software. Some advanced software systems are also likely to be sufficiently different from traditional kinds of copyrighted works that the analogical distance between the precedents and a software innovation may make it difficult to predict how copyright law should be applied to it. What copyright protection should be available, for example, to a user interface that responds to verbal commands, gestures, or movements of eyeballs?

Digital Media

The digital medium itself may require adaptation of the models underlying existing intellectual property systems.[84] Copyright law is built largely on the assumption that authors and publishers can control the manufacture and distribution of copies of protected works emanating from a central source. The ease with which digital works can be copied, redistributed, and used by multiple users, as well as the compactness and relative invisibility of works in digital form, have already created substantial incentives for developers of digital media products to focus their commercialization efforts on controlling the uses of digital works, rather than on the distribution of copies, as has more commonly been the rule in copyright industries.

Rules designed for controlling the production and distribution of copies may be difficult to adapt to a system in which uses need to be controlled. Some digital library and hypertext publishing systems seem to be designed to bypass copyright law (and its public policy safeguards, such as the fair use rule) and establish norms of use through restrictive access licensing

---

[84]Samuelson, Digital media and the changing face of intellectual property law, 20 Rutg. Comp. and Techn. L. J. 323 (1990).

agreements.[85] Whether the law will eventually be used to regulate conditions imposed on access to these systems, as it has regulated access to such communication media as broadcasting, remains to be seen. However, the increasing convergence of intellectual property policy, broadcast and telecommunications policy, and other aspects of information policy seems inevitable.

Networks

There are already millions of people connected to networks of computers, who are thereby enabled to communicate with one another with relative ease, speed, and reliability. Plans are afoot to add millions more and to allow a wide variety of information services to those connected to the networks, some of which are commercial and some of which are noncommercial in nature. Because networks of this type and scope are a new phenomenon, it would seem quite likely that some new intellectual property issues will arise as the use of computer networks expands. The more commercial the uses of the networks, the more likely intellectual property disputes are to occur.

More of the content distributed over computer networks is copyrighted than its distributors seem to realize, but even as to content that has been recognized as copyrighted, there is a widespread belief among those who communicate over the net that at least noncommercial distributions of content—no matter the number of recipients—are "fair uses" of the content. Some lawyers would agree with this; others would not. Those responsible for the maintenance of the network may need to be concerned about potential liability until this issue is resolved.

A different set of problems may arise when commercial uses are made of content distributed over the net. Here the most likely disputes are those concerning how broad a scope of derivative work rights copyright owners should have. Some owners of copyrights can be expected to resist allowing anyone but themselves (or those licensed by them) to derive any financial benefit from creating a product or service that is built upon the value of their underlying work. Yet value-added services may be highly desirable to consumers, and the ability of outsiders to offer these products and services may spur beneficial competition. At the moment, the case law generally regards a copyright owner's derivative work right as infringed only if a recognizable block of expression is incorporated into another work.[86] How-

---

[85]Samuelson and Glushko, Intellectual property rights in digital library and hypertext publishing systems: An analysis of Xanadu, Proceedings of Hypertext '91 39 (1991).

[86]See *Lewis Galoob Toys, Inc.* v. *Nintendo of America*, 780 F.Supp. 1283 (N.D. Cal. 1991) (game enhancement device not an infringing derivative work).

ever, the ability of software developers to provide value-added products and services that derive value from the underlying work without copying expression from it may lead some copyright owners to seek to extend the scope of derivative work rights.

Patents and Information Infrastructure of the Future

If patents are issued for all manner of software innovations, they are likely to play an important role in the development of the information infrastructure of the future. Patents have already been issued for hypertext navigation systems, for such things as latent semantic indexing algorithms, and for other software innovations that might be used in the construction of a new information infrastructure. Although it is easy to develop a list of the possible pros and cons of patent protection in this domain, as in the more general debate about software patents, it is worth noting that patents have not played a significant role in the information infrastructure of the past or of the present. How patents would affect the development of the new information infrastructure has not been given the study this subject may deserve.

Conflicts Between Information Haves and Have-Nots on an International Scale

When the United States was a developing nation and a net importer of intellectual property products, it did not respect copyright interests of any authors but its own. Charles Dickens may have made some money from the U.S. tours at which he spoke at public meetings, but he never made a dime from the publication of his works in the United States. Now that the United States is a developed nation and a net exporter of intellectual property products, its perspective on the rights of developing nations to determine for themselves what intellectual property rights to accord to the products of firms of the United States and other developed nations has changed. Given the greater importance nowadays of intellectual property products, both to the United States and to the world economy, it is foreseeable that there will be many occasions on which developed and developing nations will have disagreements on intellectual property issues.

The United States will face a considerable challenge in persuading other nations to subscribe to the same detailed rules that it has for dealing with intellectual property issues affecting computer programs. It may be easier for the United States to deter outright "piracy" (unauthorized copying of the whole or substantially the whole of copyrighted works) of U.S. intellectual property products than to convince other nations that they must adopt the same rules as the United States has for protecting software.

It is also well for U.S. policymakers and U.S. firms to contemplate the possibility that U.S. firms may not always have the leading position in the world market for software products that they enjoy today. When pushing for very "strong" intellectual property protection for software today in the expectation that this will help to preserve the U.S. advantage in the world market, U.S. policymakers should be careful not to push for adoption of rules today that may substantially disadvantage them in the world market of the future if, for reasons not foreseen today, the United States loses the lead it currently enjoys in the software market.

13

Biotechnology Case Study

GEORGE B. RATHMANN

I want to describe a bit of the history of the biotechnology field to give you a strong sense of the importance of this field, not just in itself but as a prelude to a new technology as it develops over the next century. I then relate that history to some questions that have been raised and finally relate my conclusions with respect to biotechnology to the objectives of the conference.

As rocky as the road for biotechnology was in the United States, what we see coming up on the world scene is much more difficult, much more serious. We desperately need a legal system to solve the problems, and it is our hope that there are ways of dealing with these issues.

The biotech era really dawned when Watson and Crick defined the structure of deoxyribonucleic acid (DNA). As with many world-shattering discoveries, this was simple and concise—a publication of one page outlining the structure of DNA (*Nature*, April 25, 1953, p. 737). They also had the vision to say it would affect not only how we looked at deoxyribonucleic acid, but how we looked at life itself and our ability to understand living systems. There would be products, there would be opportunities, and there would be new insights that would be most important. All that was recognized in a one-page article.

As important and earth shaking as that was, from the standpoint of the commercialization of biotechnology, something nearly as important occurred on June 17, 1980, when the Supreme Court ruled that live organisms could be patented. It was well recognized as important at the time, but I think few people realized how important it was for launching the commercialization of biotechnology.

In that patent, Dr. Ananda Chakrabarty, who was at G.E. at the time, claimed an organism that would digest oil. The invention was never commercialized, but it told the world that this field was going to be important and there were going to be commercial opportunities. An investment in trying to understand the biochemistry of life would pay off in the sense that the intellectual property could be protected. Within four months (October 14, 1980), the biotechnology company Genentech went public and jolted Wall Street with a rise in its stock price from $35 to $71 1/4. So it is clear that as of that date, biotechnology assumed increasing commercial importance.

At that time, in October 1980, I was looking at the opportunity to start a biotech company called Amgen and we were putting out a document that we hoped would raise $15 million. Partly because of Genentech's success, we were able to raise $19 million—with only a scientific advisory board, one employee, and promises for two future hires. So it certainly had a profound effect on whether Amgen would ever be. As a matter of fact, within a year, Amgen, Genetics Institute, Immunex, Genetics Systems, Chiron, and many others companies were formed. Within two years, more than 100 companies were formed as this era was launched.

Now, the *Chakrabarty* decision made it look simple: life forms were patentable. Genentech, Cetus and many others afterwards launched public offerings, recognizing the commercial potential that biotechnology would lead to new discoveries of valuable intellectual property, which could be protected by patents. In reality, it was not quite that simple and the launchings were not that consistent.

Venture capital funds vacillated quite a bit, although after 1980 there was a very substantial influx of venture capital (Figure 13-1). There were periods when it went down, and periods when it went up. Although these look like gigantic numbers, remember it takes about a quarter of a billion dollars to bring a pharmaceutical product to market. It probably takes more than that to commercialize something important in agriculture, food, or other areas. So this flow of venture capital was actually inadequate to keep it going. Of course, the public made the difference, but it can be seen that this was not exactly a consistent, reliable source of funds, either.

If we smooth everything out, the market value of biotechnology stocks moved dramatically from 1980, when it was literally zero, to 1991, when it was more than $35 billion (Figure 13-2). Those of us in the industry saw some very serious bumps in that curve. In 1987 some biotech companies lost 30-40 percent of the value of the company in a matter of a few days. When you finally smooth everything out, it looks a lot simpler and surer than it felt.

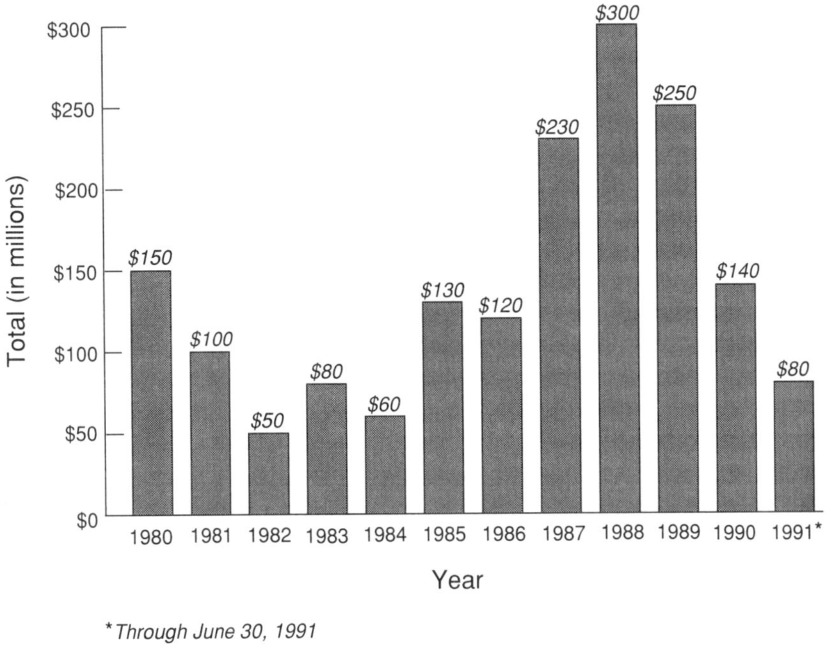

FIGURE 13-1 Venture Capital Disbursements in Biotechnology
Source: Venture Economics and Ernst and Young

Figure 13-3 shows the amount of capital raised through public stock offerings. In 1991, more money was raised in six months than for many years, and as a matter of fact, when the total figures came in for the year they exceeded $4 billion—equal to all the money that had been raised in the previous years since the launching of commercial biotechnology. Of course, the big news is $550 million in initial public offerings. Those are new companies whose survival may mean wonderful improvements to our lives around the world. At the same time they will be facing some of the rocky roads that the earlier companies faced. So we can see that it is not a steady, easy trip.

Product sales in the industry today have reached about $8 billion and are expected to reach $20 billion by the year 2000. That may be a very conservative figure. The drug industry worldwide by that time will be well over $200 billion, and biotechnology is contributing roughly half of the most important products today. By the time the year 2000 comes around, biotechnology-derived products could be even more important. Of course there should be many other parts of the biotech industry that are commercially interesting by that time.

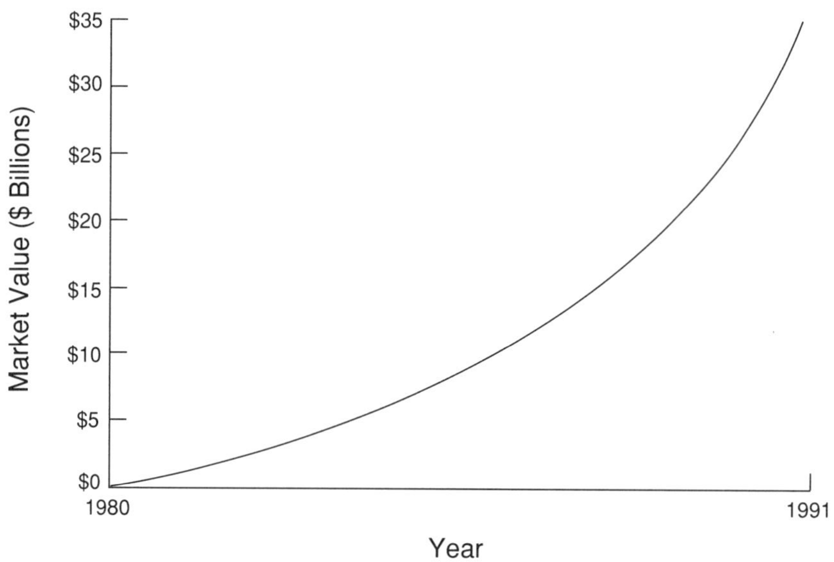

FIGURE 13-2 Market Value of Biotechnology Stocks

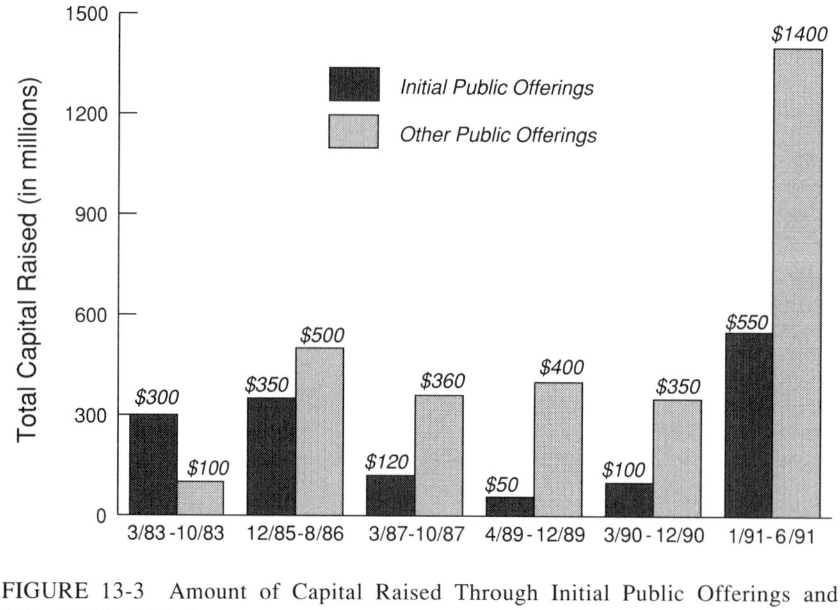

FIGURE 13-3 Amount of Capital Raised Through Initial Public Offerings and Other Public Offerings
Source: Paine Webber and Ernst and Young

So we are looking at something of great importance to the economy of the country and to international trade, which is discussed below.

I was asked by the National Research Council to address several questions. The first was, What adjustments in intellectual property rights have been made? Well, of course, the first is the allowance of claims to living organisms. The United States certainly led the way there. It was a very important opportunity that organisms that produced a pharmaceutical material could be claimed in patents. We had something tangible to claim even if the product being produced was already known or already had been defined.

One of the things that has been evolving over the last few years, and certainly in 1987 had a pretty dismal outlook, is referred to as *In re Durden*. This case implied that just because you have a novel starting material on which you carry out a process to produce another material, the process is not automatically patentable. That case was often interpreted much more severely to mean that unless the process is highly inventive, mere novelty because of novel starting materials does not make it patentable. So it was not possible in 1987 to get claims to the process that was going to produce, for example, in Amgen's case, erythropoietin by using a novel organism.

Because inventors could not claim the process, they had a very serious problem. They could not invoke any rights at all against companies who used their organism overseas, produced the product, and brought it in. They did not have a final product claim; they did not have a process claim; and there was no mechanism for protecting against the direct theft of the organism overseas—copying it, or following the teachings of the patent, and then just shipping the product to the United States.

However, an evolution has occurred since then. Certainly, a lot of process claims have now been granted. There is a bill authored by Congressman Boucher that would give guidance to the Patent Office to make sure it issues those claims. Without those claims, the organism patent is meaningless with respect to overseas competition. What if the overseas country does not issue the organism patent? The organism has only one purpose—to produce the protein, so the inventor is left with no protection against importation. Amazingly enough, the inventor is protected from infringement in the United States by U.S. companies but is unable to stop foreign infringement and U.S. importation. The trade implications are clear.

This has been a very serious problem that is now being addressed. Yet there are still concerns from people who wonder if it is really "fair" to keep foreign companies from bringing their products into the United States. They ask, "Isn't that protectionism?" This a very strange interpretation of fairness. I think these inventions are clearly being copied and misappropriated by foreign companies. Changes may or may not move smoothly, but these issues should be resolved in the next few years, and more and more companies are availing themselves of the process protection, though some opportunities have been abandoned after *In re Durden* objections.

There have been great differences in the interpretation of the scope of claims. My initial discussion is limited to the United States because global issues have really only come into play in the least five years. Even in the United States, the scope of claims has been quite a difficult issue with which to deal. The questions stated are, If the claims are too broad, doesn't it mean we are inhibiting the diffusion of technology? If the claims are too narrow, doesn't it mean that the inventor really is disadvantaged? I could say a lot about that, but in actual fact I will cite the record. A Boston court in the United States leaned toward a pretty narrow interpretation of the claims. In a Delaware court, a jury decided that the *Genentech* case should be very broadly interpreted and cover structures quite different from the ones that were defined in the patent simply because all the rest were straightforward once the patent teachings were available. So these are still issues, but I think we will move toward a pretty clear understanding over the next few years.

The effect on biotechnology advancement has not been smooth even in this country. Patent uncertainty has encouraged second entrants, who then plead that since they made such a significant investment, believing they were not going to be prevented from manufacturing the product, the terms of the claims of the patents should be relaxed. This has certainly been an expensive mistake in many cases.

Major delays in issuance of patents have prevented some innovators from pushing their products as rapidly as they could, because they feared that they might never have coverage and once they proved the success of the product, it could be duplicated relatively readily. I think many of us in the business got a lot of encouragement from the Orphan Drug Act, because that act suggested that we at least could get six years of protection if we were the first to have a product approved for an orphan indication. If we never received adequate patent protection, we still might be able to recoup our investments, which was very comforting. There has been a lot of controversy about the Orphan Drug Act and whether it should serve as a kind of substitute for the Patent Act. Nevertheless, it helped an embryonic biotechnology industry raise money and sustain its early critical momentum.

Finally, patents played a key role in attracting pharmaceutical companies' investments. These were very important for some companies in the early days. Even though the pharmaceutical companies were not the innovators, they certainly helped support many new biotechnology companies and they clearly needed the confidence of patent exclusivity.

As stated in congressional testimony by Dr. P. Roy Vagelos, Chairman of Merck & Co., "To sustain their ability to discover and develop products which form the basis of American competitiveness, U.S. pharmaceutical companies count on renewed government support . . . in strengthening international protection of intellectual property rights." We can illustrate that perhaps even more significantly in the biotech industry.

For example, in 1986 a pharmaceutical product would cost about $94 million and take somewhere between 10 and 20 years before entering the market. Some kind of protection is certainly required before that kind of investment is made. The figure today is $240 million. That number has been challenged by Congress and looked at many ways by the Office of Technology Assessment (OTA); the latest OTA study says that costs may often be that high, although sometimes they may be lower. However, it does not require a lot of arithmetic to figure this out. The pharmaceutical industry in this country alone spends about $10 billion on R&D per year, and about 30 new products—30 new molecular entities—are approved each year. That comes out to be more than $300 million invested for each success.

In fact, there are at most only four or five new therapeutic products approved each year that are important and if you divide by that, you arrive at astronomical figures for important new therapeutics. Also, all this investment is required years before you can enter the market and start to get a return. So this certainly fits the pattern of something that requires protection, and patents look like the way to do it.

In 1986 the average development time of a new pharmaceutical product was 10 years. The interesting thing is that biotechnology has compressed that time. Because of the rational design of these products, their remarkable efficiency and safety profile, and the understanding and cooperation of the U.S. Food and Drug Administration, the average development time is about four to seven years today for biotechnology products, which is a big help. However, it is still a long time and a large investment.

So let us review how biotechnology was commercialized. What happened is not particularly logical, not what anyone would have deduced sitting around a table trying to decide what was going to happen. When a biotech company decided it wanted to launch a product, it had to build a company to launch the product. All the different stages and structures had to be built—the vectors and expression systems, purifications, scale-up, manufacturing, clinical testing, regulatory submissions, and marketing. Surprisingly enough, almost all of these things were in place in major pharmaceutical companies, yet almost every single important invention was done by independent biotechnology companies. That is the fact; that is what we have to deal with. How were they able to do all this, why would they be the first to do it, and was it effective? Is it not terribly inefficient to have to create a company for each new product? That is what was done.

Small, start-up biotechnology companies were responsible for many miracle drugs. For example, Amgen developed erythropoietin, and we now know that 10 milligrams per year, one-fiftieth of an aspirin tablet, will prevent 20 or more transfusions for people that are deficient in erythropoietin—and there are many more. Chiron produced the answer to hepatitis C, which is something that has plagued society and challenged scientists for more than 30 years—a well-defined disease about which nothing could be done. Cetus discovered ways of amplifying genes. Individual inventors, individual small companies, are pioneering and finding important new molecules and insights that are changing the way medicine is practiced today. This was done in a way that perhaps was hardly predictable—small, independent companies got started and did this all on their own—but this is exactly what happened. Sometimes it occurred with the support of large companies, but none of the key innovations and developments throughout the field were made by the large companies.

As I said, it was a fairly rocky road. I think that is important. The fragility of a new technology and the need for immediate action are more critical than making long-range plans to do wonderful things over long periods of time. These companies are fragile and their viability is always in question. Their survival is in jeopardy at all times. Take 1989 as an example. Headlines blared, "Clouds gather over the biotech field." Interestingly enough, firms were stumbling on regulation and patent problems. The patent situation looked very confused at that time. It was very difficult again to get financing, and the feeling was that many companies would go out of business and some did.

If we look at the number of financings, we see what has faced this emerging technology—and will probably apply to every new technology—big financing surges, dry spells, big surges. The dry spell in 1984 and 1985 seemed to last forever. We learned it can take eight quarters before you see another chance to raise money. When 1987 came along, the stock market wilted, and 1988, 1989, and 1990—one after another—were all very bad years. Of course, 1991 salvaged a lot of companies, but those were dangerous times for fragile, embryonic businesses.

So some protection is required. There is no question that patent protection fits the need in terms of the large investment required over a long period of time. The question is always asked, however, whether keeping the inventions secret would work. Well, it doesn't. Once the gene has been described, it is trivial to produce the product. Even if the gene is not described anywhere, once the structure is out, once the product is available even in clinical trials, the structure can be determined and often easily duplicated at a much lower cost. The cost is even lower because the copier only has to copy winners. He does not have to duplicate the losers. The copier avoids the major investments that the innovator had to make.

So international protection becomes the issue today. The problems in obtaining worldwide protection are difficult. There are many countries that do not honor the patent system. Surprisingly, countries that do not have strong patent systems (e.g., China, India, Argentina, Brazil) are *not* troublesome to the biotech field, although the pharmaceutical industry has expressed concern. However, international trade competition with countries that purport to have a patent system is a very serious issue.

For example, Japan is a strong competitor. In Japan, patent flooding surrounds innovator's patents. The Japanese patent office grants narrow patents instead of broad ones. I think it is pretty obvious to those in this industry that small companies need broad patents. If you are going to try to compete in the marketplace with giants, you had better know that you have some reasonable protection against obvious duplication or partial duplication. The Japanese system has not produced many biotechnology innovations and has not produced biotechnology companies. Our problems with the Japanese system are narrow patents, sometimes taking 10 or more years to issue, and patent flooding, which surrounds the inventor's contribution and forces him to join up with a large, entrenched Japanese company to survive.

To summarize, developing countries have concerned some industries, but they have not been competitive in biotechnology. Europe has awarded strong patents that afford U.S. innovators reasonable protection. Japan has been a very serious issue. Today we see two companies in Japan enjoying the products of Amgen—two products approaching a billion dollars in sales, at prices two to four times that of the products in this country, guaranteeing high profits. It is very easy to see what is going to happen over the long term. Those companies are going to be able to invade other countries in the field of biotechnology and be very active participants in trade.

The question then is, Can the United States dictate or influence international patent practices? Well, somehow it has to. This sounds unfair to some, but it is equally unfair to have misappropriation of intellectual property.

We know the history of what happened: Japan behind, Japan even, Japan ahead. The outlook is very serious. If we think back about that 20-year period around the 1960s when U.S. patents were not being upheld, that may have been why it was easy for the Japanese to move in and take over the territory.

Now, for future challenges: The federal government's patenting of the genome was a hypothetical question until a short time ago. Would this be serious? It has now become a very practical question. The U.S. Patent Office is currently examining the NIH's application for patents on certain gene sequences. In the meantime, the Industrial Biotechnology Association has held discussions with Reid Adler of the National Institutes of Health (NIH), biotech executives and administration officials who are examining this issue. What should the NIH do with respect to all of these gene patents? A good start is to provide a forum between industry, NIH and other interested parties to see if we can understand whether these patents should be applied for, whether they should be issued, and if issued, how they should be handled.

Finally, can patents be issued faster? The U.S. Patent and Trademark Office's numbers on the average time of application pendency are very strange and not helpful. The Patent Office has always figured out ways to say it is doing things in two years when, in fact, there has not been a useful biotech patent that has taken less than four years, and usually five. If we cannot get meaningful numbers, I don't think the problem can be solved. I think the Patent Office is misleading all of us.

In terms of the conference objectives, I would like to close with these thoughts concerning a few final issues: First, with respect to international perception of the importance of intellectual property rights, the world acknowledges that the United States was the pioneer in biotechnology, and that it was done by risk capital, as well as federal support of R&D, originally. The positive contribution to human welfare is acknowledged worldwide. That does not mean that all the countries in the world want to give strong patent protection for biotechnology, which is a very difficult issue.

Second, with respect to biotechnology patents, in the United States, the road has been rocky but reasonably satisfactory. Worldwide protection will ultimately be critical. It is sad that this did not occur long ago. Because of this lack, we are seeing companies in foreign countries appropriating U.S. technology to get started.

Finally, with respect to conflict resolution, the most precious resource of a budding new industry or budding new technology is time. The solutions have to be time sensitive. Grandiose solutions that involve 60 or 70 countries, and take years and years, will mean that a lot of the companies will fail before the solutions are in place. I think people should be aware of that.

I would remind you of one last thing. This is an industry of small companies. If you look at the profile of public biotechnology companies, only 13 percent have more than 300 employees, and none have more than 2,000 employees. If we look at all biotech companies (publicly and privately held), there are only 3 percent with more than 300 employees. We are dealing with a very, very broad-based, small-company business and my remarks apply as well to my firm, ICOS, which we started within the last year, as well as to the largest biotech companies, which are still relatively small. These are the companies seeking patent protection. Strong protection can hardly "disadvantage small companies" as some critics suggest.

14

Semiconductor Chip Protection as a Case Study

MORTON DAVID GOLDBERG

Intellectual property is old. Semiconductor chips are new. I have been asked to look at how the two have worked together, and what we can learn from the experience.

First, I review the history of the technology and of the intellectual property law that Congress custom-tailored to protect it. Then the provisions of that law and what some of us see as its shortcomings are discussed by focusing specifically on three areas: how the law defines the technology, its broad exceptions to proprietor's rights, and the difficulty of internationalizing protection.

Let us look first at a brief history of the technology, just as Congress had to do before it passed the Semiconductor Chip Protection Act (SCPA) of 1984.[1] The semiconductor chip was invented in 1959;[2] and the first microprocessor chip was developed in 1971.[3] By the early 1980s, developers could fabricate chips containing more than 100,000 transistors.[4] From the 1970s through to the 1990s, the chip has become so ubiquitous that it is found in products ranging from automobiles to refrigerators to personal computers and a vast variety of "personal electronics."

---

[1] 17 U.S.C. sec. 901 et seq.

[2] H.R. Rep. No. 98-781, 2 at n. 2 (1984).

[3] Id.

[4] Id. at 2.

The story of course does not end there. Since the early 1980s, the pace of innovation in semiconductor technology has accelerated. Chips currently in production contain in excess of 1,000,000 transistors. Dynamic random access memory chips (DRAMS), which have set the pace of progress in the industry, have provided a fourfold increase in capacity *every three years*—even though each increase has required engineers and scientists to solve ever more complex problems, driving the technology to even greater heights.

The photolithographic process used to fabricate the vast majority of semiconductor chips is conceptually relatively simple. The manufacturer applies a layer of photoresist (a material that reacts to light and resists the action of certain chemical agents) to a wafer of material called a substrate. The photoresist is exposed with a predetermined pattern. After being "developed," portions of the photoresist are washed away, leaving the substrate exposed. The substrate is then treated with a chemical agent that may etch material away from the exposed part, deposit material on it, or permeate into it. The manufacturer removes the photoresist and then repeats the process for each of the multiple layers required to form the device.[5]

The photolithographic process just described has many applications beyond semiconductor chips. The substrate does not have to be silicon (or any semiconductor for that matter), and the product does not have to be electronic circuitry. Manufacturers can use photolithography with masks on a variety of substrate materials, such as glass, polycrystalline silicon, sapphire, ceramic material, superconducting material, magnetic domain material—the list goes on and on, and continues to grow.

Moreover, the resulting product does not have to be a "chip." It can be a flat-panel display, a miniature motor and gears, a thin-film recording head, or any one of a number of items that are not usually considered to be electronic circuitry. It is possible that within a few years, virtually every portion of computer hardware, from the display to the mass storage devices to the packaging for chips, will be fabricated by using some kind of masking process.

Each stage of the process, from preliminary design through fabrication, requires investment, skill, creativity, and just plain hard work. As the technology became increasingly important in the U.S. economy, additional legal protection at some stage of the process appeared to be necessary to protect this investment if innovation was to flourish.[6] Beginning in the late

---

[5]Alternatively, circuitry can be written directly on the surface of the substrate by using electron beams. This technique is not yet widely used commercially.

[6]I say "additional protection," because related aspects were already provided protection—and continue to be—under the patent and copyright statutes, as well as trademark law and the laws of trade secrets and contracts.

For example, the patent statute protects any new and useful process, machine, manufacture or composition of matter, and any new, original and ornamental design for an article of manu-

1970s, innovative semiconductor manufacturers pressed Congress to grant additional intellectual property protection for this particular form of investment in monetary and intellectual capital. The result was the Semiconductor Chip Protection Act.

From the outset, the legislative effort to fashion protection for chips focused on the "mask works." These are the series of masks that bear the circuit designs used to expose the photoresist in the fabrication process. Each mask bears the information that dictates which areas are to be exposed and which are to be covered during a given step in the process. Together, the masks describe the entire three-dimensional topography of the finished product.

However, this focus was not inevitable. If the form of protection chosen for semiconductor chip products had been derived from patent law, more emphasis might have been placed, for example, on the fabrication process or on the product than on the intermediate masks.

Initial proposals for a chip protection law called for an extension of copyright law, declaring mask works to be pictorial, graphic, and sculptural (PGS) works, notwithstanding their utilitarian purpose.[7] Ordinarily, the design of a "useful article" is protectable as a PGS work only to the extent that it incorporates aspects that are physically or conceptually separable from their utilitarian aspects.[8] The proposals would have created an exception to this limitation for mask works.

---

facture. 35 U.S.C. sec. 101, 171. The copyright statute protects original works of authorship. 17 U.S.C. sec. 102(a).

However, as indicated, some felt that the forms of protection provided by existing versions of laws such as these did not suffice. For example, because of their utilitarian purpose, the layout designs of semiconductor chip products, whether embodied in masks or in the finished product, were not protected as pictorial, graphic, or sculptural works under copyright law. See note 8 and accompanying text. Also, although some chip designs might be sufficiently novel and nonobvious to qualify for patent protection, it was felt that a great many would not. Finally, since the layout design of a chip can be observed by inspecting the final product, it would be difficult to maintain the requisite level of secrecy to qualify for trade secret protection once a chip is sold to the public. Consequently, manufacturers urged Congress either to amend the patent or copyright laws or to enact a sui generis law.

[7]See, e.g., H.R. 1007, 96th Cong., 1st Sess. (1979).

[8]17 U.S.C. sec. 101 (definition of "pictorial, graphic, and sculptural works").

It is sometimes said that a "useful article" cannot be protected by copyright and—proceeding from this premise—that the copyright protection of computer programs (if considered merely as part of a utilitarian computer) must be suspect or entitled to only a lesser level of protection. However, useful articles are defined in the Copyright Act, textually and historically, only in the context of PGS works; and computer programs are not PGS works but, rather, literary works under 17 U.S.C. sec. 102(a)(1) . See "Fallacies and Fables about 'Useful Articles,'" in Goldberg and Burleigh, Copyright protection for computer programs: Is the sky falling? 17 American Intellectual Property Law Assn. Quarterly Journal 294 at 319-322 (1989). Cf. *Lotus Dev. Corp.* v. *Paperback Software Int'l.*, 740 F. Supp. 37, 52, 71-72 (D. Mass. 1990).

These efforts failed, partly because of the limitations the drafters tried to place on copyright protection for mask works, and partly because it was unclear how provisions of the draft legislation did and did not relate to the Copyright Act as a whole. Mask works were narrowly defined by reference to the definition of "semiconductor chip product." They were to have an abbreviated term of protection—10 years—and were to be subject to special exceptions for reverse engineering. Later drafts of the legislation created a new category of works with a separate bundle of rights distinct from those generally accorded to works under section 106 of the Copyright Act.[9] Because the legislation failed to provide appropriate integration of chip protection into the copyright statute,[10] it was not adopted.

Instead, Congress chose to create a sui generis law outside of the copyright statute. The SCPA bears the basic features I have described: narrow subject matter, 10-year term of protection, and a broad exception for reverse engineering. Congress attempted to strike a balance, as the copyright and patent laws do, between innovators and imitators. It also attempted to allow a certain degree of flexibility within the narrow confines of a custom-tailored law drafted to meet the specific concerns of manufacturers of a specific product at a specific point in its technological development.

DEFINING THE TECHNOLOGY

It would seem axiomatic that in fashioning intellectual property protection for a new technology the first order of business is to determine what the technology is. The foundation of the SCPA was codified in one of its basic premises: that the technology was the fabrication of semiconductor chip products.[11] This was not a surprising choice, given the state of the technology in the late 1970s and early 1980s when the bill was drafted, and the fact that the law was passed largely at the behest of the semiconductor industry. However, time and change have quickly eroded both foundation and premises.

For example, a more relevant technology might be the fabrication of products on any substrate using masking techniques. As noted above, making a chip does not differ very much conceptually from the making of many

---

[9]See, e.g., H.R. 1028, 98th Cong., 1st Sess. (1983).

[10]H.R. Rep. No. 98-781, 9-10 (1984).

[11]A "semiconductor chip product" is defined in the SCPA as "the final or intermediate form of any product—(A) having two or more layers of metallic, insulating, or semiconductor material, deposited or otherwise placed on, or etched away or otherwise removed from, a piece of semiconductor material in accordance with a predetermined pattern; and (B) intended to perform electronic circuitry functions . . . ." 17 U.S.C. sec. 901(a)(1).

other products. Considerable skill and creativity are invested in the design of the mask works that determine the topography of those products, but this design work is easily appropriated since, in essence, each copy of the product carries its own blueprint with it.

In addition to semiconductor chip products, many other products are being made, or will be made, by using masking techniques. They include not merely devices such as microprocessors and DRAMs that are traditionally thought of as chips. They also include, for example, thin-film heads, flat panel displays, micromechanical devices, chip packaging, magnetic bubble devices, magnetic mass storage devices, optical devices, and superconducting devices. Yet because they are not formed on a semiconductor substrate, or are not intended to perform electronic circuitry functions, it may be difficult to consider those items as "semiconductor chip products" under the act. Much of this important technology would thus not be given any of the special protection that the SCPA attempts to provide.

In the future, even devices such as transistors—the heart of what is commonly known as the computer chip—may be fabricated in a way that may deny them protection under the SCPA. For example, researchers have recently made a promising speed breakthrough in lateral bipolar transistors.[12] Transistors of that kind, with their low power consumption, could power high-performance, low-power computers in the future. The devices are made of silicon deposited or "grown" in very thin layers on an insulator; they are not etched from or deposited on a semiconductor substrate.[13] Consequently, chips made in this way might not constitute semiconductor chip products within the meaning of section 901(a)(1) of the SCPA; and even though they would perform the same functions as any other computer chip, they might not be protected under the act.

To a greater or lesser extent, each of these devices shares (from an intellectual property perspective) the salient properties of semiconductor chips: considerable skill and creativity are invested in the design of the masks, and the design can be appropriated easily because it is borne on the product's surface. There is no principled reason why they should not receive the same protection, but in its sui generis approach the SCPA was too custom-tailored so as to fit a specific technology.

---

[12]Chips: IBM more than triples the speed of a kind of transistor that could power high-performance, low-power computers, EDGE: Work-Group Computing Report (December 16, 1991); "IBM triples the speed of transistor," Infoworld (December 16, 1991).

[13]Id.

PROTECTION—SCOPE AND LIMITATIONS

Protection for mask works is subject to several significant limitations, the most important of which is the reverse engineering exception. Section 906(a) of the SCPA permits

> a person to reproduce the mask work solely for the purpose of teaching, analyzing, or evaluating the concepts or techniques embodied in the mask work or the circuitry, logic flow, or organization of components used in the mask work; or . . . [to perform such analysis and] to incorporate the results of such conduct in an original mask work which is made to be distributed.

As with the technology of chip design and manufacture, the technology (and cost) of reverse engineering has not stood still since the enactment of the SCPA. Companies specializing in chip analysis can now "peel" or "strip" away a chip's various layers and provide cross sections, topological layouts, and material analyses for $10,000 to $30,000.[14] These same companies can also provide the "paper trail" that is a key element of a reverse engineering defense under the act. Computer programs can derive the logic diagram of a chip, and the logic diagram can then be used as the input for any one of a number of computer-aided chip design tools. The design can be modeled by using yet another software tool. A second comer can produce a "new" chip in this way for a very small fraction of the original producer's typical development costs for an innovative product. The second comer can use computerized optimization and can introduce some degree of variation from the original design.[15]

So, while section 906 of SCPA apparently sanctions it, this practice allows free riders to profit easily and cheaply from the success of others who make the R&D design investment and take the risks in bringing a new and innovative product to market.[16]

It can be argued, of course, that this kind of activity is beneficial. After all, it does encourage making at least incremental improvements in the original design and making products available at a lower price. That is the

---

[14]These prices may be on the high side. One company specializing in chip analysis advertises "off-the-shelf" chip reports for prices ranging from $980 to $1,880, with volume discounts for additional copies.

[15]The right to prepare derivative works is not among the exclusive rights granted in section 905 of the SCPA. Essentially, the only exclusive rights are to reproduce the mask work, to import or distribute a chip embodying the mask work, and to induce or knowingly cause another person to do one of those things. 17 U.S.C. sec. 905.

[16]Second comers of course may have to make substantial investments to produce these derivative chips. The costs of fabrication, and of the equipment and software tools used in fabrication of chips, increase dramatically as more and more circuits are crammed onto a chip. Nonetheless, copying is still cheaper than innovating.

essence of imitative competition. We must bear in mind, though, that the goal of our intellectual property laws is to give the public the benefit of innovative competition. By granting Congress the power to enact intellectual property protection to "promote the progress of science and useful arts,"[17] the framers of our Constitution intended to spur innovation, and innovation is what has made the United States a leader in high technology. Without it, there would be no high-tech products to imitate, and without meaningful intellectual property protection, innovation would lag.

The reverse engineering provisions of the SCPA are commonly mentioned in the industry as a primary reason that chip developers have brought few lawsuits under the act against copiers of their chip designs. Section 906(a) is widely viewed as the exception that swallows the rule of protection for mask works. Although difficult to prove, the perception strongly suggests that the SCPA may not provide meaningful protection.

INTERNATIONALIZING PROTECTION

Another major weakness of the SCPA is the absence of any workable effective international protection. Had protection for mask works or chip topography been integrated into an existing form of intellectual property protection—as noted above, copyright was the original candidate—it could have been internationalized by means of existing treaties. The sui generis treatment of chip protection under U.S. law, by contrast, has been extremely difficult to project into the international arena.

A sui generis law requires a sui generis treaty—a treaty that must be negotiated without any international consensus on what sort of regime for protection is appropriate. The primary multilateral effort to date was the Washington Treaty,[18] prepared under the auspices of the World Intellectual Property Organization (WIPO). The process was strongly influenced by developing countries hostile to intellectual property protection generally, and it produced a treaty so flawed that not one single major chip-producing country could support it.[19] Chief among the problems that the United States

---

[17]"The Congress shall have Power . . . To promote the Progress of Science and useful Arts, by securing for limited Times to Authors and Inventors the exclusive Right to their respective writings and Discoveries" (U.S. Constitution, Art. 1, sec. 8, cl. 8).

[18]Treaty on Intellectual Property in Respect of Integrated Circuits, as opened for signature on May 26, 1989.

[19]In addition to the United States and Japan, the European Community declared that it would not sign the Washington Treaty. 4 World Intellectual Property Report 140 (1990). As of May 25, 1990, only eight countries had signed it: Egypt, Ghana, Guatemala, India, Liberia, People's Republic of China, Yugoslavia, and Zambia. 4 World Intellectual Property Report 240 (1990). Members of WIPO or the United Nations, and certain intergovernmental organizations can still become parties to the treaty by depositing an instrument of accession. Id.

found with the Washington Treaty were the inadequate term of protection, the lack of specific protection for mask works incorporated in a finished product, broad provisions for compulsory licenses, and excessively permissive treatment of so-called innocent infringers.

There has been an effort to cover protection of chip topographies in the language of the Trade Related Intellectual Property Rights (TRIPS) agreement proposed in the General Agreement on Tariffs and Trade (GATT).[20] Section 6, article 35 of the TRIPS draft distributed by GATT Director General Arthur Dunkel on December 20, 1991, provides that TRIPS parties will provide protection in accordance with what are in effect the substantive provisions of the Washington Treaty, as supplemented by further provisions in section 6. The additional provisions endeavor to remedy the most significant shortcomings of the treaty.[21] It is far from certain, however, that the TRIPS effort will come to fruition.[22]

The only means remaining under the SCPA for internationalizing chip protection is through reciprocity with countries that protect (or are taking steps to protect) U.S. mask works on substantially the same basis as the SCPA.

The SCPA provides two means for giving foreign nationals reciprocal treatment. The first is a presidential proclamation under section 902(a)(2). This provision permits the president to extend protection under the act to foreign nationals for mask works that are first commercially exploited abroad. This extension can be made only after a finding that in the foreign country, U.S. mask works receive national treatment or treatment equivalent to that given under the SCPA. To date, there have been no presidential proclamations under section 902(a)(2).

The second reciprocity provision, section 914, was an eleventh-hour amendment to the draft law.[23] Section 914, International Transitional Pro-

---

[20]GATT is a multilateral treaty covering a broad group of trade issues. TRIPS is the agreement on trade-related aspects of intellectual property rights (including trade in counterfeit goods), that is to form a part of GATT if the current five-year-old negotiations yield a viable agreement.

[21]For example, article 38 provides for a 10-year minimum term of protection (as opposed to an 8-year term under the WIPO treaty), and article 37 provides for royalty payments by "innocent infringers" (as to certain stock in hand or previously ordered) after receiving notice of the infringement.

[22]The TRIPS text covers a number of areas of intellectual property protection. Failure to reach agreement on areas that are not directly related to chip protection may doom the effort. Of even greater importance is the fact that TRIPS is a part of the current round of GATT negotiations. These negotiations have deadlocked a number of times over such contentious trade issues as agriculture. At this writing, it is quite possible that there will be no TRIPS agreement because—for reasons entirely unrelated to intellectual property—the Uruguay Round of GATT may fail to produce an agreement.

[23]It was adopted too late to be included in the final version of the House Report on the act.

visions, permits the Secretary of Commerce to issue temporary orders that extend the benefits of the law to nationals of countries that are "making good faith efforts and reasonable progress toward" entering a bilateral or multilateral treaty with the United States or enacting domestic legislation along the lines of the SCPA. The "transitional provision" was originally scheduled to expire in 1987, but has been extended twice and is currently scheduled to expire on July 1, 1995.[24]

Although originally meant only as a stopgap measure to be used until other countries met the requirements of section 902, section 914 has been the sole vehicle for internationalizing protection of semiconductor chips. Nineteen foreign countries (including Japan, Germany, Canada, and the United Kingdom) have been granted interim protection under section 914 orders issued by the Secretary of Commerce.

By employing this "arm-twisting" method of inducing foreign countries to adopt our own approach, the United States has achieved some positive results in internationalizing the protection of chip topography, but the means is hardly ideal for international comity. No sovereign nation appreciates another nation dictating its laws to it, and resentment of this approach may be one of the factors hindering cooperation in multilateral efforts to harmonize the protection of chip topography worldwide.

Moreover, it is unlikely that the United States can rely forever on temporary section 914 orders to induce countries to make permanent changes to their laws. These countries must recognize that the current status of indefinite temporary reciprocity cannot be permanent, that is, unless the legal fiction of perpetually "making good faith efforts and reasonable progress" is stretched to the breaking point.

CONCLUSIONS

The SCPA is a law that was custom designed to address narrow concerns. It freezes in the law definitions that relate to only a portion of a specific technology—as that technology existed when the legislative effort began, several years before the law's passage. It also freezes in the law a balancing between innovative and imitative competition that was based on a competitive environment vastly different from what exists today (e.g., as discussed above, the substantial decline in the cost of reverse engineering relative to the cost of developing an innovative product).

Apart from the narrowness of its subject matter, the protection afforded is thin indeed. Establishing a paper trail can be as simple as ordering from a catalog and can be a successful defense to infringement under the reverse

---

[24]Pub. L. No. 102-64 (1991).

engineering provisions if a defendant introduces some—it is impossible to say how much, since we have no judicial guidance[25]—variation on the original design.

It is possible that, in time, Congress will enact a legislative "fix" that will remedy at least some of the law's shortcomings. However, as the technology continues to develop, will Congress, or the law itself, keep pace? Unlike the Copyright and Patent Acts, the sui generis SCPA has neither a flexible and expansive subject matter nor a historic body of principles and precedents for adapting to changes without repeated congressional action.[26]

Moreover, the solutions in the United States, be they judicial, legislative or administrative, may differ greatly from the solutions abroad, since for a sui generis scheme of protection there is no common body of copyright or patent principles or multilateral treaty to guide the courts, legislatures, and administrative agencies in other countries.

ACKNOWLEDGMENTS

I acknowledge with gratitude the substantial contributions to this chapter by Jesse M. Feder, Esq., my colleague at Schwab Goldberg Price & Dannay. In the technical discussion, I am heavily indebted to the expertise of Drs. Praveen Chaudhari, Tak Ning, and Webster Howard, of the IBM Corporation, for an appreciation of the present state of the semiconductor industry and its possible future directions. In all cases, however, I am solely responsible for the interpretation that appears in this chapter and for any error or omission it might contain.

---

[25]Neither of the decisions rendered in *Brooktree Corp.* v. *Advanced Micro Devices, Inc.*, 705 F. Supp. 491 (S.D. Cal. 1988) (order denying preliminary injunction) and 757 F. Supp. 1088 (1990) (order denying defendant's motion for judgment notwithstanding the verdict) illuminate this issue.

[26]Although it is beyond the limited scope of this case study, I think it is instructive to contrast the experience of chip protection under the SCPA with the experience of computer program protection under copyright law. Congress amended the Copyright Act in 1980 essentially only to confirm the explicit congressional recognition in 1976 of computer programs as copyrightable literary works. Pub. L. No. 96-517, 94 Stat. 3015, 3028. In 1990, Congress granted a rental right to computer program copyright owners, as it had already done to owners of copyrights in sound recordings. Pub. L. No. 101-650, 104 Stat. 5089, 5134, 5135. No further recourse to Congress has been necessary.

Since the passage of the 1976 act, the technology of software engineering and the shape of the software industry have evolved considerably. Many new and difficult issues have arisen. Notwithstanding concerns expressed by some commentators (see Chapter 11) the courts have met these challenges successfully by applying traditional copyright principles. Moreover, there is a growing international consensus that computer programs are literary works protected under the Berne Convention.

15

Optoelectronics

EUGENE I. GORDON

For the purpose of this exposition, intellectual property rights (IPRs) in science and technology consist of two key components: inventions and proprietary information. The basic purpose of patent and patent license law, and the body of law and precedent surrounding protection of proprietary information, is to encourage investment in research and development, and in manufacturing and marketing activities that utilize IPRs. The basic intent is to encourage commercial enterprise. Viewed from the simplest perspective, intellectual property is a property and subject to the protection of the law. A key issue in new technologies such as software and biogenetics, is what constitutes property and for how long it should remain property. A second key issue is the variation in protection available in different countries. The extreme importance to the United States of a revitalized manufacturing capability makes reexamination of the legal and practical aspects of protecting IPRs desirable.

This chapter differs from most of the others in two important aspects: First, it presents the view of an engineering practitioner rather than a legal strategist. Second, it focuses on proprietary information as well as patents.

Optoelectronics has become pervasive in all aspects of high technology, including manufacturing technology. Products based on optoelectronics are prominent in such fields as information, communication and entertainment systems, medicine, R&D, education, and defense. It would be hard to imagine life without products utilizing optoelectronics. Slackening of investments in activities involving optoelectronics would stall progress in many aspects of commerce, improvements in the quality of life, and our ability to defend ourselves. Optoelectronics is interdependent with electronics, communication and information technology, and computer technology, and shares strategic importance with them.

Optoelectronics is not a new field. Around 1880, Alexander Graham Bell patented the Photophone, a system for voice communication utilizing amplitude-modulated sunlight as the electronic signal carrier. During the 1920s, the initial experiments on television were carried out at the AT&T Bell Labs, and extended and carried into commercial reality by the RCA David Sarnoff Labs and others during the 1930s. The advanced cathode-ray tubes (CRTs) developed during that period at Bell Labs were crucial to the early development of radar, so important for success in World War II. Advanced CRTs were used as the memory elements in the first electronic telephone switching service introduced by AT&T late in the 1950s. The laser was invented at Bell Labs in the mid-1950s, and in 1960 the first working laser was demonstrated at Hughes Aircraft Company. In 1970, the first low-loss fiber-optic transmission medium was demonstrated by Corning Glass, and the first room-temperature, continuously operating semiconductor laser was demonstrated at Bell Labs and in the Soviet Union. These form the heart of the current long-haul terrestrial and oceanic transmission systems for voice and data. For example, the first transoceanic underwater communication cable was installed in the mid-1950s. It utilized low-frequency, electron tube repeater and wire cable technology, and had only a few tens of channels. The first optoelectronic cable, TAT-8, based on semiconductor laser repeaters and fiber-optic transmission paths, was installed in the Atlantic Ocean between the United States and England and France, and operated first in December 1988. It has 40,000 voice channels in a cable the diameter of a finger, about the same size as the original electron tube cable. TAT-9, to be installed in 1992, will have more capacity than all previous cables combined. The semiconductor laser is also used in compact disc recording.

In 1969, the first charge coupled device (CCD), and in 1970 the first miniature color video cameras based on CCDs, were invented and developed at Bell Labs. The CCD has become the sensing element basis for most fax machines, the Camcorder, and many important military applications used in the recent Gulf War. It is often used in commercial television and in surveillance applications. The film-based still camera will eventually be supplanted by a CCD-based camera, and computer-based editing and transmission will change the nature of photography for advanced amateurs and news reporting.

As the name implies, optoelectronics is a part of the broader technological field known as "electronics." In optoelectronics, light rather than electrons plays a role in moving energy or information from one point to another. Actually, all energy and information in electronics is conveyed in the form of electromagnetic waves. The real distinction is the frequency of the waves or the energy of the photons involved. Electronics implies photon energies associated with microwaves, radio frequency, low frequency, and ultimately direct current with vanishingly small photon energy. Optoelectronics implies that the photons involved in the interaction have energy associated with infrared, visible, or ultraviolet radiation.

Optoelectronics has become the preferred means for broadband, long-distance signal transmission for shore-to-shore or transoceanic communication. As an electronic path it is the preferred means to distances as short as meters. In the next decade this may be reduced to centimeters and ultimately millimeters. This will have a substantial impact on the speed and compactness of electronic systems. Ultimately optoelectronics will provide transoceanic communications without repeaters that will be of much lower cost than terrestrial long-haul equivalents and easier to implement because the right of way is so unconstrained. This will have a profound effect on the nature of international business and politics.

In many systems, optoelectronics is key to the information display essential for humans involved in the system. Sensing is often based on light, and optoelectronics consequently plays a key role in sensing components in many manufacturing systems and is the basis for most computer-integrated manufacturing.

In my opinion, the reason for special attention to optoelectronics in this forum lies in the advanced nature and rapid pace of the technology; its high vitality despite its long history; and its pervasive, ubiquitous influence. I have not perceived anything special about optoelectronics other than its vitality and pervasiveness in virtually any modern system. The ubiquitous and growing presence of optoelectronics in business offices, stores and markets, manufacturing facilities, financial institutions, and military arsenals, and its basic role as the articulation that binds this country and the world, are relatively new.

In most cases, optoelectronic technology is a component technology. Optoelectronic components are included in larger electronic systems. Conventional electronics is complex, but the basic elements are relatively stable. Both electronics and optoelectronics are practiced as a business in which components are made by specialists and assembled into systems or equipment by others. There are exceptions, but even very large companies such as AT&T and IBM purchase many of the components they use. The distinction I make is that manufacture of many optoelectronic components is typically not at the scale of electronic components, and there are many more small suppliers. The manufacture of advanced integrated circuits is becoming international in nature because of the enormous investments in facilities required. Optoelectronics is not yet at that point and is not likely to get there except in highly specialized situations.

The elements of vitality and relative smallness pose special problems associated with stimulating entrepreneurial activity and investment. I would like to focus on two of these: protection of proprietary information and patent law.

PROTECTION OF PROPRIETARY INFORMATION

The vitality and pervasiveness of optoelectronics provide enhanced opportunities for entrepreneurial and small-business activities. Frequently, new businesses are started by individuals who have left the employment of a large, well-established company. Alternately, the individuals leave the employment of the large company and join a smaller company engaged in aspects of the same business. Both situations occur frequently in rapidly growing areas of commerce, and optoelectronics is especially distinguished in this respect. Thus, there are special problems in optoelectronics not found in well-established, stable technology areas.

The loss of individuals to another company generally causes some concern to a large company, which may have a significant market (or potentially significant market) in a given area and may have invested substantially in R&D and manufacturing facilities in that area. The company has invested significantly in the hiring and training of individuals in the technology area. The individuals leave gaps that must be filled by relatively untrained and unskilled new employees and possibly by new managers. These individuals working for a smaller company may attract more of the large company's employees to the other company. They carry with them proprietary information that potentially can be used to the advantage of the smaller company. Although the large company possesses economy of scale, the smaller company can usually move with greater alacrity, boldness, economy, and freedom than the large company. It can capture niche markets, utilize the knowledge of the market gained from new individuals, and capitalize on the proprietary information and the associated technology without having incurred the R&D cost. The smaller company can cause considerable damage to the actual and potential market of the large company.

The law provides substantial protection for companies holding proprietary information. Companies can approach other companies that they believe may be using their proprietary information unlawfully, and request discussions and development of means to give them reassurance that their information is protected. Indeed, in the absence of a demonstrably aggressive approach to protection, they will lose the legal protection of their proprietary information. Not only must they scrupulously guard it internally to maintain its status as proprietary information, but they must be tough and aggressive with others who they suspect may be violating it.

Companies that refuse to cooperate are subject to injunctions and litigation. When damages can be demonstrated, the awards are often trebled. Thus the legal consequences of the unlawful use of proprietary information can be painful for the company found guilty. Proving guilt is another matter. Litigation is expensive, distracting, and time-consuming, and the results are not predictable. Large companies are not anxious to use legal protection, but it poses no major downside risk unless the company engages in demonstrably illegal behavior in the name of protecting its proprietary information. However, for the small company, litigation as the respondent can be disastrous. The upside is minimal, and the downside is the potential for loss or major damage to the company and major legal expenses.

Large companies sometimes use the threat of litigation or actual litigation effectively as a major weapon in their attempt to protect their proprietary information. Unfortunately, they may also use it for other purposes that are not legal. The law provides protection from wholesale raiding of the employees of one company by another. That is a separate issue and is not covered here. The loss of employees may be viewed as significant to the company even when the numbers are sufficiently low that no clear case for raiding can be made. Hence, companies sometimes use protection of proprietary information as a screen for what in reality are attempts to limit loss of key employees (when the number lost is lower than that protected against by conventional legal practice).

Ostensible protection of proprietary information can be used to hinder the start-up or small company so that it cannot compete effectively in several ways: by tying its management up, by diverting its funds to legal expenses, by inhibiting its ability to raise additional capital, by casting doubt on its ability to deliver a product unencumbered by legal difficulties, and by limiting its ability to hire new people who may be concerned about the future of the company. Indeed, litigation or the threat of litigation casts a deep shadow on new or small companies, and effectively hobbles them. It further protects the large company, whose own employees are intimidated by what they see and are loathe to leave to join the target company or other potential target companies. It also subtly intimidates employees of the large company from planning other start-ups.

The point is that although companies have a right and an obligation to protect their proprietary information, they may also use protection of proprietary information in violation of antitrust law. One may argue that individuals in small companies or aspiring to new businesses have a right to earn a livelihood using their intellect, experience, training, and acquired knowledge from previous employment. It may be further argued that a first company, trying to prevent individuals from joining a second company or practicing their profession in the area of their expertise on behalf of that company, because of the alleged use or inevitable use of the proprietary information of the first company, is acting in restraint of the rights of the individuals involved to earn a livelihood. The boundary between experience and acquired knowledge on the one hand and proprietary information is not well defined, however, and is probably not amenable to good definition. The courts have a difficult time with these issues, but they manage.

In some cases, however, the first company merely charges "inevitable use of their proprietary information" by virtue of the close association of the individuals involved with that proprietary information and the associated technology. The threat of an injunction or lawsuit is sometimes enough to allow the first company to gain unfair advantage. The implication is that you are "guilty because of the potential for committing a crime." Later, I describe such a circumstance, and also try to draw some conclusions and make some recommendations.

PROTECTION OF PATENTS

With recent changes in the patent law, the process of obtaining a patent for an invention has been vastly improved. Getting a patent is time-consuming and expensive (on the order of $10,000-$20,000), but the process is straightforward. Most patents are defensive in nature; that is, the purpose is to preserve the right of the assignee to practice an invention without interference from others, or to prevent others from using the technology and harming the business of the assignee. Some patents are offensive in nature; their purpose is to make money by selling or licensing the patent. Either way, the granting of a patent is not a guarantee of a valid patent.

Although many patents are strong, many other patents are issued that are not valid. Sometimes they are not distinguishable from existing patents or disclosures in the public domain. It is not clear what the source of the problem is. Possibly the examiners are overloaded and do not search or distinguish carefully. In some cases, because of overload, examiners are pressed into working in areas in which they have limited experience. Possibly the Patent Office has chosen to let the courts make the tough decisions.

It is sometimes said that no patent is valid until it is tested in court. The cost of defending a patent under challenge, however, can be extremely high, for both the challenger and the defender. When the court is placed in the position of determining the validity of a patent relating to sophisticated technology like that usually involved in optoelectronics, neither the judge nor the jury may be sufficiently knowledgeable or experienced to make a reasoned judgment. Thus, the litigant with larger financial resources may have the advantage.

The alternative is a much tougher and more thorough scrutiny of patent applications by the examiners. This would make patents more difficult to acquire and probably would act to inhibit submission of applications, putting the burden on the inventors and their companies to be more discriminating. Although this would reduce the number of patents issuing, I suspect that the quality would be vastly increased, and the number of tests in court would be reduced, making the issuing of a patent more meaningful. A short case history is presented later, in which the author served as an expert witness in a patent interference trial between two well known laser companies, which illustrates some of the above points.

PROTECTION OF PROPRIETARY INFORMATION: AT&T VERSUS LYTEL

During 1982-1983, AT&T prepared for divestiture. In early 1983 the eventual founder of a semiconductor laser company announced his intention to take early retirement from AT&T Bell Labs after 26 years of service, in turn, as a member of technical staff, supervisor, department head, and finally laboratory director. During his last four years he had led a laboratory developing and introducing into manufacture devices for use in fiber-optic communication systems.

After his retirement in July 1983, this individual started an active consulting career and worked regularly for AMP, Inc. of Harrisburg, Pennsylvania. AMP was a well-known connector company interested in moving into fiber-optic interconnection technology. While consulting, the individual in question prepared a business plan for starting a company to manufacture semiconductor lasers and other products that would be of interest to AMP. In November, he completed the plan and incorporated LYTEL, a Delaware corporation, to be the start-up company. He also presented the plan to AMP management, who indicated strong interest in being an investor. He informed AT&T of his plans to start a laser company. Its main concern, as expressed to him, was that he not use AT&T proprietary information and not raid its employees or hire too many of them. The number six was suggested as a maximum tolerable loss.

AMP's main concern was that LYTEL not use any proprietary information belonging to AT&T or do anything that would upset them; AT&T was an important customer for AMP's products. Those concerns were major ones for the founder as well, because he valued his association with AT&T and felt that using its proprietary information would be unethical. At this point the initial management team for LYTEL came together to plan the company and refine the business plan. One of the main activities was establishing process instructions for the products to be manufactured that would use only processes defined in the published literature. The team also tried to establish what information AT&T would be able to legitimately claim as proprietary information. They concluded that there was virtually nothing, for two reasons: (1) AT&T had a policy of open and timely publication of research results. This was dictated by the public utility commissions that authorized the license contract fee (LCF) as a legitimate operating company expense. The LCF was the source of the research funds used by Bell Labs. (2) Although AT&T played a key role in the early research on semiconductor laser diodes (SLDs), it had made some decisions during the decade of the 1970s that put it well behind in the long-wavelength SLD technology that LYTEL planned to manufacture. For example, the long-wavelength SLDs for TAT-8 and the continental fiber-optic systems implemented during the decade of the 1980s were made for AT&T by Hitachi. AT&T was manufacturing only short-wavelength SLDs. Thus LYTEL was confident that it would not compromise AT&T.

The term sheet for the purchase by AMP of LYTEL shares was signed in February 1984. The final stock purchase was scheduled for May, and the initial group of managers assembled to begin work as of April 1, 1984. It consisted of two newly resigned employees of AT&T Bell Labs and three former employees, none recently employed by AT&T, one of whom had never worked on SLDs. LYTEL was confident that it had done nothing that could antagonize AT&T. The company was surprised when several days later it received a formal letter from AT&T asking LYTEL to voluntarily accept an injunction not to go into the laser business or anything associated with it. The basis for the request was the claim that "LYTEL would inevitably compromise AT&T's proprietary information." It was implied that if LYTEL did not accept the injunction, AT&T would sue. It appeared, in retrospect, that AT&T's response the previous November had simply given it time to put together a plan of action.

LYTEL's reaction at the time was one of surprise, whereas AMP's reaction was one of dismay. As a result, there was a long delay in completing the stock purchase while LYTEL tried to negotiate an arbitration agreement that would assure AT&T that its proprietary information was not being compromised. During that time the founder's complete energy was dedicated to dealing with the AT&T action. LYTEL ran out of money advanced by AMP and salaries were at risk. Finally, in August, LYTEL reached an agreement on arbitration with AT&T. The AMP stock purchase was completed shortly thereafter.

The arbitration agreement was onerous, from LYTEL's point of view. LYTEL had to disclose its business plan and all the details of its processes. AT&T produced a list of items that it claimed were proprietary information. LYTEL was not allowed to see the list, ostensibly to protect AT&T proprietary information—AT&T claimed LYTEL might be reminded of something it had forgotten. LYTEL was not allowed to talk to AT&T lawyers about the claimed proprietary information. LYTEL had to work through its own lawyers, who could not tell it what AT&T was claiming. Its lawyers could deal with the list only by asking LYTEL engineers leading questions. The arbitrator looked for overlap in the list. After the overlap was established, discussions could ensue over specific items. Once the arbitrator made a decision on overlap, and LYTEL agreed to remove it from its process, AT&T would be allowed access to the facility to monitor compliance and look for violations. In addition, LYTEL had to agree not to solicit any AT&T employees during the period of arbitration and could hire only if it were directly approached.

Two years of intense effort, and a half million dollars of LYTEL's initial capital expended in legal fees, resulted in a finding of no overlap. LYTEL suffered an extreme setback from which it has not yet completely recovered, although the company is now making good progress. It is now a wholly owned division of AMP. All but one of the original group are gone. Retrospectively, it is fairly clear that LYTEL's entry into the market was delayed. That delay was destructive because a crucial market window was lost. LYTEL's ability to hire key employees was severely compromised. As I see it, the freedom of choice of AT&T's own employees was compromised by fear of the consequences of joining LYTEL. This fear was exacerbated by internal documents distributed to AT&T employees describing the situation with LYTEL.

As I see it, AT&T's strategy was consistent with allowable legal practice and accomplished its purpose. In my view, however, although its position was defensible, it was not on the high ground. Given the current state of legal practice in this area, LYTEL made a major error in informing AT&T of its plans, thus allowing AT&T to act before LYTEL's funding was completed. With funding in place, LYTEL would have had a stronger negotiating position and might not have been forced to accede to an onerous agreement. It seems unlikely that AT&T would have been willing to test the idea of inevitable use of its proprietary information in court, and in my view, if it had come to litigation, AT&T probably would have lost.

The ability to force a small company into an agreement that adversely affects its fortunes by virtue of its inability to sustain litigation is a severe deterrent to the ability of individuals to leave their employment and engage in entrepreneurial activities. It is also a deterrent to joining another company to work in the same area. Legislation to restore some balance by increasing the risk and burden of proof for larger companies in such cases would greatly enhance the willingness of individuals to engage in entrepreneurial activity. It would promote diffusion of knowledge by freeing potential job switchers from the fear of unfounded legal action. Most certainly, this would lead to increased jobs and wealth generation for the country, and an enhanced technology base, as it has in the past.

RECOMMENDATION ON PROPRIETARY INFORMATION

I believe that a simple expedient, if it were legally available, would have avoided the entire problem. It would not have been unfair to AT&T, or compromised its proprietary information, if LYTEL could have obtained simply and quickly an injunction requiring AT&T to wait until LYTEL produced its first product before taking any action. Certainly, violation of AT&T's proprietary information was of little consequence while LYTEL was simply *preparing* to manufacture and sell. At the point of actual production an arbitration agreement could have been negotiated without duress and means could have been established to protect AT&T. In my view, it would have been a much fairer, more balanced, and less expensive agreement.

I believe that if that legal recourse were commonly available, it would eliminate or greatly mitigate the heavy-handed practices of companies who take advantage of their size and the limitations of the legal system to act in restraint of trade or to limit the freedom of choice of individuals earning a livelihood. In any case, it would be desirable to make unlawful any claim of inevitable crime.

PROTECTION OF PATENTS: SPECTRA PHYSICS VERSUS COHERENT RADIATION

In 1984 I was asked by attorneys of Lyon and Lyon of Los Angeles to serve as an expert witness in a patent interference trial between Spectra Physics and Coherent Radiation, both manufacturers of various kinds of gas lasers. Lyon and Lyon represented Spectra Physics. The interference involved a patent held by Coherent Radiation, who charged that Spectra Physics was using the Coherent Radiation patent in manufacturing a high-power argon ion laser for use at ultraviolet wavelengths. The technology involved the use of a series of metal disks to define the linear gas discharge path in an argon ion laser. The metal disks were more resistant than previous technology to the deleterious effects of a high current discharge, especially sputtering, and effectively carried heat away from the discharge region. The prior technology for visible wavelengths involved the use of quartz and ceramic tubes to define the discharge path, which had limited current capability. The ultraviolet lasers, which represented an important market, required extremely high current to operate, so the metal disk technology was a good solution.

Among the other expert witnesses supporting the Spectra Physics effort were former colleagues from Bell Labs and Hughes Aircraft Company, all of whom had worked on the argon ion laser following its discovery in early 1964. The argon ion lasers in question were first demonstrated and described by Ed Labuda and myself of Bell Labs in collaboration with Bill Bridges of Hughes Aircraft Company. Improvements were described later by Ed Labuda and myself. A particular improvement patented by Bell Labs and described in the open literature was the use of metal disks to define the discharge path. After studying the Coherent Radiation patent and the issues of the case I became convinced that the Coherent Radiation patent was invalid. I agreed to serve as an expert witness.

The events as I understood them were that Spectra Physics had embarked on the metal disk program following the teachings of the Gordon-Labuda paper. It made some improvements on the original idea but concluded that a new patent was not possible. The development work stalled for reasons not relevant here. One of the individuals involved in the development left Spectra Physics and joined Coherent Radiation, which started a similar program shortly thereafter. Spectra Physics reinstated its program and eventually marketed a new laser family based on the metal disk technology. In the meantime, Coherent Radiation developed an almost identical technology and applied for a patent that later issued. It did not reference the earlier Gordon-Labuda paper or the Bell Labs patent. It also introduced a competing metal disk laser product line.

Coherent Radiation then proceeded to claim that Spectra Physics had infringed its patent and sought relief, after failure to reach agreement on a settlement, by bringing suit against Spectra Physics. Eventually, the case was heard in San Jose, California. The jury ruled that the patent claims were invalid, and Spectra Physics was able to move ahead on its commercial offering. In my view, justice was served. However, many millions of dollars were spent as a result of the initial dispute and the litigation that followed. It drained energy and emotion unnecessarily over a period of several years. Both companies have had financial difficulties for other reasons. The trial added an unnecessary burden. In my opinion, all of this would have been avoided if the patent examiner had done a thorough job of researching the patent and technical literature, or if Coherent Radiation had properly referenced the earlier work in its application.

RECOMMENDATION ON PATENTS

The response of the Patent Office to applications is quite rapid. The office works efficiently, and in my opinion, patent law serves its intended purpose well. In my view, however, patent examination must be raised to a higher standard. The examiners must do a more thorough job of researching past patents and the literature to discover prior art. If mistakes are made, they should be on the side of caution. More thorough examination will certainly limit the number of patents that the office can process, but it will also limit the number of applications. Truly useful patents, rather than patents that can harass and impede, will result. The benefit to commerce is hard to estimate, but it would be significant.

SUMMARY

Optoelectronics, although not new, is a frontier technology area. It has been a source of commercial vitality worldwide and a significant element in the quality of life. It may be viewed as a strategically important technology. The United States has made virtually all the key inventions and technology developments. Almost all of the key products and devices, however—including displays, imaging devices, copiers, printers, facsimile terminals, lasers for communications, audio compact discs, optical memories, product scanners for point of sales terminals, and devices for many military applications—are made mainly overseas. It will not be easy to turn this situation around unless investment in R&D and manufacturing in electronics, particularly optoelectronics, is increased. Obviously, there are many options for increasing domestic investment and all must be explored, but one factor we can control is to improve the ability of issued patents to hold up under litigative scrutiny, thus making investment in new products more attractive. Given that protection of proprietary information is important if investments in R&D are to be made and that entrepreneurial activity must also be encouraged, it is important to find the correct balance of IPR law that encourages investment in both large and small companies.

Discussion

Conference discussion of the chapters in this section focused on the choice between traditional intellectual property paradigms and alternative approaches to protecting intellectual property in new technologies. Increasingly, important new technologies seem to fall in the interstices between the traditional paradigms. Perhaps the fundamental issue can be stated as, What should be done about nonpatentable, noncopyrightable innovation?

Concern was expressed that some of the most economically important technological changes in the twenty-first century will consist of incremental innovation that is easily reproduced. In biotechnology, for example, many of the most important future developments may not be patentable because technically they will not meet the criterion of nonobviousness. The cure for cancer may also be unpatentable because of the nonobviousness criterion. This problem, which is very difficult to deal with, is likely to be serious for biotechnology and other advanced technologies in the twenty-first century.

Opinion on the wisdom of sui generis approaches was split. In one view, sui generis systems can be used to protect things that do not involve an inventive step but yet have economic value. This view suggests that other values besides inventive step or expression, such as lesser discoveries, deserve protection. Paradigms exist for this kind of protection, in the German copyright law, the law of petty patents, or the old German *Gebrauchsmuster*.

On the other hand, concern was expressed that sufficient attention has not been given to the problems inherent in sui generis statutes. Critics of such approaches commented that courts are needed to interpret even the most seemingly obvious statutes. The more an intellectual property right (IPR) statute departs from traditional patent and copyright law, the more years it will take for the courts to complete the initial round of interpretation. By then, it may be time to reconsider whether the statute has been overtaken by changing technology.

A member of the audience who was involved in drafting the Semiconductor Chip Protection Act (SCPA) of 1984 underscored the problems with sui generis law that arise from not being in the mainstream of intellectual property law. The uncertainty of developing new intellectual property law leads legislators to be very cautious and makes them reluctant to speculate about possible future new technologies.

A staff member of the House Judiciary Committee offered a series of contrasting perspectives on the SCPA of 1984. He noted that the sui generis approach to semiconductor chips was proffered by the intellectual property community, which feared "distortion by shoehorn." He recalled that the reverse engineering provisions in the law were offered by the semiconductor industry itself. He argued that the reciprocity provisions of the SCPA have succeeded in achieving bilateral relationships with all other semiconductor-producing countries. Finally, in his view, the Washington Treaty (for semiconductor mask work protection) was not really a failure; it led to increased discussion and refinements that are now a part of the IPR discussions in the Uruguay Round of the General Agreement on Tariffs and Trade (GATT).

One of the concerns expressed about sui generis approaches is that they would lead to a piecemeal approach to legislation. Clearly, technology-by-technology sui generis laws would not be helpful. Such an approach is not necessarily the only alternative, however. A law professor in the audience argued that the legal profession has had 200 years of experience with cases concerning the protection of technologies that do not meet the requirements of the major IPR paradigms. This is arguably sufficient to derive a conceptual basis for a new IPR paradigm for nonpatentable, noncopyrightable innovation.

Another issue concerned the principles that should guide the choice between sui generis or existing statutory provisions. Different principles arise, depending on whether the goal is to protect the property of the generator of new technology or to increase social benefit from the new technology. One person gave as an example the protection of computer program interfaces, which leads to a lack of standardization and reduced value to end users of much of the new technology. In the traditional formulation of the IPR policy question as a trade-off between the innovator and imitators, sight may be lost of the interests of the public. It was suggested that with respect to protection of the user interface and perhaps more broadly, intellectual property law should balance the rights of the consuming public, initial innovators, and competitors. It was further noted that it is very difficult to address these kinds of issues within existing legal and theoretical frameworks.

The question was raised whether sui generis approaches and international harmonization of IPR laws are contradictory, particularly given that the United States and Japan did not accept the Washington Treaty. One response to this concern was that sui generis approaches are somewhat at odds with harmonization, but that there are going to be problems in harmonization even if the traditional paradigms are used, because courts in different countries may not reach the same decision. Professor Barton's recommendation that IPR reform be handled through standing study groups or commissions might alleviate this problem. International agreements on sui generis IPRs could be sought by establishing such study groups and making them international in composition.

Another response was that the ability of the European Community (EC) to negotiate a software directive has demonstrated the possibility of attaining multilateral protection for sui generis rights. (Observers making this argument view the EC software directive as a sui generis approach, although it is formally a copyright approach.)

Chapter 11 raises concerns about the effects of IPRs on small businesses. Would narrower patent claims be antithetical to small businesses? Would large companies dominate international commissions of the type envisioned by Professor Barton? To assess the first question, it is necessary to balance the scope of the claims of a small business against that of patents already issued. One view offered at the conference was that, given the costs of litigation, the disadvantage to a small business of broader claims held by others would be greater than the benefit of its own broader claims, which it might not have the capability to exploit. With respect to the second question, although it is possible that large companies might dominate, at least in the commission context activities would be much more visible.

Another point that came out in discussion is the politically controversial nature of IPR issues and the roots of that controversy in strong economic interests. Any body of law in the IPR area significantly affects the rights of various players. For example, if the law is changed to give strong protection to computer interfaces, those who control large computer networks are given significantly greater power and economic rents. There is probably no way of avoiding this. The question for public policymakers is, How far should the monopoly go?

One audience member commented that legal cultures and traditions have evolved worldwide over hundreds of years and in different cultural contexts. The challenge as viewed by this person is to put in place over several decades a worldwide system that would be amenable to all countries. To date, efforts have consisted mainly of propagating a Western cultural view of IPRs as broadly as possible. Some non-Western countries have voluntarily adopted Western-style IPR laws in the race to modernize, but the West cannot count on this continuing. The negotiating positions of Brazil and India in the GATT talks on IPRs prefigure this change. If the goal is a lasting and stable global order, nations must move beyond imposition or adoption of one particular country's model to reach a consensus about a model or variety of models that respect cultural differences.

Another broad question from the audience was whether the IPR legal system has to solve all resource allocation problems generated by new technology. The implication was that too much of a burden may be put on the IPR system, and concomitantly not enough emphasis on other areas of public policy, to deal with the economic issues arising from new technology. This question leads to the topic of Section VI, which attempts to put IPR issues in a global perspective.

VI

Global Intellectual Property Rights Issues in Perspective

Introduction

This section represents the final session of the conference, which addressed the question, What next? A distinguished, eclectic panel was asked to think about certain major features of the international intellectual property right (IPR) regime and the interaction between science and technology and IPRs. In addition to synthesizing some of the more significant themes emerging from the conference, the panelists offered personal—and, in some cases, contrary—views on the future direction of the global IPR issue.

In Chapter 16, the first speaker, **Robert E. Evenson**, professor of economics at the Economic Growth Center at Yale University, addresses the concept of IPRs as infrastructure, specifically with reference to their role in developing countries. He argues that international IPR conventions have not worked well for the poorest countries because they do not have exporters' interests to protect. Evenson contends on this basis that it may not be realistic for the United States and other advanced industrialized countries to expect to use trade laws, such as the General Agreement on Tariffs and Trade (GATT), to convince developing countries to join the "club," because they simply have different economic and technological development objectives, perspectives, and capabilities. He suggests that the newly industrialized countries (NICs) and "near NICs" have used IPRs to facilitate competence and capacity building, but that IPRs have not been their major policy instruments. Evenson recommends that the discussion of IPRs in developing countries emphasize the stimulation of local R&D as opposed solely or primarily to facilitating technology transfer from other countries.

David C. Mowery, associate professor of business and public policy in the Walter A. Haas School of Business at the University of California at Berkeley, examines the investment and trade effects of changes in the IPR regime. He points out that it is impossible to judge whether the worldwide strengthening of IPRs would be in the U.S. national interest, because adequate empirical information on the costs and benefits is lacking. Mowery notes that the intersectoral trade-offs being negotiated within GATT to gain greater IPR protection may hurt some U.S. industries. He adds that a GATT settlement on IPRs may have only modest effects on foreign investment because so many factors play a role in the globalization of markets. Mowery stresses that, for the United States, much more needs to be learned about the domestic economic effects of the strengthening of U.S. domestic intellectual property protection that has taken place over the past decade.

The third panelist, **Michael Borrus**, is codirector of the Berkeley Roundtable on the International Economy, also at the University of California at Berkeley. He focuses on regional asymmetries in the cost of and access to technology development and their implications for the future use of IPRs. Borrus comments that the rapid development and global diffusion of technology have reduced the ability of U.S. industry to appropriate know-how through non-IPR means (e.g., by maintaining lead time), which has motivated the United States to call for stronger protection. He notes that technologies in which the United States is strong seem to be particularly vulnerable to this phenomenon and adds that increased protection of IPRs does not address the underlying problems of asymmetrical access to accumulation of knowledge that exist among nations and are being exacerbated by the emergence of regionalized clusters of economic and technological competence. Borrus suggests that perhaps the time has come to consider complementary alternatives to the strict focus on IPR protection, which might include efforts to strengthen the nation's and firms' capacities to cycle technology more rapidly, to appropriate know-how developed elsewhere, and to use and diffuse new technological innovation more effectively.

Robert W. Lucky, the former executive director of the Communications Science Research Division at AT&T Bell Laboratories, discusses trends in technology development and the future assertion of IPRs. He expresses concern about the fact that corporate R&D centers such as Bell Labs are finding it increasingly difficult (from the standpoint of corporate profitability) to justify investment in esoteric lines of fundamental research, particularly when many competitors do not make a similar investment and yet are able—through a variety of mechanisms (including the standard-setting process, the open literature, and conferences)—to access much of the intellectual property that a company such as AT&T produces. Lucky notes that researchers do not respond to the patent incentive; indeed, their motivation is to make their results public as rapidly and completely as possible. He notes a number of global trends in the conduct of R&D that are exacerbating the problem.

The last panelist, **Eugene B. Skolnikoff**, professor of political science at the Massachusetts Institute of Technology, addresses the relationship between technology and sovereignty, and the need for new mechanisms to resolve international IPR issues and conflicts. He argues that as global economic competition replaces older security concerns, strengthening IPRs may be seen as nationalistic and protectionist. Skolnikoff predicts that frictions between developed and developing countries will likely continue, because of the lack of developed country concern for the welfare of developing countries, and that the interest of many developing countries in some form of national and international protection is likely to parallel the extent to which knowledge spreads locally and indigenous technological capability is enhanced. He adds that he expects the bargaining power of developing countries to increase because their cooperation on global issues (e.g., climate change) is increasingly required to achieve additional progress. Skolnikoff also argues that it is unrealistic to expect to create one integrated, global IPR system that would provide adequate dispute resolution and be capable of keeping up with technological change. He acknowledges that although this is contrary to the view expressed by many at the conference, he believes it is a "fact of life." At the same time, Skolnikoff emphasizes that whether or not the United States likes it, the United Nations and other international organizations will remain essential for resolving IPR problems, and argues on this basis that the United States will continue to use the United Nations and to participate in the formulation of new international agreements, because there is little choice.

16

Global Intellectual Property Rights Issues in Perspective: A Concluding Panel Discussion

ROBERT E. EVENSON

The current movement toward unified, global intellectual property rights has gained considerable momentum. U.S. government agencies are leading the movement, and most market-oriented countries are supporting the U.S. position in the General Agreement on Tariffs and Trade (GATT) negotiations and other venues. Developing countries, on the other hand, are resisting, both formally in international forums and informally through less-than-aggressive administration of their own intellectual property right (IPR) legislation. As a result, considerable international tension and animosity exist between most developing countries and many developed countries—notably the United States—over bilateral trade law actions and the GATT negotiations.

The traditional Paris, Berne, and other international conventions are functioning quite well to achieve IPR compliance between developed countries. They are not functioning to achieve compliance between developing and developed countries, however. Lax administration of IPR laws and acts of "piracy" largely were overlooked until several years ago, when U.S. interest groups brought them into the policy domain. The resulting shift of IPR issues into the domain of trade law and policy has had important consequences for both developed and developing countries. These changes have been achieved at some diplomatic costs. A GATT agreement will lower these costs.

Yet is it realistic for the United States to hope that enforcing IPRs through trade law will solve the problems inherent in the older system? Is it really the case, as with education and perhaps some trade concessions, that developing countries will join the "club" in support of unified, global IPRs, or are there factors that constitute such real differences in economic interests between developing and developed countries as to threaten this goal?

As one considers these issues for developing countries, it is important to remember that IPRs are not designed merely to facilitate the transfer or export of technology from one country to another. They are designed to stimulate R&D and inventive activity in all countries. It is relevant to ask whether they actually do this in developing countries. They also are designed to stimulate the removal of secrecy from ideas so that those ideas can facilitate and stimulate other inventions. Finally, IPRs are limited rights. The seller (exporter) of IPRs should not expect full capture of all the economic returns associated with an IPR. Claims of losses by exporters of IPR-protected items to developing countries should be assessed accordingly.

As a basis for discussing these issues further, summary data comparing relevant economic variables across groups of countries are presented in Table 16-1. Data are reported for six types of developing economies. It should be obvious that developing countries encompass many types of economies. The six categories in the table are based on work by Weiss (1990). These categories of technology capacity are not intended to be "stages," although they do reflect different levels of institutional development. They are basically differentiated by this capacity to develop and implement technology.

These classes offer a broader and better sense than is usually provided of the range of economies encompassed within the term *developing economies*. Developing countries range from traditional economies (1a) through economies that are regarded as nearing newly industrialized status (2b), and those newly industrialized countries (NICs) that are regarded as being on the threshold of global technological competitiveness (2c).

It is important to note that all of the Stage 1 countries (roughly 60 to 70 countries) for the most part do not operate intellectual property systems of any real substance. Most, however, have some form of intellectual property system. Bangladesh, for example, has a patent law and a patent office, and it administers a patent system. It has only recently added a second domestic examiner, however, which is an index of its limited capacity to stimulate or examine technology agreements. Most of the countries in Stage 1 do not have adequate staffing or court systems to administer IPR laws according to the standard expected by the United States and GATT negotiators.

The situation changes for the Stage 2 countries. Table 16-2 provides several indicators to illustrate this. The indicators are organized by using

TABLE 16-1 Representative Countries at Various Stages of Scientific and Technological Development

Technology Capacity Classes	Asia and Near East	Latin America and Caribbean	Sub-Saharan Africa
Stage 1: Emerging Islands of Modernization			
1a. Traditional technology-based economy	Yemen, Laos	Surinam	Equatorial Guinea
1b. First emergence	Nepal, Papua New Guinea	Haiti, Guyana	Ethiopia, Burkina Faso
1c. Island of modernization	Sri Lanka, Tunisia, Indonesia	Jamaica, Peru	Kenya, Ivory Coast, Zimbabwe
Stage 2: Struggle for Mobilization and Mastery			
2a. Mastery of conventional technology	Iran, Malaysia, Turkey	Colombia, Argentina	
2b. Transition to newly industrialized country	India, Thailand, Hong Kong	Mexico	Republic of South Africa
2c. Threshold of technological competitiveness	Singapore, Taiwan, South Korea	Brazil	

the same classifications as in Table 16-1, as well as categories representing the recently industrialized Mediterranean countries (e.g., Spain) and mature, developed countries in the Organization for Economic Cooperation and Development (OECD). The extraordinary growth performance of countries in categories 2b and 2c, particularly their industrial growth, is apparent. They have outperformed poorer countries and mature OECD countries by a large margin.

Turning to R&D intensities in the public sector (i.e., R&D spending relative to production value), note that agricultural experiment stations have been established throughout each of the Stage 2 countries and, in fact, have been quite effective. The poorer countries do have R&D capacity in public sector agricultural fields.

The Stage 1 countries, however, have virtually no industrial R&D capacity. Some have a small capacity in the public sector, but there is little evidence that it is very effective. Private sector R&D begins to be important in the Stage 2 countries. The NICs spend roughly 1 percent of the value of their industrial product on R&D, which is well below the OECD standard of 2+ percent. It is also below the standard of the recently industrialized Mediterranean group. R&D personnel are relatively low priced in these economies, however, so the ratio of scientists and engineers to industrial product is comparable to that in developed countries.

Most of the Stage 2 countries do not make large investments in science, and the poorer countries make almost none. Data on inventions are instructive, even if inventions are not strictly comparable among countries. Inventions per inventor are clearly highest in the NICs, which have about twice as many inventions per inventor as do OECD countries. They are lowest for the poorer developing countries.

It is also clear from Table 16-2 that less developed countries do not export technology. They are large *importers* of technology. Some of the advanced NICS, such as South Korea, Taiwan, and Brazil, do export some technology, but the poorer countries have no such export capability. This has implications for the way traditional intellectual property conventions have functioned.

For all practical purposes, the international IPR conventions have not worked well for any of the developing countries. I have argued elsewhere (Evenson, 1990) that this is largely because they do not have exporters' interests to protect. They also have little to gain from secrecy removal because foreign inventions are already public. They are not threatened by loss of rights in other countries because they simply do not have such rights. Most of their modest invention is local, imitative, adaptive, and well suited to their own economies. This is a very important type of invention, but conventional patent systems often do not provide satisfactory protection because of their high "inventive step" requirements.

A primary condition for successful IPR piracy is that the pirating country has competence. This rules out most Stage 1 economies, which aspire to be pirates. After all, IPR piracy is closely associated with development success. Stage 1 countries do not, however, have the engineers and scientists to reverse engineer and copy complex inventions, although they can engage in simple counterfeit production. Most of the countries in categories 2b and 2c do have the competence to pirate more complex technologies. If a country is competent and has few exporters' rights to protect, it can engage in some pirating of IPRs under the traditional IPR conventions, which have few sanctions to punish piracy.

It is important, however, to make a distinction in this regard between copyrights, trademarks, and inventions. For developing countries, inventions are more important to economic growth than designer clothes or other goods protected by trademarks and copyrights. Developing countries have a vital interest in using IPRs to stimulate domestic invention. They have a lesser stake in the stimulation of local trademarked goods.

TABLE 16-2 Growth and R&D Indicators by Type of Economy

	Economy Type							
	Traditional (1a)	First Emergence (1b)	Island of Modernization (1c)	Mastery of Conventional (2a)	Transition (2b)	NIC (2c)	Recently Industrialized Mediterranean	Mature OECD
REAL GROWTH								
1980s (Average annual percent increase in GNP)								
Agriculture	2.20	2.20	2.50	3.80	2.80	3.10	2.50	1.40
Industry	2.00	2.10	2.30	5.10	6.00	8.10	3.50	2.20
Total	2.50	2.60	2.80	4.70	5.30	7.10	3.00	3.00
GNP per capita (Average annual growth rate 1965-1990)								
Total	0.50	0.50	1.50	2.40	2.50	6.10	2.80	2.50

				R&D INTENSITIES				
Public Sector								
Applied R&D								
Agriculture	0.40	0.40	0.40	0.60	0.70	0.80	0.80	1.50
Industry	0.10	0.10	0.10	0.15	0.40	0.30	0.25	0.30
Basic R&D								
Science	0.02	0.02	0.03	0.04	0.10	0.20	0.25	0.40
Private Sector								
Applied R&D								
Agriculture	0.00	0.00	0.01	0.05	0.10	0.20	0.50	1.50
Industry	0.00	0.00	0.02	0.05	0.05	1.00	1.20	2.30
				INVENTION INDICATORS				
Inventions/Inventor[a]	—	—	—	0.05	0.10	0.50	0.30	0.20
Import Share[b]	—	—	0.90	0.95	0.81	0.64	0.80	0.31
Export Ratio[c]	—	—	0.00	0.00	0.05	0.10	0.20	1.70

[a]Average ratio of patented inventions per scientist and engineer engaged in R&D.
[b]Proportion of domestic patents granted to foreign inventors.
[c]Ratio of patents granted abroad to patents granted in home country.

Have developing countries actually used their own IPRs to good purpose in this regard? The answer would appear to be that almost all of them have failed to use IPRs in their own best interests. They have allowed conflict over the terms of importing technology to overshadow the possible benefits that they might realize by developing domestic inventive competence and capacity. Their response all too often has been to have laws on the books, and a nominal system of enforcement with little funding, in the hope that this will enable them to avoid paying "unfair" licensing fees. For easily pirated, counterfeit goods, weak IPRs in the past allowed certain domestic groups to gain.

The NICs and the near-NICs have used IPRs to facilitate the development of domestic inventive competence and capacity, but IPRs have not generally been their major policy instruments for this purpose. Most important, they have maintained a trade policy regime that has stimulated exporters of technology to sell to them, and they have purchased huge amounts of technology at low prices. They also have engaged in some piracy, although the extent of their piracy is usually overstated.

What next? Is the enforcement mechanism associated with recent initiatives, including the current GATT round, going to work better than past mechanisms in developing countries? There is certainly going to be much more concern, much more policy dialogue on intellectual property rights, than previously. The real interest of countries without exporters' rights to protect, however, is likely to be as limited as it has been in the past. The shift to using trade rights as a means to protect IPRs is probably going to be the new enforcement regime for most developing countries.

Will developing countries move aggressively to use IPRs to their own advantage? Very few policymakers in developing countries are asking whether IPRs actually stimulate domestic invention and capacity development or whether they improve the ability of developing countries to buy technology on better terms.

There is also a related question about whether the willingness of foreign technology suppliers to provide technology to developing countries is improved by stronger IPRs. The literature does not show strong correlations between direct foreign investment and the strength of IPRs.

The natural comparative advantage of most developing countries is to imitate—in the direction dictated by that country's resources. In the agricultural sector, the scope for adaptive invention is large. Almost all agricultural biological inventions (e.g., plant varieties) are, in fact, so location specific that there is little importing of technology. Utility models are a way of encouraging imitative inventions.

Current developments in IPRs do pose some threats to public sector R&D institutions. Agricultural research systems in the United States are perceived to be threatened by the expansion of IPRs in the plant and animal area, particularly in the area of protecting naturally occurring genetic resources. The patent protection of parent material and genes in plants—as well as, to some degree, the finished varieties of crops—could block public disclosure of research results and the free flow of genetic resources.

Plant breeders' rights, the system used in the United States to protect plant varieties, have by and large avoided this problem. Experience has shown that agricultural scientists have accommodated themselves to the use of breeders' rights. However, the real threat comes from the growing pressure to provide full patent protection to plants and animals, and to block some of the genetic resource exchanges that occur in standard scientific and research systems. For the poorest countries (1a), the only capacity they have for R&D (other than what they purchase through technology contracts) is in the agricultural sector. It is important that their researchers not be cut off from vital flows of genetic resources.

As developing countries recover from the "surprise attack" launched by the United States and converted into the GATT initiative, they may be able to develop a more positive set of tactics and programs than they have managed to date. For the short run, they could bargain aggressively with the developed countries for trade concessions, in return for strengthened domestic treatment of foreigners' intellectual property. Developing countries have not been very good at this. Their stance has been one of continued resistance to strengthening IPRs and including them in the GATT. They would probably do better by acknowledging that IPRs are going to be part of trade laws and policy, and then proceeding to bargain for concessions.

For the longer run, developing countries will have to face up to the question of what represent optimal IPRs from their national perspective. The current debate and conflict are over IPRs that serve the interests of developed countries. There is a great deal of latitude for the better use of existing IPR instruments, the development of new IPR instruments, and more effective administration of IPR systems.

Consider the existing IPR regime. The natural comparative advantage of developing countries depends on where they are in the technology capacity classification. Stage 1 economies can do some adaptive inventing, but they are not internationally competitive. They should be developing the capacity to purchase technology more effectively, learn from experience, and develop minor modifications. The utility model (petty patent) will suit this purpose well. It should be strengthened in most developing countries to reward "blue-collar" invention and to stimulate the incentive to seek new products and process improvements systematically.

Many of the developing countries currently purchase technology through complex turnkey arrangements with foreign firms. They not only pay a high implicit price for the technology they have purchased but often, in the process, lock themselves into that technology setting for a long period of time. Stronger IPRs probably would improve the terms of contracting for firms operating in category 1 economies. That is, the recognition of IPRs might clarify the terms of contracting and lower the cost of the entire package.

For the Stage 2 economies, where inventive capacity exists, stronger IPRs should stimulate the development of this capacity. Again, the utility model can be strengthened to make it, in effect, a national IPR that is almost as strong as the conventional patent. Countries might seek to define an "imitation" patent, in which a foreign and a national pair of corporations shares the rights. Such inventions may emerge from foreign investment partnerships.

Developing countries will have to invest more in their IPR systems than is now the case. They will have to rely on international search services (i.e., to determine what technology has been protected) to a greater extent than they now do, and they will have to bring IPRs into their industrial policy mix in a positive and aggressive way if they are to realize their potential benefits. The current state of affairs is not optimal from a developing country point of view. Yet the system that is in the best interest of the United States and other developed countries is not also in the best interests of the developing countries unless the latter can gain large trade concessions in return for stronger IPRs for foreigners.

The best response to the pressures placed on a developing country by bilateral trade and GATT negotiations will vary according to the extent of the country's scientific and technological capacity. For many countries that now have low capacity, it is unlikely that their IPR systems will change much over the next two decades, but more advanced countries will make IPR changes. If they can free themselves from their fixation on the international dimension of IPR policy, they can realize considerable gains.

DAVID C. MOWERY

I have been asked to discuss the effects on investment and trade of a change in the IPR regime. My focus is on a few topics raised within that context and on the effects of a successful completion to the Uruguay Round package that included Trade-Related Aspects of Intellectual Property rights (TRIPs) and the other provisions being dealt with by the many negotiating groups at work in Geneva. Three reasons are discussed for suspecting that the near-term effects of an IPR agreement may be quite modest, particularly with respect to foreign investment and foreign investment flows among industrialized nations. Those three reasons are followed by a more speculative discussion of an additional issue—the likely welfare consequences for the United States of stronger domestic and international property rights protection. The IPR agenda of the Uruguay Round and the U.S. leadership in bilateral and multilateral negotiations concerning IPRs complement recent initiatives to strengthen the rights of patent holders in the domestic economy as well.

Predicting the effects of an IPR agreement, particularly on investment, requires that one rely on recent historical or empirical data. As Edwin Mansfield and others have suggested in this volume, these data are highly equivocal in the conclusions they support. Certainly, the existing empirical studies support the conclusion that within industrialized economies, and the United States in particular, the effects of IPRs on different industries vary quite substantially. By the same token, the importance of intellectual property protection, particularly patents, to different industries varies quite considerably, according to a survey conducted by Mansfield (1986) and an inquiry by a group of Yale economists (Levin et al., 1987) who surveyed industrial research managers on the importance of patents as a means of capturing value from new technologies.

This evidence suggests that patent protection and other forms of formal intellectual property protection are most important in regard to pharmaceutical and chemical-related technologies. One implication of this finding is that some of the "side payments" necessary to reach the Uruguay Round trade agreement are likely to be made by U.S. industries that derive very little benefit from an IPR agreement. The textile industry, for example, does not rate highly in terms of the importance of patent protection. Yet the U.S. textile industry is likely to be required to provide additional market access in a Uruguay Round package. The distribution of interindustry benefits and costs associated with a Uruguay Round agreement that includes stronger IPRs will, therefore, set the stage for a lively domestic political debate over ratification of the package.

The available data on the differential importance of IPRs across different industries are quite limited and quite dated in several important ways. As noted earlier, the United States is "flying blind" in regard to the costs, benefits, and overall effects of stronger IPR protection.

What are the limitations of these data? Many of these studies focus on relatively large, multiproduct firms. They rarely devote comparable attention to small start-up firms in asking managers about the importance of items such as patents. The authors of the Yale survey have suggested that if more small firms had been included, formal instruments of IPR protection might have been given greater weight. If you ask start-up entrepreneurs about their critical assets, they will in many cases respond that their patent applications or the patents they have received are the key assets of the firm.

Another problem with the results of the surveys is the fact that a patent that is not completely "airtight" may still assist in the creation of a more smoothly functioning market for intellectual property. The cross-licensing that has been quite widespread in the semiconductor and computer industries has been facilitated by patent protection. Patents reduce some of the frictions and simplify some of the negotiations over licensing and cross-licensing transactions so that even if a patent does not completely exclude others, it may still support the development of markets in technology.

The third reason for caution about these surveys is the fact that they draw on managers' reactions during the late 1970s and early 1980s. This period predates the significant strengthening of the rights of patent holders in U.S. domestic courts that occurred with the creation of the Court of Appeals for the Federal Circuit, as well as other statutory and enforcement changes. Thus, the surveys reflect perceptions formed during an era of weaker domestic protection for IPRs. Arguably, were these surveys administered today, the results might differ because of the stronger enforcement and strengthening of patent holder rights that have occurred. This hypothesis is speculative, but it points up a very important research need, namely, the need to update and extend these surveys, both to cover a broader array of firms (software, for example, scarcely existed as an industry at the time the surveys were conducted) and to bring the changed environment more centrally into the responses of managers.

Another area in which the evidence on IPRs is highly equivocal has also been mentioned both by Mansfield (Chapter 5) and by Primo Braga (Chapter 6), who noted the weak nature of the evidence that links investment flows, or even domestic R&D investment, and growth of gross domestic product to the strength of domestic IPR protection. This again reflects a number of problems in the available data. For one, it is not possible to disaggregate different types of foreign investment or to distinguish between R&D and marketing-related foreign investment, and so on, which makes the empirical evidence here extremely weak.

A second reason to expect that the near-term effects of an IPR agreement may be quite modest is the likelihood that a Uruguay Round TRIPS settlement will have fairly modest effects on direct foreign investment, particularly among industrialized economies. Although it is possible to argue that stronger protection for patents and copyrights will strengthen international markets for technologies, reduce frictions and transaction costs, and so on, these markets will still be afflicted with a number of problems that stem from fundamental uncertainty about the characteristics of new technologies themselves. Licensing negotiations, for example, may involve small numbers of people in an environment with many possibilities for opportunistic behavior and withholding of information. These problems will not necessarily be eliminated by the creation, or enforcement, of stronger IPRs.

In addition, the forces driving growth in direct foreign investment, particularly among industrialized economies and between industrialized and newly industrialized economies, are much broader and are certainly subject to a much wider array of influences than those flowing solely from the international IPR regime. The point made by Armstrong in Chapter 8 concerning globalization really concerns the growing importance of access to markets, access to sources of technology, and proximity to customers. Most of these motives operate independently of the IPR regime, and few of them will be altered significantly by the strengthening of IPRs. Some shift may occur in the mix of destinations for certain types of foreign investment, along the lines suggested in Chapter 5, but I think that on the whole the effect will be modest.

A final reason to think that a Uruguay Round agreement covering IPRs may have modest effects on direct foreign investment is the fact that these effects will be heavily influenced by the results of other negotiating groups on closely related issues, particularly those affecting direct foreign investment. The negotiations over Trade Related Investment Measures (TRIMS), antidumping policy, and rules of origin, all are likely to have effects on direct foreign investment flows that are at least as significant as those exerted by an IPR agreement. Stronger IPRs could place smaller firms in a more advantageous position with respect to foreign exploitation of their intellectual property because of the possibility that this stronger protection could facilitate technology licensing, but that is an extremely speculative hypothesis.

Let me say a few words about how a GATT-based IPR agreement might affect the thrust and some of the modalities of U.S. policy. Not only are the effects of a multilateral GATT-based IPR agreement likely to be modest, but also the United States will likely continue to rely on bilateral (gentle and not so gentle) forms of persuasion.

As suggested by Gorlin in Chapter 7, the existing provisions covering IPRs in the "Dunkel draft" contain very important advances, but they also contain major loopholes, particularly with respect to enforcement and the speed with which individual developing countries will adopt the provisions. One can pledge to observe these commitments and still invest very little in enforcement. As a result, the United States will likely continue its bilateral vigilance and pressure, and other industrial economies probably will also.

A second and more important challenge to a GATT-based IPR agreement is an issue raised by David in Chapter 2 that arises throughout this report. Any IPR regime is subject to constant challenge from the evolution of technology. These pressures have operated with considerable force in the U.S. domestic intellectual property system. Sui generis and other forms of protection have been invented, extended, stretched, or otherwise recut to accommodate—sometimes perfectly, sometimes imperfectly—the ever-changing demands of technology.

The adaptation of IPRs to new technologies has not been easy in the U.S. domestic policy arena, in Congress, or in the courts. I would submit that this adaptation process is going to be even more difficult in a GATT-centered, multilateral IPR regime. Challenges will be raised continually by the evolution of technologies, and those technologies will challenge the boundaries of existing forms of protection and place great demands on the consensus-based, multilateral system of GATT decision-making. The Uruguay Round may produce a stronger dispute settlement mechanism, but it will still be barely sufficient to the task. These challenges and problems will continue to force the United States and probably the European Community to rely heavily on bilateral and plurilateral methods to enforce and extend IPRs.

For all of these reasons, the effects on direct foreign investment of an IPR agreement in the Uruguay Round of the GATT are likely to be modest. Moreover, the effects of an IPR agreement will be heavily influenced by Uruguay Round agreements on other issues. I would hedge such a prediction, however, by underlining the great uncertainty that stems from the very imperfect evidence on which any predictions must be based.

Finally, I want to comment briefly on the question raised by Professor Zvi Griliches as to whether tighter IPRs are beneficial for the domestic U.S. economy. I suggest that there are grounds for qualified skepticism about the effects of stronger intellectual property rights on U.S. economic or technological performance. As Ergas (1987) and others have argued, U.S. postwar technology policy has focused heavily on creating new technologies, that is, funding research but devoting very little attention to the adoption and application of these technologies. This policy contrasts with those of other industrialized economies, including Japan and Germany. Indeed, some evidence (see Edquist and Jacobsson, 1988) suggests that U.S. performance, relative to Japan and Germany, is particularly weak in the adoption and application of new technologies within manufacturing. Other evidence (Mansfield, 1988) suggests that U.S. firms have particular difficulties in sourcing or absorbing technologies from outside the firm.

The efforts of the past decade to strengthen domestic and international protection of intellectual property do very little to address this "downstream" adoption weakness that is particularly important to the U.S. economy. Without advocating weaker IPR protection or enforcement, I think that at a minimum, the United States should retain, and perhaps even restore, some balance to the agenda of its technology policy. We certainly need to learn much more than we currently know about the domestic economic effects of the strengthening of domestic intellectual property protection that has taken place over the past 10 to 12 years.

In conclusion, let me just repeat my concern that we are flying blind in this important area of policy. I do not think that we, as a nation, have a complete picture of the costs and benefits of stronger domestic or international IPR protection.

MICHAEL BORRUS

The previous chapters have provoked several observations and thoughts that I want to share. First, there has been a clear premise underlying this report, articulated by Armstrong (Chapter 8) and many others, that the science and technology world and its associated industrial activities are somehow globalizing. Now it is certainly true that the costs of technology development are rising substantially. At the same time, the useful life of products that embody a particular technology is getting shorter and shorter. The result is a need to establish and maintain some kind of global market position to sustain the pace of development by earning sufficient economic returns to keep the game going.

A few data points are useful to illustrate this problem of rising costs and declining time for new technology development. Almost all high-technology companies agree that, based on their experience and spending patterns, developing new technology has become increasingly expensive. The ubiquitous silicon microchip provides an illustrative case. Half a decade ago, a leading-edge chip design could cost $1 million to $2 million; the associated process technology, $50 million to $100 million to develop; and the manufacturing facility, an additional $100 million. Today, chip designs are in the $10 million range, process technology ranges upward of $250 million to $500 million, and manufacturing is about the same. In short, development of new silicon technology costs on the order of 5 to 10 times its cost only five years ago.

Moreover, the breadth of technologies that must be mastered seems to be widening at similar rates, which adds dramatically to the expense. Consider, for example, just the manufacture of leading-edge electronic products, such as a hand-held television-videocassette recorder or a notebook computer. Manufacturers have to master semiconductor, display, battery, packaging, precision mechanical and magnetic component, sensor, software, microdesign, and systems integration technologies, among others—a far more complex task than was required in producing the televisions or personal computers of just a decade ago. Pharmaceutical companies similarly have to master biotechnological approaches in addition to chemistry, a dramatic broadening of technological possibility and requirement.

Simultaneously, the useful lives of high-tech products are declining dramatically, thereby reducing the available time to recoup the accelerating expense of technology development. Pharmaceuticals again provide a good example: injectable cephalosporins for bacterial infections were first introduced in 1965; 12 years later, sales of their second-generation successors finally surpassed those of the first generation, but the fourth generation began to overtake the third in only a year (Nevens et al., 1990). Similarly, electromechanical-turned-electronic products (e.g., typewriters, telecommunication switches, cameras, and automotive subsystem controls) used to have a useful life measured in decades; today, such products barely last three to five years before successors overtake them. New workstation generations now appear roughly every two years, a far cry from the decade-long useful life of the mainframe computational capability they are subsuming. Also, Japanese automakers have cut the time of new automobile development and manufacture from eight to three years.

These constraints mean that major developers of new high-tech products have to move fast to market, and attain global scale and position, just to have a chance of recouping and continuing their accelerating investments. The very same constraints push toward increased cooperative activity to defray the accelerating costs of new technology development and of global market positioning. These constraints also create the need for rapid, responsive, and worldwide regimes of intellectual property protection because innovators have so little time to stake their claims and recoup their costs before the next innovation lays waste to their invention.

Although this discussion has presumed that these very real pressures equate to globalization, I think the premise is worth examining. I would first observe that in most technology-based industries, particularly electronics and informatics, "globalization" is in some sense the reciprocal of the massive erosion of the once-predominant position of U.S. science and technology, of U.S-based production activities, and of U.S.-owned companies. In place of U.S. predominance, there are now major players, substantial leading-edge know-how, and new technologies resident in places outside the United States.

These new capabilities and players seem to be clustering geographically, due to political and economic forces that range from the North American Free Trade Agreement to the trade and investment patterns of Japanese industry in East Asia. There seems to be regional clustering: a Japan and Asian cluster, an American cluster, a European cluster, and so on. The reality of regionalization, rather than the image of symmetrical globalization, seems to me to have very different implications for intellectual property protection—under common constraints of collapsing time and increasing cost—than the paradigm that has been assumed during much of this volume.

For one thing, the know-how and technology that are increasingly resident in each of these regions are not equally accessible between regions. The relevant institutions of science and technology are quite different in their degree of openness and their accessibility from one region to the next. The ways in which domestic economies and industries are organized are quite different from one region to the next. The science and technology practices, and related production activities, are quite different from one region to the next. Not least, as the discussion has emphasized throughout, the intellectual property regimes (and enforcement) in each of these regions are quite different.

Technological know-how is much easier to access in some regions than in others. Compare, for example, the United States and Japan. In the United States, relevant institutions such as universities and national laboratories are quite accessible. The social networks of technical peers and engineers are relatively easy to enter. Skilled personnel, in particular, leave companies; they go from one to another. Short-term capital market constraints in the United States often force new companies—small companies with good ideas—to license technology quickly. Firms that embed certain kinds of technology and know-how can be acquired outright.

Compare that with the situation in Japan, where the relevant mechanism for technology development is largely the corporate laboratories. They and the relevant social networks are much harder to penetrate. The same kinds of capital market constraints do not exist and, therefore, do not create the same kinds of motivations. Firms cannot be acquired as easily. People do not change jobs with the same frequency or ease. Technology thus tends to accrue locally in the United States, but it diffuses very rapidly. It tends to accrue locally in Japan also, but it does not diffuse nearly as rapidly and is not nearly as accessible.

Equally important, in addition to being asymmetrically accessible, the technological know-how that is clustering in these different regions is also asymmetrically appropriable. In electronics and informatics, the kinds of capabilities that still exist at the leading edge in the United States (e.g., software, design, and architectural skills) are more easily appropriated by others. For example, software can be pirated with relative ease. Compare that ease of appropriation with the high-volume, flexible, rapid-cycle-time manufacturing skills that are embedded in the practices and the people of a Japanese corporation such as NEC. Those kinds of skills are much harder to appropriate.

The differences in accessibility and appropriability are complemented by another major regional difference—the capacity of firms and economies to cycle the know-how to which they have access and to accumulate it over time into advances in science and technology, new development and new production capabilities. Again, there is substantial variation in the ability of firms, operating within different economic circumstances in different regions, to accumulate technology. Faster accumulation is likely in Japan and Asia, for example, than in the Americas.

These regional differences exist at the same moment that time and cost constraints create the need to have closer collaborative relationships between regions. Thus, there are asymmetrical access, asymmetrical appropriability, and asymmetrical accumulation of technological know-how in different regions at a time when there is an increased need for sharing. Now, it seems to me that if your know-how is harder to appropriate, harder to access, harder to keep up with, and you are entering into shared relationships with companies from other regions that have know-how that is easier to appropriate, easier to access, and easier to keep up with, you are going to have a much greater capacity to appropriate returns from the technology in ways other than strict intellectual property protection. That is, your more exposed partner is likely to be far more interested in strict intellectual property protection than you are.

Indeed, one of the main reasons for the recent U.S. interest in a strict international IPR regime is precisely this: as the relative competitive abilities—production skills, for example—of the relevant U.S. industries have eroded in international markets, so too has their relative ability to appropriate know-how in ways other than through strict intellectual property protection. What is left to protect—information, software, ideas—really can be accomplished only through a strict IPR regime.

Given the asymmetries in regional technology access and appropriability at a time when there is a push toward shared activities, the most likely future result is increased international disputes. The issues are going to be phrased in terms of trade and investment, and they will incorporate intellectual property concerns. No matter how the Uruguay Round turns out, there will be many more disputes as these regional asymmetries begin to play out in real business relationships. The attempt to incorporate intellectual property concerns into the GATT can be seen, then, as a precursor to this future of increased conflict. However, there are going to be increasing pressures to deal with intellectual property disputes in venues and in ways that lie outside the traditional intellectual property mechanisms because existing intellectual property mechanisms appear to be ill-equipped to deal with this conflict.

The bottom line is this: Inevitably, increased protection will always lag the pace and costs of technological advance and the controversies that regionally asymmetrical access and appropriability are going to generate. For that reason, the U.S. approach seems to me to be quite defensive, trying to hold ground by increasing intellectual property protection. Given the potential controversies that are at stake, the need for shared research and development, the increased costs, and the decreasing cycle times for technology development, perhaps the United States ought to consider a different approach by looking at it from a different perspective. Perhaps the United States ought to consider complementary alternatives to the strict focus on increased IPR protection. Such a complement might include strengthening the nation's and firms' capacities to cycle technology more rapidly, to appropriate know-how that is developed elsewhere, and to diffuse and use new technological innovation more effectively. It may be somewhat heretical, but it is surely worth echoing one of Paul David's points in Chapter 2, namely, that it is not obvious whether an economy derives greater long-term benefits from stricter IPR protection that rewards innovation or from protecting less and choosing to favor the more rapid exploitation and use of technology.

Perhaps a preferable model might be to favor exploitation and use, rather than stricter protection, that could be accomplished through a liberal licensing regime, with reasonable royalties and without the ability to use blocking injunctions as a remedy. Such a liberal IPR regime might also include some degree of public subsidy for innovation. This is a model much like the one under which AT&T Bell Laboratories operated from 1956 through 1984. That model just might be preferable to the elaborate, and increasingly complex, system of sui generis protection of intellectual property that some have advocated in the report.

ROBERT W. LUCKY

My role is to tell you about the trends that I see in technology development that are likely to affect the environment for intellectual property in the future. I would like to say that I am a working researcher, but that would be self-flattery. I am a research manager, and my goal—my point of view—is to find ways to create incentives for investment in research. I am really very, very concerned about this.

To the degree that IPR protection can help in that effort, I am all for it. To the degree that it creates disincentives, I am not. After thinking about this a fair amount, my conclusion is that it would be wrong to put the burden for creating all those incentives on intellectual property alone. Intellectual property rights are only part of a much bigger fabric of government and corporate policies, which involves tax laws, trade barriers, competency of management, the market, and so on.

I drive to work in the morning and, as I approach my labs at AT&T, I go past a water tower with three legs. It is shaped like a transistor. I drive underneath it, and it reminds me of AT&T's historic past. The state of New Jersey has placed historic markers along the road proclaiming that, from this site, the first signals were transmitted to a satellite; the first signals from outer space were received and radio astronomy was born in 1927. The discovery of radiation from the "big bang" won a Nobel Prize for two of my colleagues at Bell Labs.

As I drive in, I think of what we are doing now at Bell Labs. I feel the burden as I think that I just closed down radio-astronomy research. People said, "What is in it for AT&T?" There was a letter to the company newspaper in my in-box yesterday from an employee who asked why we are investing money in future technologies, such as computer-generated environments, when we are laying off people. My having to answer this is not an enviable position.

I go to meetings, and the business people have taken over— largely what we call "bean counters." They say, "Why are we spending money on research, explain this to me again? This is an investment. We can put our money in research, or we can put it in something else. Now, explain to me how putting the money in research is an investment?"

We tell them, "You are going to get a lot of money back; all the studies show that." They say, "Gee, that is really nice, and when do we get this back?" We say, "Well, in 20 years, give or take a bit." They say, "That does not sound so good." Then they ask, "Who gets the money back?" and we are forced to admit, "Well, probably not you." Those are the facts. You cannot get around them.

From the standpoint of AT&T, two threats—one from outside and one from inside the United States—occupy a lot of my attention. The international threat, epitomized by the Japanese—although it certainly does not involve them alone—is the speed with which they are able to capitalize on invention. I do not blame them for it, but we in the United States just have not been very good at this. We have got to get our own act together. I think of something like fiber optics, which has a front end of invention (the basic R&D) and a back end (the embodiment of research in a product). AT&T built up its fiber optics business over several decades—built it up to the point at which it suddenly became a good business—and then the rules of the game were changed. It becomes a question of business processes and manufacturing competencies, and that is where the payoff is. The problem is that the payoff from the front end—the research that led to the fiber-optic technology—does not seem to exist. So that is one threat.

The other threat comes from within the United States, from other companies that are attempting to compete with AT&T. They are not all small, start-up companies; one has about 35 percent of the market. Yet they undertake no—zero—research. They are a living demonstration that you do not need to do research to be successful in this business. When the bean counters look at it, they say: "We have to compete against this company. Let's add up the bottom line. Where are we spending? What is the cost structure in these two businesses? What is this line, research, with nothing over there on the other side and why we are doing this?"

I think about this again and again. One of our competitors has, on occasion, indicated that it is even proud of the fact that it has undertaken no research. In fact, the chief executive officer of that company reportedly has said that Bell Labs was "AT&T's expensive hobby. You do not need this to succeed." I think about this and how they get the benefits of AT&T's research without investing in it and about what would be fair.

I do not mean just access to technology. There is much more than that out there. There is this whole seething, boiling, international community of researchers and developers who are building a knowledge pool of how to do telecommunications. There are the standards bodies, the literature, the conferences, and all the get-togethers and meetings through which people accumulate wisdom. I have no quarrel with companies that contribute to the knowledge pool, but what happens when one or more companies opt out? This is, in fact, the case today.

Let me address the issue of motivation. I am fascinated with the idea that IPRs are intended to provide incentives to inventors to make their results publicly available in return for exclusivity. When you deal with actual researchers, however, you realize that the overwhelming majority are not motivated by this at all. Indeed, their motivation is to publicize what they have. Period. This is the world they live in. In fact, the thing that upsets researchers most is when they submit a paper for company approval for publication, and it is turned down because it may have patent significance. They want to convene review boards, and then they say, "Maybe I should work on something else, something that is not so applicable, rather than something that has patent significance if I am not allowed to publish." Publishing is their world, the classic motivation of researchers.

I convened a focus group of researchers the other day, and I asked them about intellectual property and what it meant to them, and whether it was an incentive or disincentive. One said, "I don't know; if it went away, it would be no big deal." I am less concerned about them than I am with business leaders and how they see intellectual property. In our company, it is a defensive posture. We cross-license everybody. It is just a way of opting out of this whole system, isn't it?

Let me, finally, review a series of trends that I see occurring in technology that make a difference in the research environment. *Globalization* is the first. AT&T Bell Labs used to be a unique place; we were alone in the world. Now technology is strewn about the world, and competence is everywhere. Further, we are constantly forming consortiums and joint ventures. As a result, we are never quite sure who are our friends and who are our enemies. It is "mix and match" in business today, and it is happening all over.

A second trend is *collapse of the time scale*. There is not time to do anything any more. In fact, outside of research, people stop writing things. We are fast becoming an oral culture, and in terms of getting a profit, you have got to get in and get out. It has little or nothing to do with this large flywheel of research.

A third trend is *rising complexity*, for example, programs with multimillion lines of code, networks that one does not understand anymore. There is an accumulation of complexity that makes it much harder to make progress. Things have gone beyond the understanding of individual people.

A fourth trend is that *the physical world is changing to the virtual world*. No one in the telecommunications industry cares about hardware devices any more. Don't invent another transistor; I don't want it. Tell me about software and things like that. So all of the telecommunications companies are cutting back on their physical sciences work.

A fifth trend is *instant access to information everywhere*. Networks exist for every kind of thing. Since the disappearance of limitations on information flow, fashions sweep the world of technology, and there is instant information access everywhere. One day it is high-temperature superconductivity; the next day it may be cold fusion; then it is something else, and everybody flits from here to there at the same time.

A sixth trend is that *the world is now run by standards and open architectures*: "Let us get together and develop this, and then we will sort out who gets what." I do not know how this kind of world works at all. I do not understand why I put 30 years of work into picture processing, to have it standardized and given to everyone, with the returns going to the people who now can manufacture the best in that system.

A final trend is *the cost of research*. It is rising much faster than the cost of living. In the world of electronics, where one now speaks in terms of pentaseconds and gigabits and things like that, the potential dollar costs are enormous. So it is a world full of complexity, difficulties, and mystery; and as a research manager, I have a problem.

EUGENE B. SKOLNIKOFF

I want to step back a bit from the details and think about the IPR issue as it fits into a larger international framework. First, there is the obvious point (but worth repeating) that change is genuinely a constant. Armstrong (see Chapter 8) and others have made the point earlier in this report. Technology changes; it is dynamic, and it is hard to anticipate. I would generally agree with the list of changes given by Robert Lucky, but let me review them and add some.

Aside from the dynamic nature of technology, one must recognize that knowledge and competence in science and technology are spreading rapidly and growing all over the world. The United States is no longer the dominant power in every field, although it probably has the greatest technological breadth. In some areas it remains ahead, whereas in others it no longer leads, and that has become a very important factor in the IPR debate. It has been suggested by an earlier speaker that unauthorized access to technology is the natural result of the spread of competence and knowledge; there is likely to be much more of it in future years. As a result, the incentives and need for, and the purposes of, intellectual property protection are going to change as technology changes, as indigenous capabilities change, and as growth in competence continues to spread around the world. While unauthorized access to knowledge will continue to grow, the motivations of many countries, particularly the NICs, regarding intellectual property will change. There is likely to be growing interest among many countries in some form of national and international protection that will parallel the spread of knowledge and indigenous capability.

Yet, we in the United States can expect from that a continuous battle between IPR protection and unauthorized use, and continual fighting over the details of agreements and over what ought to be included. I believe we should be very wary of thinking that we can ever bring the battle to closure, to some sort of an international agreement that could settle the issue. It is an unrealistic and inappropriate goal, which cannot be accomplished. Although we may want to seek agreements that rationalize a variety of the IPR problems we face today, it is unrealistic to expect that we could create one integrated system somehow having an adequate dispute resolution mechanism while being able to keep up with technological change. I realize that this assertion flies in the face of some of the views expressed in this report, but I believe that it is a fact of life.

In the political and economic context, we have to keep in mind the increasing globalization of many factors, not only technology but also markets; companies; social, economic, and political relationships; and environmental and other issues. We are living in the midst of a rapidly changing scene in which international economic competition, especially in high-technology trade, is emerging as a major—perhaps *the* major—international political issue for the future. It is replacing security competition. Yet while we see the very welcomed loss of Cold War confrontation, ethnic, national, and local forces and concerns have been unleashed that will create their own security dimensions for many years to come. Thus, ironically, we see a rise in nationalism accompanying globalization. In fact, I believe nationalism in the coming years will probably be the strongest political force with which we have to contend.

The United States, too, is engaged in an economic competition with strong nationalist tendencies. We obviously are not doing terribly well in this competition, for a host of reasons. It is resulting in rising pressure to protect the products of our huge R&D investment. In that framework, IPRs are viewed as one of the elements of this international trade competition, one of the elements of protectionism. They will sometimes be used as a barrier to the entry of foreign goods.

International trade itself, the larger context in which IPRs must be seen, is now very different than it was. Michael Borrus discusses the asymmetries between the United States and Japan earlier in this chapter, and I am in close agreement with his comments. I want to put this in a slightly different context, however; that is, international trade issues, which were seen in the past as matters of export subsidies and of tariffs and duties of various kinds, now go much deeper into society. International trade issues now involve cultural issues, different forms of economic structure and tradition, and even questions about the national support of R&D, which is in effect a kind of subsidy. Moreover, international trade questions are going to get very much more difficult in the years ahead, whether or not the Uruguay Round is concluded successfully. Intellectual property rights will be embedded as one piece among many in international trade negotiations, negotiations that are subtle and complex and that will go to the heart of the sociocultural structure of a country.

Let me mention some other issues relevant to the international scene. There has been much talk about developing countries and the North-South relationship. I can only foresee a continuing conflict of interest between the North and the South, one that is moderated by the growth and transition of some developing countries into NICs whose views of the world—and of their own self-interest—will change and move closer to that of developed countries. I think that the developed countries' disinterest, if I may put it that way, in the problems and difficulties of developing countries, which has become apparent over the past few decades, is likely to continue, at least with respect to the poorer nations.

Several factors may change, however. One is that developing countries will have increased bargaining power against the North because of the emergence of global issues that require their cooperation. Climate change is an example of an issue that will require trade-offs, and the countries of the South are likely to attempt to draw linkages among quite disparate issues as a means of leverage to obtain resources, transfers of technologies, and other benefits. This will require a long time to work out, because I see little indication that the United States or other industrialized nations are yet seriously willing to put substantial resources into a North-South bargain. They will do so only if they have no other choice.

The last area I want to mention is the United Nations and international organizations in general. They obviously are essential for dealing with IPRs and the other problems discussed in this report. It remains true that the United States prefers bilateral pressure and bilateral relationships to working through multilateral organizations. I do not think that this policy preference is likely to change soon, even with a change of administrations. In the effort to get broader participation in international agreements, multilateral organizations inevitably end up being slow and unwieldy, which has the effect of reducing the power and influence of the United States in that context.

We will continue to use the United Nations because we have no choice. We will participate and develop positions as necessary, and there will even be considerable momentum toward new international agreements. However, those agreements will be limited and hard to reach, and they will usually lag behind technology. For these reasons, the United States (and probably other developed countries) will be reluctant rather than enthusiastic participants in international solutions to IPR problems.

Finally, I must take note of a relevant aspect of the American political process. I would argue that despite all the strengths of this nation, we are increasingly fragmented; our political process makes it difficult to come to grips with complex issues. Divergent and fragmented interests have so many different sources of leverage in the formulation of policy that it will continue to be very hard, especially on contentious issues such as IPRs, to reach agreement and to implement those policies consistently in the international arena. I believe that we are not in a good position to handle this aspect of our role.

REFERENCES

Edquist, C. and S. Jacobsson. 1988. Flexible Automation. Cambridge, England: Blackwell.

Ergas, H. 1987. Does technology policy matter? in Technology and Global Industry, H. Brooks and B. Guile, eds. Washington, D.C.: National Academy Press.

Evenson, Robert. 1990. Intellectual property rights, R&D, inventions, technology purchase, and piracy in economic development: An international comparative study. In Science and Technology: Lessons for Development Policy, R.E. Evenson and Gustav Ranis, eds. Boulder, Colo.: Westview Press.

Levin, R.C., A.K. Kleverick, R.R. Nelson, and S.G. Winter. 1987. Appropriating the returns from industrial research and development. Pp. 783-820 in Brookings Papers on Economic Activity. Washington, D.C.: Brookings Institution.

Mansfield, Edwin. 1986. Patents and innovation: An empirical study. Management Science 32:173-181.

Mansfield, Edwin. 1988. Industrial innovation in Japan and the United States. Science (September 30):1769-1774.

Nevens, T. Michael, Gregory L. Summe, and Bro Uttal. 1990. Commercializing technology: What the best companies do. Harvard Business Review (May-June):155.

Weiss, Charles, Jr. 1990. Scientific and technological constraints to economic growth and equity. In Science and Technology: Lessons for Developing Economies, Robert Evenson and Gustav Ranis, eds. Boulder, Colo.: Westview Press.

Discussion

During the conference, the panel moderator, Arden L. Bement, Jr., invited members of the audience to offer comments and to ask questions of the panelists.

A copyright attorney in the audience expressed sympathy with Robert Lucky's comment about the establishment of technology standards and the inability of R&D-intensive companies to capture fully the benefit of many years of investment when the technology it has developed on a proprietary basis suddenly becomes the international standard. She noted, however, that there is a problem in developing consensual regulations when, for example, proprietary software is being included in a standard. Some would advocate an approach whereby, if proprietary material is to be brought to the negotiating table, it must be labeled as proprietary and a prior agreement must be reached to license it under certain terms and conditions. This differs from Lucky's view that technologies can simply be appropriated as part of the standard-setting process.

Robert Lucky: I did not mean to give the impression that the typical pattern is for an individual and company to work on something for 30 years and then have it expropriated as a standard. It does happen, but the more usual practice is that the standard arises out of an international collaboration that is very deep and where there is a real process of invention taking place in the standards bodies. I recall a meeting at which the chief scientist of AT&T met with researchers and said, "Whether

> you people like it or not, the future systems are being designed by the standards committees."
>
> I also want to comment on the software issue and whether the software would be put on the table and would be proprietary. What is put on the table are *algorithms*, rather than software. I share the feeling of many that the real essence of software is the algorithms, really the unpatentable part of it. That is where the real genius and the real invention are, not clothing the algorithms in the code itself. If the algorithms are put on the table, people think they are mathematics and they are available to the world, but that is what the work and investment went into.

Another participant commented on Robert Lucky's frustration about his company's inability to recapture fully its R&D investment and speculated whether it was representative, in microcosm, of U.S. unwillingness to recognize that it had lost its technological hegemony. He suggested that this raised again the question of whether the United States should support a differentiated or undifferentiated international intellectual property right (IPR) system, and he asserted that a differentiated system runs counter to the U.S. view of the world, circa 1945 or 1950, and that a differentiated system may well make more sense in terms of the realities of the current world.

The speaker suggested looking at the evolution of IPR issues over time in the context of trade negotiations. He used Robert Evenson's country categories and asserted:

> Trade preferences have absolutely no value for the Bangladesh's, the type 1 countries of the world. They were important for a short period of time for the newly emerging countries, and they became quite irrelevant very quickly once the Taiwans and Koreas made it in the export world. When they were withdrawn, there was not a whimper. A differentiated IPR system, which might involve letting the developing countries of the world have a different kind of patent protection and possibly longer periods for differential treatment in the length of patents and so on, is now being discussed in Geneva and elsewhere.
>
> But the NICs are now realizing, and not just under pressure from the United States, that it is in their own interest (particularly if they are investing elsewhere) to worry about a homogeneous kind of IPR protection. So, one can have differentiation without great cost to the innovation process in the advanced countries and still provide incentives for investment in the LDCs (less-developed countries). Unlike some of the panelists, I do not believe that there is a necessary conflict here. I think these are global issues that do have positive-sum games, if we are just a little more flexible in our own attitude and realize that the world is changing and we have to march with the times.

A government official in the audience questioned whether Robert Lucky's example of companies that compete with AT&T in telecommunications' markets, but do not invest in R&D, was something analogous to the case of secondary drug manufacturers. That is, is it a case of a generic type of manufacturer attempting to create a niche to make money on an older technology that is in the public domain, or is it really stealing current technology?

Robert Lucky: I really did not say that they are stealing, but maybe they are. It may be wrong to equate the other company in my example with AT&T. We really are two different kinds of companies. The other company is more equivalent to one of the business units within AT&T, but that business unit happens to contribute a great deal to research, which is not duplicated by the efforts of AT&T's competitors. So, I believe that this is a process of international knowledge pooling, and here you have a company that is not contributing to the knowledge pool but is living off the work contributed by others.

Another government official stated his sense that IPRs could be important, but that many other things affect technology flows and international competitiveness, and they are more difficult to get at. He asked whether some of the attention to the problem of IPRs both internationally and within the U.S. government is simply a way of dealing with something more tangible because we are impotent with regard to the other factors.

David Mowery: In my limited and perhaps somewhat heretical view, intellectual property reform bears a close relationship to antitrust reform in that these are gestures that do not have on-budget costs. The consequences are uncertain, but they are political gestures that certainly suggest that something is being done. In recognizing the uncertainty of the ultimate benefits and costs in both of these policy areas, they are actions that the federal government can take for which it does not bear the whole cost.

Eugene Skolnikoff: Let me add that while I agree with David Mowery, one ignores at one's peril the momentum of government. It is worth recalling that the patent system has been around since 1790 and there are people in Washington very much concerned with it. A whole industry exists, and it is very natural for the government to want to focus on it. Thus, I do not think that we focus on IPRs because we do not want to do something about other problems. Nevertheless, the net effect is that we are not doing very much about those other problems.

An audience member offered four brief comments on what has been going on in industry with respect to IPRs. He noted that uncertainty in the IPR area is worse than wrong decisions, so it is important to do whatever can be done to accelerate the process of stabilizing the rules. His second comment was that failing to deal with IPRs because they may not be the *central* competitiveness problem will just make things worse, so we should do what we can even though the IPR community does not have any direct influence on capital formation, which is a major business problem.

The speaker's other comments offered a longer-range perspective. One had to do with the viability of giving an IPR monopoly to a sequence of things (e.g., numbers or genetic code) that is found in nature. He argued for the need to think seriously about what we are doing (or not doing) with sequences of things that are put together in a novel way by human beings. He suggested that as the world moves into a global information age, a lot of thought is needed about that simple construct or question.

Finally, he took note of the assertion made earlier that the historical and cultural roots of our monopoly abilities, grants or whatever, are really different in other cultures than they are in the Western World. If true, this point will require attention from the IPR community before international rules can be solidified.

An industry representative commented that he had hoped to hear more about IPR problems in the context of the aerospace sector. He focused particularly on the effort to develop a national aerospace plane by three airframe manufacturers and two engine manufacturers, who have formed a consortium to build within about a decade an airplane that takes off like a plane, goes to space, and returns to earth. They have a contract that involves two departments of the U.S. government. The industry executive raised a question about the technology implications and rights to data that bear upon industry consortiums, particularly in areas of high risk.

Robert Lucky: The whole business of consortiums has suddenly come of fashion. AT&T wants to do it. I am in the midst of forming three consortiums with different groups, including other companies and universities. In every case, we have submitted proposals that say we will work out the intellectual property rights if you give us the money. Everybody knows that if you bring in the IPR people, it will kill the whole thing. This is a very fundamental issue. For example, in the case of the consortium on optical networks we have with a bunch of competitors, we are asked, "When are you working for the consortium and when are you working for your company?" We just cannot resolve it, and yet we want to pursue this kind of cooperative behavior.

A final commentator noted that there has been a significant paradigm shift underway regarding the way in which R&D is conducted. He suggested that it used to be undertaken in a serial fashion, that is, first "R," then "applied R," then "applied D," then "D," and so on. However, in the past decade or so, this has shifted gradually into a more parallel paradigm in which research, market input, design input, and manufacturing all proceed in parallel. He speculated that the paradigm shift may have been brought about by radical changes in the time scale in which new technology approaches the marketplace, as well as by the ability to ask marketing and engineering questions in a more scientific way. One implication may be that research funds could become easier to obtain when the work goes on in a parallel, rather than a serial, mode.

Moderator's Summation

ARDEN L. BEMENT, JR.

I would like to pass on some reflections on this interesting panel discussion. First, it has been clear throughout the conference that there are many dialectics and dichotomies in this field that are driving changes. We can expect that this evolution in thinking will continue.

Second, there clearly is a need for additional research and intellectual content. It is like one economist addressing another and saying, "Well, if this works in practice, will it really work in theory?" If we are going to do research, it really does need to be interdisciplinary; researchers from law, business, management of technology, and public policy are needed so the IPR problem can be examined in a much broader context.

Third, in listening to the various presenters, it seems that ideology is giving way to pragmatic reality. This may be due to external pressures in the global context that are causing many nations to redefine their self-interest and to trade-related internal pressures stemming from entrepreneurial activity in the development of a domestic economy. It seems that those countries in which political, economic, and cultural ideologies are less strongly coupled probably will be able to adjust more rapidly than those countries where these ideologies are much more closely linked.

Fourth, we talk about strong versus weak IPR systems in terms of high stimulation and high diffusion. It occurs to me that the LDCs are caught somewhere in the middle. One might observe that diffusion rates are becoming less and less dependent on the strength of the IPR regime in terms of concurrence of discovery. There is greater investment around the world in R&D in general, but also a leveling of technical capability and R&D infrastructure. So it does not necessarily follow that a weak IPR regime equates with higher diffusion rates.

Further, based on comments made during the conference, it seems that three conditions must be met if a country is to benefit from a strong IPR system. One is a strong national science and technology infrastructure. A second is industries that are skilled in developing globally competitive strategies, as well as the management of technology in the global context. The third is strong protection mechanisms. Having a strong regime without a strong protection mechanism is much like having a national border that is not defended.

Moreover, the United States is not strong in all three areas. We certainly have a strong science and technology infrastructure, but we are not uniformly strong across our industries. We do not possess all the capabilities necessary to develop effective global competitive strategies and the management of technology in terms of not only the acquisition of technology but also its adaptation and integration for competitive advantage.

Fifth, during the conference we discussed congruency in international IPR systems, which it seems to me is going to be more and more important because of the collapse in time and costs, as Robert Lucky points out in Chapter 16. If you look at the time span from the application for intellectual property protection, to the discovery of an infringement, to final litigation and eventual resolution, one could go through—over that time span—three or four product generations or product cycles; perhaps as many as 30 improvement patents; and in the case of biotechnology, perhaps three or four progeny generations.

In looking to the future, it seems to me that we are now getting close to an IPR analogue of quantum theory: namely, the "attempt rate" against IPR barriers will be much greater and will accelerate. Given enough time, the opportunities to circumvent or tunnel through the competitive barriers that are established by IPRs will become greater. It also occurs to me that there is an uncertainty principle at work as well—that time and uncertainty or error rate seem to be equal to a constant. The more you try to compress time in the protection of intellectual properties, the greater is the error rate, and that error rate equates with the litigation costs. So we could be facing a point of diminishing returns in trying to push that too hard.

Sixth, in the developed countries, much more so than in the LDCs, invention tends to be more "want oriented" than "need oriented." Creative marketing is really translating wants into perceived needs so that market invention is becoming almost as important as product invention. Whereas in the past, necessity was the mother of invention, more and more invention is becoming the mother of necessity.

So again, the ability to establish markets as a competitive strategy will become increasingly important. The significance of that is if I can flood the market with patents, if I can be the first to commercialize and perhaps the first to dominate in global market share, and if I can do that before I have to face litigation, then I can certainly afford to be fairly expansive, fairly generous when it comes to possible infringement. This is where the time constants become very important, especially with regard to relatively strong IPR regimes.

Finally, we have discussed sui generis IPR approaches in almost every session of the conference. This strikes me as the "je ne sais quoi" part of IPRs; namely, it deals with "intrinsic beauty," where embedded intelligence is going to become more and more the way in which value will be created in the future. I expect that this will be an expanding area of IPR protection, and the degree to which other parts of the IPR regime will be able to adapt to it is uncertain.

Coda

Issues for Future Research

As might be expected on a topic as complex and multifaceted as intellectual property rights (IPRs), a variety of rich and promising avenues of future research have been discussed in this volume. The purpose of the conference, however, was not to produce a set of recommended priorities for future IPR research. Nevertheless, given the important intellectual explorations that are clearly needed with respect to both the domestic and the international aspects of the IPR problem, we present below an *unprioritized* list of issues, derived from ideas raised in this volume, that may warrant further investigation in the short term. We leave it to others to evaluate the merits and relative priority of these research issues and to formulate appropriate strategies for responding to them.

Research Issue: The introduction of IPRs throughout the world has involved propagating as broadly as possible a *Western* cultural view of the concepts of ownership and rights. Some non-Western countries have voluntarily adopted Western-style IPR laws in the process of modernization. Western countries cannot necessarily count on a continuation of this pattern of adoption, however, as the negotiating positions of Brazil and India in the Trade Related Aspects of Intellectual Property Rights (TRIPs) negotiations appear to reveal. Other cultures and legal traditions, including those in Asia and throughout the Islamic world, may have different concepts of optimal ways to encourage creative participation in society. These alternative cultural traditions and practices must be better understood in building a new global IPR paradigm.

Research Issue: The argument that weak forms of IPRs or high levels of piracy have possibly negative effects on innovation and economic growth must be taken very seriously. This line of thought must be contrasted with the view expressed by Paul David in Chapter 2 that, under some conditions, IPRs can have seriously detrimental consequences for the process of innovation. There are currently few data on the effects of IPRs on invention and innovation under different conditions that might help resolve this debate.

Research Issue: The effects of high levels of IPR protection on the economies of developing countries have been little studied because the field of economics has begun to devote serious attention to the IPR problem relatively recently. Development theory previously assumed that the principal route to development was through capital formation.

Research Issue: No clear consensus has been reached in this volume on the superiority of a uniform, high-protection, global IPR system over a differentiated system, which is determined by individual national interests. Here again, adequate data do not exist to substantiate either view. Moreover, analyses of the short- and long-term benefits to developing countries of one approach versus the other are almost entirely lacking.

Research Issue: The lack of good data and information on the benefits and costs of strong IPRs to developing countries will likely affect the outcome of the current General Agreement on Tariffs and Trade (GATT) negotiations on TRIPs. How far can the United States expect to push developing countries to strengthen their IPR systems when it cannot be shown that the current level of protection is too low or that stronger protection would be in their interest?

Research Issue: The United States has been able to make headway with the newly industrialized countries (NICs) on IPR issues through the use of bilateral negotiations and the threat of trade retaliation. It is debatable, however, whether the United States will have much further success with this strategy. Is there evidence that losses due to IPR infringement have declined in those countries? Is there evidence that stronger IPR protection by the NICs has stimulated technology transfer or indigenous innovation?

Research Issue: In Chapter 5, Edwin Mansfield suggests three types of studies that might help to estimate the size of the effect of stronger IPR protection on the promotion of indigenous technological innovation activities in developing countries:

- a study to determine the effects of stronger patent protection on the size and composition of the R&D expenditures of firms located or headquartered in selected developing countries and on their rate of commercialization of new products and processes;
- a study to explore the costs and benefits to developing countries of modifying their patent systems; and

- a study to estimate the effects of stronger IPR protection on the size and composition of R&D expenditures by multinational firms in developing countries.

Research Issue: How precisely should IPR laws attempt to define and focus on specific technologies? The process of scientific and technological advance is changing in ways that challenge the effectiveness of IPRs in stimulating economically valuable innovations. Is the current IPR system capable of adequately handling new technologies? If not, is it preferable to modify existing IPR forms or to examine alternatives? What might be the nature of these alternatives?

Research Issue: One of the concerns expressed about sui generis approaches to IPRs is that they would lead to piecemeal legislative solutions. Also, it is not clear whether sui generis approaches and international harmonization of existing IPR laws are compatible. An alternative to sui generis laws that is less frequently mentioned is the development of a fundamentally new IPR legal approach that would be, in effect, a new paradigm. What would be included in the basic outlines of such an approach?

Research Issue: Bryan Harris notes in Chapter 6 that IPR harmonization for its own sake cannot be justified without a basic understanding of "the relationship between the economic interests of intellectual property owners and intellectual property users." The push by the United States for harmonization in GATT has moved forward without a full understanding of possible negative impacts on some sectors of U.S. industry. More research is needed to elucidate the effects of strong versus weak IPR protection on the use and development of new, protected technologies.

Research Issue: What are the practical effects on corporate competitive strategies of the fundamental differences in patent law and practice between the United States, which requires that an inventor demonstrate that he/she is the "first to invent," and other advanced industrialized countries, which adhere to a "first-to-file" approach?

VII

Appendixes

APPENDIX

A

CONFERENCE AGENDA

Global Dimensions of Intellectual Property Rights in Science and Technology

NATIONAL ACADEMY OF SCIENCES
Washington, D.C.

Wednesday, January 8, 1992

8:00 a.m. **Registration**

8:30 a.m. **Welcome:** Gerald P. Dinneen, Foreign Secretary, National Academy of Engineering

8:35 a.m. **Opening Remarks**

Conference chair: Albert R.C. Westwood, Martin Marietta Corporation

8:45 a.m. **History and Theory of Intellectual Property Rights**

MODERATOR: Albert R.C. Westwood
PRESENTER: Paul A. David, Stanford University

9:20 a.m. Questions

9:30 a.m. **Comparative National Approaches to Intellectual Property Rights**

MODERATOR: Karl F. Jorda, Franklin Pierce Law Center
COMMENTATORS:
James E. Armstrong III, Armstrong, Nikaido, Marmelstein, Kubovcik, and Murray

Bryan Harris, International Consultant, European Community
Carlos A. Primo Braga, The World Bank
Deepak Nayyar, Jawaharlal Nehru University

10:10 a.m Open Discussion

10:40 a.m. Break

11:00 a.m **Convergence and Divergence in Intellectual Property Rights, Technology, and Global Relationships**

MODERATOR: Herbert C. Wamsley, Intellectual Property Owners, Inc.
PRESENTER: John A. Armstrong, IBM Corporation

11:35 a.m. Discussants:

John T. Preston, Massachusetts Institute of Technology
Bruce Merrifield, University of Pennsylvania
George W. McKinney III, Beacon Venture Management Corporation

12:35 p.m. Open Discussion

1:00 p.m. Lunch

2:00 p.m. **Definition of Adequate and Appropriate Protection of Intellectual Property: Opposing Visions**

MODERATOR: Anne W. Branscomb, Harvard University

The Argument for a Single Worldwide System:

Robert M. Sherwood, International Business Counselor

The Argument for Differing Levels of Protection:

Claudio Frischtak, International Consultant

2:40 p.m. Questions

2:50 p.m. **Update on International Intellectual Property Rights Negotiations:**

Jacques J. Gorlin, The Gorlin Group

3:10 p.m. Open Discussion

3:35 p.m. Break

3:55 p.m. **The Effects of Unauthorized Use of Intellectual Property**

MODERATOR: Gustav Ranis, Yale University
PRESENTER: Edwin E. Mansfield, University of Pennsylvania

4:30 p.m. Open Discussion

5:00 p.m. **Adjourn, Reception**

Thursday, January 9, 1992

8:00 a.m. **Registration**

8:30 a.m. **The Impact of Technology on Intellectual Property Rights**

MODERATOR: Albert R.C. Westwood
PRESENTER: John H. Barton, Stanford University

9:05 a.m. Open Discussion

9:20 a.m. **Case Studies**

MODERATOR: John T. Preston, Massachusetts Institute of Technology

Software: Pamela Samuelson, University of Pittsburgh

9:40 a.m. Questions

9:50 a.m. **Biotechnology:** George B. Rathmann, ICOS Corporation

10:10 a.m. Questions

10:20 a.m. Break

10:35 a.m. **Semiconductor Chips:** Morton David Goldberg, Schwaab, Goldberg, Price and Dannay

10:55 a.m. Questions

11:05 a.m. **Optoelectronics:** Eugene I. Gordon, New Jersey Institute of Technology

11:25 a.m. Questions

11:35 a.m. Open Discussion

12:00 noon Lunch

1:00 p.m. **Intellectual Property Rights and Competitive Strategy**

MODERATOR: Jacques J. Gorlin

PANEL:

Otto A. Stamm, CIBA-GEIGY AG
Michiyuki Uenohara, NEC Corporation
W. L. Keefauver, Consultant
Antonio Medina Mora Icaza, ANIPCO

2:00 p.m. Open Discussion

2:30 p.m. **What Next?**

MODERATOR: Arden L. Bement, Jr., TRW Inc.

DISCUSSANTS:

Robert E. Evenson, Yale University
David C. Mowery, University of California at Berkeley
Michael Borrus, University of California at Berkeley
Robert W. Lucky, AT&T Bell Laboratories
Eugene Skolnikoff, Massachusetts Institute of Technology

3:45 p.m. Open Discussion

4:15 p.m. **Moderator's Summation**

4:25 p.m. **Chair's Comments**

4:30 p.m. **Adjourn**

APPENDIX

B

Biographies of Contributors

Albert R.C. Westwood, Conference Chairman

Albert R.C. Westwood is Vice President—Research and Technology for Martin Marietta Corporation. He received his B.Sc., Ph.D., and D.Sc. degrees in metallurgy and materials science from the University of Birmingham, England, and joined Martin Marietta Laboratories (then RIAS) in 1958, becoming its Director in 1974. Subsequently, he became Corporate Director of R&D in 1984, Vice President—Research and Development in 1987, and Vice President—Science in 1990. He assumed his present position in August 1990.

Dr. Westwood has published some 120 technical papers, mostly concerned with environment-sensitive mechanical behavior or R&D management, and his scientific contributions have been recognized by a variety of awards and fellowships, including the Beilby Gold Medal (1970), fellow of the Institute of Physics (1967), of the American Society for Materials International (ASMI) (1974), of the American Association for the Advancement of Science (AAAS) (1986), and of The Minerals, Metals, and Materials Society (TMS) (1990); and election to the U.S. National Academy of Engineering (1980) and the Royal Swedish Academy of Engineering Sciences (1989). He has also served as Campbell Memorial Lecturer (ASMI, 1987), Henry Krumb Lecturer (American Institute of Mining, Metallurgical, and Petroleum Engineers (AIME, 1988), and American Society for Mechanical Engineers (ASME) Distinguished Lecturer (1989-1990).

His current professional responsibilities include: Trustee of AIME; Past President of TMS-AIME; Past President of the Industrial Research Institute; Member of the Board of Directors of Martin Marietta Energy Systems; Member of the Visiting Committee to the National Institute for Standards and Technology; Member of the National Critical Technologies Panel (Office of Science and Technology Policy); Chairman of Advisory Panel to the National Science Foundations's (NSF) Directorate on Science, Technology and International Affairs; and Chairman of the National Research Council's (NRC) Commission on Engineering and Technical Systems.

James E. Armstrong III

James E. Armstrong III, Senior Partner, Armstrong, Nikaido, Marmelstein, Murray and Kubovcik, Washington, D.C., has been involved in the practice of patent law for more then 36 years. He received a Bachelor of Science degree in chemical engineering in 1951 from Michigan State University and an LL.B. from the University of Maryland School of Law in 1957. He was registered to practice before the U.S. Patent and Trademark Office in 1956 and was admitted to the Maryland bar in 1957 and the District of Columbia bar in 1970.

Mr. Armstrong is fluent in Japanese and has spent two to four months in Japan each year for the past 21 years on patent and licensing matters.

Mr. Armstrong is a member of the American Bar Association, the American Intellectual Property Association, the U.S. Trademark Association, the Licensing Executives Society, and the American Group of the Association Internationale pour la Protection de la Propriété Industrielle (AIPPI).

His publications include *The Japanese—Successful Patent, Know-How and Joint Venture Relations,* with Levine, Richman and Seward, Patent Resources Group (1973, 1974), *The Thought Process—Essentials for the Drafting of U.S. Patent Specifications and Claims*, by Armstrong and Nikaido, Japan Group AIPPI (1975); Revised Editions (1980), (1986) *Fundamentals of Technology Transfer*, Pegan and Armstrong (1991) and several articles in *les Nouvelles* (the Journal of the Licensing Executives Society) on the cultural impact of Japanese-American licensing negotiations.

Mr. Armstrong has served as a lecturer and instructor with the Patent Resources Group (Directed by Professor Irving Kayton of George Mason University) since 1973. He has also been a guest lecturer at the World Trade Institute in New York on antitrust and licensing matters on several occasions since 1978, and has presented several lectures and workshops at meetings of the Licensing Executives Society. Over the past 21 years, Mr. Armstrong has given more than 100 lectures and seminars (usually in Japanese) to numerous Japanese corporations, Japanese patent associations, and professional societies on a wide variety of topics relating to patents and licensing. During 1986-1989, Mr. Armstrong was a guest lecturer at Peking University and several other Chinese universities, where he taught fundamentals of technology transfer.

John A. Armstrong

John Armstrong, IBM Vice President, Science and Technology, was born in Schenectady, New York, in 1934. He received an A.B. in physics from Harvard College in 1956, and his Ph.D. in 1961 from Harvard University for research in nuclear magnetic resonance at high pressures. He then, as a Research Fellow, switched to lasers and nonlinear optics, working with Professor N. Bloembergen.

In 1963 he joined IBM Research as a Staff Member. He spent 1967-1968 at the IBM Research Laboratory in Zurich and returned to Yorktown as Manager of Quantum Optics. Between 1976 and 1980, as Director of Physical Sciences, he was responsible for a major part of the physics, chemistry, and materials science at IBM Research. In 1980 he joined the IBM Corporate Technical Committee headed by the IBM Chief Scientist. In 1981 he was made manager of materials and technology development at the IBM East Fishkill development laboratory, working on advanced bipolar technology and associated packaging. In 1983 he returned to the Research Division as Vice President, Logic and Memory, in 1986 he was named Director of Research, and in 1987 was elected IBM Vice President and Director of Research. In May 1989 he was elected a member of the Corporate Management Board and named to his current position in which he is responsible for ensuring IBM's technological excellence and leadership in research. He also has management responsibility for the research division, technical strategy development, technical journals and professional relations, and technical personnel development.

Dr. Armstrong is author or coauthor of more than 50 papers on the subjects of nuclear resonance, nonlinear optics, the statistical properties of laser light, picosecond pulse measurements, and the multiphoton laser spectroscopy of atoms.

He was Chairman of the Advisory Committee for physics of the NSF. He is a Fellow of the Optical Society of America, a Fellow of the American Physical Society, a Fellow of the AAAS, a Fellow of the American Academy of Arts and Sciences, and a Fellow of the Institute of Electrical and Electronic Engineers. He is a member of the National Advisory Committee for Semiconductors and cochairman of its Working Group on Technology, a member of the Massachusetts Institute of Technology Physics Visiting Committee, and a member of the Policy Board of the National Nanofabrication Facility. He is also a trustee of Associated Universities, Inc., and a member of the Policy Steering Committee of the Governor's Conference on Science and

Engineering Education, Research and Development: Developing New York State's Action Plan for the 1990s. In 1990 he was elected to the Board of Overseers at Harvard University and to the Board of Advanced Network and Services, Inc.

In 1987 Dr. Armstrong was elected a member of the National Academy of Engineering and a foreign member of the Royal Swedish Academy of Engineering Sciences. In 1989 he was awarded the George E. Pake Prize of the American Physical Society. In 1990 he received an honorary Doctor of Science degree from the State University of New York at Albany.

John H. Barton

John H. Barton, the George E. Osborne Professor of Law at Stanford Law School is director of that school's International Center for Law and Technology. He teaches international business transactions, law and high technology, and international environmental law. He consults extensively to the international agricultural research community on intellectual property and biosafety regulatory questions in developing nation agricultural biotechnology. He organized a May 1992 meeting on Pacific Basin technology issues in the post-Uruguay Round world, in cooperation with the University of Hong Kong Law School.

He is coauthor of a leading international business law casebook, *The Regulation of International Business*, and has published and spoken widely on biotechnology, genetic resources, and international technology transfer issues. His most recent writings include "Catch-up Strategies for Technologically Proficient Developing Nations," presented at XVI Simposio Nacional de Pesquisa de Adminstracao em Ciencia e Tecnologia, in Rio de Janeiro in October 1991, and "Patenting Life," *Scientific American* 264:40 (March 1991).

He served as a member of the NRC Committee on Managing Global Genetic Resources and is now a member of the National Institutes of Health Recombinant DNA Advisory Committee. His undergraduate degree is from Marquette University (1958) and his law degree is from Stanford (1968).

Arden L. Bement, Jr.

Arden L. Bement Jr. is Vice President, Technical Resources of TRW, Inc. Before joining TRW in 1981, he was the Deputy Under Secretary for Defense for Research and Engineering. From 1976-1979 he was the Director of the Material Science Office for the Defense Advanced Research Projects Agency. Before that he held positions at General Electric, Battelle Memorial Institute, and the Massachusetts Institute of Technology.

Dr. Bement serves on the Statutory Visiting Committee of the NIST

Visiting Committees for Oak Ridge, Lawrence Livermore, and Argonne National Laboratories. He is a member of the Visiting Committees for MIT, Carnegie Mellon University, and John Hopkins University, and of the Advisory Committees for the University of Michigan, Ohio State University, and Howard University.

Dr. Bement has participated in a number of international bilateral exchanges including the U.S.-USSR Bilateral Exchange Program in Magnetohydrodynamics, and bilateral exchange programs with Japan, Canada, Sweden, Denmark, Norway, and the United Kingdom. He has also served as an United Nations Scientific Advisor on Atomic Energy.

He is the author or editor of four books and author of over 90 journal articles on material science, energy, and defense technology. He has received numerous awards including the Engineering Citation Award, University of California, Los Angeles; Distinguished Civilian Service Medal, U.S. Department of Defense; Outstanding Achievement Award, Defense Advanced Research Projects Agency; and the Outstanding Achievement Award, Colorado Engineering Council.

Dr. Bement is a member of the National Academy of Engineering, a Fellow in the American Society for Metals, a Fellow of the American Nuclear Society and a Fellow of the American Institute of Chemists.

Michael Borrus

Michael Borrus is a Director of the Berkeley Roundtable on the International Economy at the University of California, Berkeley; he teaches in the joint School of Engineering-Business program on the Management of Technology. A member of the California state bar, Dr. Borrus has worked on high technology and trade issues for the last decade.

He regularly consults on technology policy and business strategy issues with various governments and firms in the United States, Asia, and Europe—including, most recently, the National Advisory Committee on Semiconductors. His recent works include *The Highest Stakes: Technology, Economy and Security* (Oxford University Press, 1991); *Competing for Control, America's Stake in Microelectronics* (Harper and Row, 1988); "High Technology in the Pacific Basin: Analysis and Policy Implications," a paper prepared for the U.S. State Department; and "Information Networks and Competitive Advantage: The Issues for Government Policy and Business Strategy (Brussels and Paris, EEC-OECD, September 1990).

Paul A. David

Paul A. David, professor of economics and professor of history, by courtesy, was appointed William Robertson Coe Professor of American Economic

History at Stanford University in 1977. He is also Director of Research for the High Technology Impact Program at the Center for Economic Policy Research at Stanford. He is an elected fellow of the International Econometrics Society and of the American Academy of Arts and Sciences, and has served as Vice President and President of the Economic History Association.

Paul David was born in New York City and was educated at the High School of Music and Art. He majored in economics as an undergraduate at Harvard College, where James Duesenberry was his honors thesis advisor and where, in 1956, he received an A.B. in economics Summa Cum Laude. After two years at Cambridge University as a Fulbright Scholar and research student under the supervision of Peter Mathias and R.C.O. Mathews, he returned to Harvard University for further graduate study in economics. There he became a member of Alexander Gerschenkron's Economic History Workshop. Joining the Stanford Faculty in 1961, he was promoted to the rank of Associate Professor of Economics in 1966 and Professor in 1970.

Professor David, whose research and teaching have covered a wide range of subjects, is known internationally for his contributions to the development of the "new economic history," using the theoretical and statistical tools of modern economics to reconstruct and analyze economic life in the past, and studying its connections with the present. A recent paper on this theme is "So, How Would It Matter if 'History Mattered'?: Path-Dependence in Economics and Its Long-Run Implications," *Working Papers in Economic History*, (The Australian National University, July 1991; forthcoming in G.D. Snooks, ed., *The Role of Longrun Analysis in Economics*).

Recent work in the area of the economics of technology and innovation include "Performance-Based Measures of Nuclear Reactor Standardization," (with G. Rothwell) CEPR Publication No. 247 (Stanford University, June 1991); "Learning from Disaster?: Changes in the Distribution of Operating Spell Durations in U.S. Nuclear Power Plants After Three Mile Island" (with G. Rothwell and R. Maude-Griffen) CEPR Technical Paper No. 248 (Stanford University, May 1991); "Technology Diffusion, Learning Spillovers, and the Optimal Duration of Patent-Based Monopoly" (with T.E. Olsen) CEPR-HTIP Working Paper (Stanford University, July 1991); forthcoming in *International Journal of Industrial Organization*.

Robert E. Evenson

Robert Evenson is a Professor of Economics at the Economic Growth Center at Yale University. After receiving a Ph.D. from the University of Chicago in 1968, he was appointed an Associate Professor at Yale. In 1974, he became an Associate of the Agricultural Development Council of the Philippines. He returned to Yale as a full professor in 1977.

Dr. Evenson has written and edited a number of books and monographs including *Science for Agriculture* (with W. Huffman; Iowa State University Press, 1991); *Research and Productivity in Asian Agriculture* (with Carl Pray et al.; Cornell University Press, 1991); *Research, Productivity and Incomes in Brazilian Agriculture: A Study of the EMBRAPA Program* (with E.R. da Cruz, J. Strauss, M.T.L. Barbosa, and D.Thomas; EMBRAPA, Brasilia, 1991); *Science and Technology: Lessons for Development* (editor, with Gustav Ranis; Westview Press, 1990).

Claudio Frischtak

Claudio Frischtak is a consultant to the World Bank based in Rio de Janeiro, Brazil. Formerly principal industrial economist at the World Bank, he has a strong interest and has published in the areas of industrial policy, regulatory regimes and reform, technology strategies and institutional support systems. His most recent publications include "The Competitive Potential of National Producers of Equipment and Systems," in P. Evans, P. Tigre, and C. Frischtak, eds., *Brazilian Informatics in Transition: Government Policy and International Trends in the 1990's* (University of California Press, 1992); "Introduction" in C. Frischtak and R. Newfarmer, eds., *Transnational Corporations: Market Structure and Industrial Performance* (Routledge and Sons, 1992); "Banking Automation and Productivity Change: The Brazilian Experience," *World Bank Industry Series Paper No. 46* (December 1991); "National Systems Supporting Technical Advance in Industry: The Brazilian Experience" (with C. Dahlman), in R. Nelson and N. Rosenberg, eds., *National Technical Systems Supporting Industry* (Oxford University Press, 1992); "Specialization, Technical Change and Competitiveness in the Brazilian Electronics Industry," *OECD Development Centre Technical Papers no. 27* (October 1990); "The Protection of Intellectual Property Rights and Industrial Development in Brazil," in F. Rushing and C. Ganz, eds., *Protection of Intellectual Property Rights in Science, Technology and Economic Performance: International Comparisons* (Westview Press, 1990); "Competition Policies for Industrializing Economies," *World Bank Policy and Research Series No. 7* (1989). Mr. Frischtak did his graduate work at the University of Campinas (Sao Paulo, Brazil) and at Stanford University and, while at the World Bank, was an adjunct Professor at the Department of Economics, Georgetown University (1987-1990).

Morton David Goldberg

Morton David Goldberg, a partner in the law firm of Schwab, Goldberg, Price and Dannay, has been a copyright practitioner, writer, and lecturer for 30 years, with increasing attention to software intellectual property. Mr.

Goldberg is an Honorary Trustee and Past President of the Copyright Society of the United States and a former Section Chairman and Section Delegate of the American Bar Association's Section of Patent, Trademark and Copyright Law; and he has served on Boards of Directors of the Computer Law Association, U.S. Trademark Association, and American Intellectual Property Law Association. Mr. Goldberg's most recent paper (coauthored with David O. Carson) was prepared for the World Intellectual Property Organization and is entitled "Copyright Protection for Artificial Intelligence Systems."

Eugene I. Gordon

Eugene Gordon is a Distinguished Research Professor in the Department of Electrical and Computer Engineering at the New Jersey Institute of Technology in Newark, New Jersey. After receiving his Ph.D. in physics from the Massachusetts Institute of Technology, he joined AT&T Bell Labs where he worked until 1983, when he founded LYTEL, Inc. He later worked for Hughes Aircraft Co., and was the Chief Executive Officer and President of Photon Imaging Corporation from 1988 to 1990. He is an active member of the Institute of Electrical and Electronics Engineers (IEEE).

Dr. Gordon participated in important demonstrations of the use of lasers in surgery, including the first use of argon lasers in treating diabetic retinopothy (blindness in diabetes). He is the co-inventor of a charge coupled device (CCD) and was leader of the laboratory doing the initial feasibility demonstrations of monochrome and color cameras, and line scanners using CCDs. He was the leader of the group that made the first visible helium neon lasers, the first continuous argon laser, the first acousto-optic light modulators and deflectors, and the first electron beam lithography machine for mask making, and that developed reliable laser diodes for submarine cable light wave systems.

Jacques J. Gorlin

Jacques J. Gorlin has been a consulting economist since October 1982. He provides advice and conducts economic analyses on a broad range of trade, high-technology, and intellectual property-related issues for Fortune 500 as well as smaller U.S. and foreign companies, and U.S. government agencies.

Since March 1986 he has served as the economic consultant to the Intellectual Property Committee, an ad hoc coalition of 13 major U.S. corporations. Mr. Gorlin also serves as consultant to the Task Force on Intellectual Property of the President's Advisory Committee on Trade Policy and Negotiations. He is a member of the Industrial Functional Advisory

Committee on Intellectual Property Rights for Trade Policy Matters, a private sector group that advises the Secretary of Commerce and the U.S. Trade Representative on trade policy.

Prior to entering the private sector, Mr. Gorlin served in a number of senior positions in both the executive and legislative branches of government. From 1972 to 1977, he was a senior international economist in the Department of the Treasury and in the Office of the U.S. Trade Representative, where he specialized in international trade and Middle East finance. As a senior economic advisor to Senator Jacob K. Javits (R-NY) from 1977 to 1981, he headed the Senator's economic staff and served as his liaison with the New York business and banking communities. In 1980-1981, Mr. Gorlin served on President Reagan's transition team, focusing on the State Department's economic functions; and in 1981, he assumed the position of Executive Assistant to the Under Secretary of State for Economic Affairs. In 1982, he served as the Senior Economic Advisor to the Administrator of the Agency for International Development.

Mr. Gorlin has lectured on the subject of international trade, high technology, and intellectual property rights before corporate, legal, and academic groups, and been a guest commentator on "All Things Considered" (National Public Radio) and other public affairs programs. He is the author of a *Trade Based Approach for the International Copyright Protection for Computer Software.* He has also contributed to numerous collections on intellectual property protection, including *Global Competition: The Role of Intellectual Property* and *Intellectual Property Rights and Capital Formation in the Next Decade.* Mr. Gorlin's monograph, "Foreign Trade and the Constitution," appeared in *Foreign Policy and the Constitution* (American Enterprise Institute). His articles have appeared in the *New York Times*, the *Wall Street Journal* (European edition), the *Journal of Commerce*, and other international publications.

Mr. Gorlin received an A.B. in history from Columbia College (1965), an M.A. from the Johns Hopkins School of Advanced International Studies (1967), and a Ph.D. from the Johns Hopkins University (1971).

Bryan Harris

Bryan Harris is an independent consultant on European Economic Community (EEC) problems to commercial and professional firms and associations. He is an Adjunct Professor (EEC Law) and Member of the Executive Board, Franklin Pierce Law Center, in Concord, New Hampshire. He writes and lectures on legal, institutional and political aspects of the European Community.

He is editor of the monthly newsletter (Monitor Press) "Competition Law in the European Communities." He is also the author of "The Law of the European Communities" (supplement to *Halsbury's Laws*, Third Edition); "The Common Agricultural Policy" (in *Halsbury's Laws*, Fourth Edition); *Franchising in the European Community* (Longmans); and *Lobbying in the European Community* (Macmillan, 1992).

From 1973 to 1983 he was Head of the Intellectual Property Division in the Commission of the European Communities, Brussels. He was the leader of the Community's delegation at the diplomatic conference on the revision of the Paris Convention for the protection of industrial property and was the initiator of the commission's Green Paper on Copyright, which was eventually published in 1988.

William L. Keefauver

William L. Keefauver is a consultant in intellectual property matters. He was formerly Vice President, Law, of AT&T with responsibility for all intellectual property legal matters and Vice President and General Counsel of AT&T Bell Laboratories. He was also a member of the board of directors of NCR until its recent merger with AT&T.

He is currently a member of the Advisory Commission on Patent Law Reform in the U.S. Department of Commerce and a member of the advisory committee on intellectual property matters of the United States Trade Representative for the current round of General Agreement on Trade and Tariffs negotiations.

Mr. Keefauver is President of the International Intellectual Property Association—the U.S. group of the AIPPI—and is a past chairman of the American Bar Association Section of Patents, Trademarks and Copyrights. He is also Vice President of the Board of Trustees of the McCarter Theatre in Princeton, N.J.

Robert W. Lucky

Robert Lucky is the Vice President of the Applied Research Group at Bellcore, Inc. and is a leading expert, author, and commentator on the state and future of data communications technology. He was formerly the Executive Director of the Communications Sciences Research Division at AT&T Bell Laboratories, where he led AT&T's research into methods and technologies on future communications systems, including optical fiber technology, data networks, mobile communications, image processing, and broadband communications services. At Bell Labs, Dr. Lucky invented the "adaptive equalizer," a revolutionary technique for correcting distortion in telephone signals that is used in all high-speed data transmission today. He authored one of the most heavily cited textbooks on data communications and the popular book *Silicon Dreams*, which analyzed the ways humans and computers deal with information. Dr. Lucky has also appeared on a number of network television shows, including Bill Moyers' Public Broadcasting Station program "A World of Ideas," to discuss the impacts of future technological advances.

Edwin Mansfield

Edwin Mansfield is Professor of Economics and Director of the Center for Economics and Technology at the University of Pennsylvania. A graduate of Dartmouth College, he received his M.A. and Ph.D. degrees from Duke University, as well as the Certificate and Diploma of the Royal Statistical Society. Before joining the University of Pennsylvania faculty, he taught at Carnegie-Mellon, Yale, Harvard, and the California Institute of Technology. He has been a consultant to many industrial firms and government agencies, and has been a member of the Advisory Committee of the U.S. Bureau of the Census and the AAAS's Committee on Science, Engineering, and Public Policy. He has been chairman of the Visiting Committee at Rensselaer Polytechnic Institute. He received the Certificate of Appreciation from the U.S. Secretary of Commerce and in 1984 was appointed to the National Technology Medal Committee.

Professor Mansfield has been elected a fellow of the American Academy of Arts and Sciences, the Econometric Society, and of the Center for Advanced Study in the Behavioral Sciences, and he has held Fulbright and Ford Foundation fellowships. He is a member of the board of directors of the American Productivity and Quality Center. He has served as U.S. chairman of the U.S.-USSR Working Party on the Economics of Science and Technology, and was the first U.S. economist to be invited to visit and lecture in the People's Republic of China under the 1979 Sino-American agreements.

He is the author of 170 articles and 25 books. His textbooks on economics, microeconomics, and statistics have been adopted at more than 700 colleges and universities, and have been translated for use abroad. He has been an editor of six journals, including the *Journal of the American Statistical Association*, and is general editor of a series of books on technological change published by the University of Wisconsin Press. In 1984 he received the Publication Award of the Patent Law Association.

George W. McKinney III

George W. McKinney III is currently a Managing Director of Beacon Venture Management Corporation, a specialized firm focused on technology commercialization in new companies. He also is President and Chief

Executive Officer of Environmental Quality Corporation, a new venture in the area of pollution prevention.

Prior to joining Beacon Venture Management, Dr. McKinney was a Managing Partner at American Research & Development (ARD), the Boston based venture capital firm. While at ARD he was the founding President and Chief Executive Officer for American Superconductor Corporation (AMSC), an emerging company producing wires and coils from oxide superconductors. He continues to serve on the Board of Directors of AMSC.

Dr. McKinney spent 18 years at Corning, Inc., prior to entering the venture capital field. At Corning he was responsible for corporate planning and business development and was Secretary to Corning's Management Committee. He also served as the primary business representative on Corning's internal patent committee.

Dr. McKinney holds an S.B. in Industrial Management from the Massachusetts Institute of Technology (MIT) and a Ph.D. in Business from Stanford University. His article "Corporate Strategic Partnerships" regularly appears in *Pratt's Guide to Venture Capital Sources*. He has served on the Board of the New England Venture Capital Association.

Antonio Medina Mora Icaza

Antonio Medina Mora Icaza has worked as a programming analyst and researcher at numerous computing institutions. At present, he is the President of Company Grupo Telos, S.A. de C.V., a Mexican software firm. He is also the President of the Mexican Software Association, ANIPCO. He studied at the Faculty of Engineering and the Instituto de Investigación en Matematicas Aplicadas y Sistema at the Universidad Nacional Autónoma de Mexico (UNAM), the Carnegie Mellon University, and the Centre Mondial pour la Informatique et les Resources Humaines in Paris. He later returned to teach at UNAM, and has lead workshops on communications networks in Mexico and Chile. Mr. Medina Mora is also a participant in the negotiations for the North American Free Trade Agreement on the Board of Intellectual Property and Telecommunications.

Bruce D. Merrifield

Bruce D. Merrifield is a graduate of Princeton University and holds masters and doctoral degrees in physical organic chemistry from the University of Chicago. Dr. Merrifield has since been in research, research administration, and new venture management, and during the Reagan administration was Assistant Secretary of Commerce for Productivity, Technology, and Innovation. Currently, he holds the Walter Bladstrom Chair for Professor of Management at the University of Pennsylvania Wharton School of Business. He also is a consultant for the American Electronic Association and has been active with the Greater Minnesota Corporation. He is a director of a number of companies and, formerly, was Vice President of Technology and Venture Management for the Continental Group.

At the Department of Commerce, Dr. Merrifield's office spearheaded landmark legislation to modify the antitrust laws (the Cooperative R&D Act of 1984) and the Technology Transfer Acts of 1984 and 1986. The office has developed the R&D Limited Partnership concept and has catalyzed the formation of more than 100 cooperative R&D consortia, involving over 1,000 companies, and many innovation centers in states and local communities.

He is a former director and president of the Industrial Research Institute, and is both a former Trustee of the American Management Association and Chairman of its Research Council. Currently, he is a member of the Directors of Industrial Research, a member of Sigma XI Honorary Society, and is a Fellow of both the AAAS and the Institute of Chemists.

Dr. Merrifield is a past member of the Advisory Board for the Binational Research and Development Foundation with Israel and is a current member of similar boards with India and France. He has served as Science Adviser to the Jordanian government and as a member of the Visiting Committee for Research at MIT and Boston University.

Dr. Merrifield also has been active in community affairs as a former Chairman of the Board of Education, a trustee of several schools, as National Chairman of the Princeton Alumni Association, a board member of several foundations, and a member of the Executive Council of the Episcopal Church.

David C. Mowery

David Mowery is Associate Professor of Business and Public Policy in the Walter A. Haas School of Business at the University of California, Berkeley. He received his undergraduate and Ph.D. degrees in economics from Stanford University and was a postdoctoral research fellow at the Harvard Business School. Dr. Mowery taught at Carnegie-Mellon University, served as the Study Director for the Panel on Technology and Employment of the National Academy of Sciences, National Academy of Engineering, and Institute of Medicine, and served in the Office of the U.S. Trade Representative during 1988 as a Council on Foreign Relations International Affairs Fellow. He has served on a number of NRC panels, including those on the Competitive Status of the U.S. Civil Aviation Industry, the Causes and Consequences of the Internationalization of U.S. Manufacturing, and the Federal Role in Civilian Technology Development. His research deals with the economics of technological innovation and with the effects of public policies on innovation; he has testified before congressional committees and served as an adviser for the Organization for Economic Cooperation and Development, various federal agencies, and industrial firms. Dr. Mowery has written or edited several books, including *Technology and the Pursuit of Economic Growth*; *Alliance Politics and Economics: Multinational Joint Ventures in Commercial Aircraft*; *Technology and Employment: Innovation and Growth in the U.S. Economy*; *The Impact of Technological Change on Employment and Economic Growth*; and *International Collaborative Ventures in U.S. Manufacturing*.

Deepak Nayyar

Deepak Nayyar is a Professor of Economics at the Center for Economic Studies and Planning, Jawaharlal Nehru University in New Delhi, India. He was a Rhodes Scholar from India at Oxford from 1967 to 1969, where he received his Ph.D. He was a Research Fellow at Oxford, and later a lecturer in Economics at the University of Sussex and a Professor of Economics at the Indian Institute of Management in Calcutta. He has recently completed one year and a half as the Chief Economic Advisor to the Government of India, and Secretary, Ministry of Finance, New Delhi.

Deepak Nayyar has written several books, including *Migration, Remittances and Capital Flows: The Indian Experience* (Oxford University Press, Delhi, 1992), and has edited *Industrialisation in India: The Debate on Growth and Stagnation* (Oxford University Press, Delhi, 1992).

John T. Preston

John Preston is the Director of the Technology Licensing Office at the Massachusetts Institute of Technology. As Director, he manages the patenting and licensing of the MIT, Lincoln Laboratory, and Whitehead Institute inventions and software. He is a member of the Board of Directors of Molten Metal Technology, Environmental Bioscience and Ergo Computing, Inc., and is Chairman of the Technology Transfer Advisory Panel for the Strategic Defense Initiative of the U.S. Department of Defense.

Mr. Preston received his B.S. in physics from the University of Wisconsin and his M.B.A. from Northwestern University. His professional activities have been directed toward technology transfer, and specifically toward issues related to starting new high-technology companies. He has founded, or assisted in founding, companies that are currently worth several hundred million dollars. In addition, about 40 companies, mostly spin-offs of MIT, have been started, in part, through the efforts of the Technology Licensing Office during his tenure.

Carlos Alberto Primo Braga

C.A. Primo Braga is an economist with the International Trade Division of the World Bank. Before joining the World Bank, Mr. Primo Braga was an Assistant Professor of Economics at the University of Sao Paulo and Senior Researcher at the Fundacao Institute de Pesquisas Economicas (FIPE), Sao Paulo, Brazil. Since 1988 he has also been a Visiting Professor at the Johns Hopkins School of Advanced International Studies, Washington, D.C.

C.A. Primo Braga received his M. Engr. degree from the Institute Tecnologico de Aeronautics (ITA) in 1976. He also received a master's degree in Economics from the University of Sao Paulo in 1980, and an M.Sc. (1982) and a Ph.D. (1984) in Economics from the University of Illinois at Urbana Champaign.

Since 1984, he has served as an economic consultant to many private companies, multilateral agencies, and government institutions in Brazil and abroad—including the World Bank and the Organization of American States. During 1987-1988, he was joint coordinator of the Brazilian team working on a Rockefeller Foundation project on the Multilateral Trade Negotiations and the Developing Countries. He has served as a member of the editorial board of several academic publications—such as *Revista de Estudos Economicos*, *Informacoes FIPE*, *Revista Brasileira de Comercio Exterior*, and *Hemisfile*—and as a member of the board of directors of the Fernand Braudel Institute of World Economics, Sao Paulo, Brazil. He has also testified before different committees of the U.S. Congress on U.S.-Latin American economic relations and his "op-ed" essays have appeared in *Folha de Sao Paulo*, *O Globo*, and *O Estado de Sao Paulo* among other newspapers.

His main research interests encompass international economics (multilateral institutions, U.S.-Latin American economic relations, foreign debt, trade, and development); economics of science and technology; intellectual property rights; industry studies (steel, frozen concentrated orange juice); and economics of education.

Dr. Primo Braga's main publications include *Brasil e a Rodad Uruguai* (coeditors C.L. Martone and E.R. Pelin; Sao Paulo, IPE/USP, forthcoming); "The Threat of a Cold Trade War and the Developing Countries," *SAIS Review*, 11:53-67 (Summer-Fall 1991); "The North-South Debate on Intellectual Property Rights," in Murray C. Smith, ed., *Global Rivalry and Intellectual Property* (Halifax: Institute for Research on Public Policy, 1991); "Il Pioano Collor o je Ambizione di una Riforma," *Politica Internazionale*, XIX:65-71 (January-February 1991); Chapters 3, 5, and 7 in Wolfgang E. Siebeck, ed., *Strengthening Protection of Intellectual Property in Developing Countries: A Survey of Literature* (Washington, D.C.: The World Bank, 1990); "U.S. Policies and the Prospects for Latin American Economic Integration," in W. Baer and D.V. Coes, eds., *United States Policies and the*

Latin American Economics (New York: Praeger, 1990); "Brazil," in P.A. Mosserlin and K.P. Sauvant, eds., *The Uruguay Round: Services in the World Economy* (Washington, D.C.: The World Bank, 1990); "U.S.-Latin American Trade: Challenges for the 1990's" *Economic Impact*, 67(2):51-55 (1989); "The Economics of Intellectual Property Rights and the GATT: A View from the South," *Vanderbilt Journal of Transnational Law*, 22:243-264 (2/1989); *Brasil 1980: Os Desafios da Crise Economica* (coeditors: C.A. Roacca, M.C. Cacciamali, and M.C. de Castro; Sao Paulo, IPE/USP, 1988); "Brazilian Public Sector Disequilibrium" (coauthors J.H. Welch and P.T.A. Andre) *World Development* 15:1045-1053 (August 1988).

George B. Rathmann

George B. Rathmann has been Chairman of the Board of Directors of the ICOS Corporation since January 1990, and President and Chief Executive Officer since September 1991. Dr. Rathman previously held top executive positions at Abbott Laboratories, Inc., and at Amgen which he co-founded in 1981. During his tenure at Amgen as President, Chief Executive Officer, and Chairman, two of the most significant biotechnology products were developed, EPO and G-CSF, hormones that control red and white blood cell growth, respectively, in bone marrow. Dr. Rathmann received his Ph.D. in physical chemistry from Princeton University.

Pamela Samuelson

Pamela Samuelson is a Professor of Law at the University of Pittsburgh School of Law. She has written and spoken extensively on intellectual property and other legal issues affecting new technology fields such as computing. She is a Contributing Editor for Communications of the ACM, and author of its regular "Legally Speaking" column. During 1985 and 1986, she was the Principal Investigator of the Software Licensing Project at the Software Engineering Institute at Carnegie Mellon University, which advised the Defense Department on needed changes to its software acquisition policy. She practiced law with the New York law firm Willkie Farr & Gallagher before becoming an academic. She is a 1976 graduate of Yale Law School and a 1971 graduate of the University of Hawaii at Honolulu.

Robert M. Sherwood

Robert M. Sherwood, an international business counselor based in Washington, has practiced law on Wall Street and as an international corporate attorney. He leads a multi-industry group of companies dedicated to better understanding of intellectual property protection in Brazil. He has spent 20 weeks since 1987 in Brazil and Mexico, researching the influence of intellectual property protection at the grass-roots level. His book titled *Intellectual Property and Economic Development* was published by Westview Press in 1990. A Brazilian edition will be published by Editora da Universidade de Sao Paulo soon, and a Spanish edition is in preparation.

He visited the Soviet Union in 1990 and will visit China this year to discuss intellectual property with officials there. He has conferred with World Bank officials regarding his research results over the last three years. He has published and lectured on Latin American debt, technology transfer, and intellectual property protection, and has taught at the graduate level. He holds degrees from Harvard College, Columbia University, and Harvard Law School.

Eugene B. Skolnikoff

Eugene B. Skolnikoff is Professor of Political Science at MIT. He was the Director of the Center for International Studies at MIT from 1972 until 1987. Originally educated as an electrical engineer at MIT, he studied economics and politics at Oxford on a Rhodes Scholarship and later received a Ph.D. in political science at MIT. He served on the White House staff in the Science Adviser's Office in the Eisenhower and Kennedy administrations, and was a Senior Consultant to President Carter's Science Adviser. In addition, he has been a consultant or adviser to several U.S. government departments, companies, and international organizations, and Chairman of the Board of Trustees of the German Marshall Fund of the United States, an American foundation. His work in government, research, and teaching has focused on science and public policy, especially the interaction of science and technology with international affairs, covering a wide range of industrial, military, space, economic, and futures issues.

Otto A. Stamm

Otto Stamm, a citizen of Switzerland, is Head of the Patent Department of CIBA-GEIGY AG, Basel, Switzerland.

Dr. Stamm was born in Switzerland in 1930. He received his doctorate from the University of Basel. After further study abroad, he was appointed lecturer in the Department of Chemistry at the Federal Institute of Technology in Zurich in 1964. In 1967 he joined the Patent Department of J.R. Geigy AG in Basel, and in 1969 gained his patent attorney's diploma after further studies at the Law School of the University of Strasburg, France.

Dr. Stamm is chairman of the Committee of Intellectual Property of the Swiss Federation of Commerce and Industry as well as a member of the working groups of many international industrial associations and is a member of the Executive Committee of the Swiss Group of the International Association for the Protection of Industrial Property.

Michiyuki Uenohara

Michiyuki Uenohara was born on September 5, 1925, in Japan. He received a B.E. degree from Nihon University in 1949 and M.S. and Ph.D. degrees from the Ohio State University in 1953 and 1956, respectively. After serving as a research assistant and instructor at both universities, he joined Bell Laboratories in 1957. In 1967, he returned to Japan and joined NEC Corporation and managed the Central Research Laboratories. He was elected a member of the Board of Directors in 1976 and was responsible for corporate R&D and engineering until June 1989. He is now Executive Adviser and holds the concurrent position of Chairman of the Board of Trustees, NEC Research Institute for Advanced Management Systems and Chairman, NEC Research Institute, Inc., in New Jersey. He is a Japanese member of the High Level Advisory Panel based on the U.S.-Japan agreement on cooperation in research and development in science and technology, and he serves on various government councils, including the Higher Education Council of the Japanese Ministry of Education, Science and Culture. He is a member of the Engineering Academy of Japan, a foreign associate of National Academy of Engineering in the United States, and a foreign member of the Royal Swedish Academy of Engineering Science.

Index

A

F

J

M

N

O

P

S

T

V

W